Methods in Molecular Biology™

Series Editor
John M. Walker
School of Life Sciences
University of Hertfordshire
Hatfield, Hertfordshire, AL10 9AB, UK

For other titles published in this series, go to
www.springer.com/series/7651

Viral Applications of Green Fluorescent Protein

Methods and Protocols

Edited by

Barry W. Hicks

US Air Force Academy, Department of Chemistry, CO, USA

Humana Press

Editor
Barry W. Hicks
Department of Chemistry
2355 Fairchild Drive
US Airforce Academy CO 80840
USA
barry.hicks@usafa.edu

ISBN: 978-1-4939-6143-6 ISBN: 978-1-59745-559-6 (eBook)

DOI: 10.1007/978-1-59745-559-6

Softcover reprint of the hardcover 1st edition 2009

Cover illustration: Front cover illustration from Figure 6 in Chapter 5.

Printed on acid-free paper

springer.com

Preface

This is the second volume I have edited on techniques utilizing green fluorescent protein (GFP) for the *Methods in Molecular Biology* series. When I began the first, there were only about 1,500 total publications, so it was relatively easy to get a broad survey of the fields making use of GFP; I read nearly 1,000 abstracts covering more than half of everything that had been published. When Humana contacted me to do another volume, I knew I would not be able to do that. Searching the SciFinder database for GFP now produces nearly 50,000 hits. Refining that search to GFP and virus in the past 5 years still produces almost 8,000 hits. Thus, even the topic of *Viral Applications of Green Fluorescent Protein* is too broad to thoroughly cover in a single volume, but I hope that, collectively, these chapters can aid anyone wanting to expand viral GFP methods into their own research. Numerous examples of both construction and application of GFP viruses are provided, and unlike the primary literature, the additional detail available can help newcomers get productive results more rapidly.

Just how important are fluorescent proteins to biomedical research today? It is almost impossible these days to pick up a recent issue of *Science*, *Nature*, *Cell* or the *Proceedings of the National Academy of Science USA*, and not find one, or more commonly, several, articles utilizing GFP. Comparing the number of publications and the rate of increase in publications utilizing GFP with some of the other very important topics (DNA microarrays, Prions and RNAi) is informative. As a reminder, Prions received the 1996 Nobel Prize in Physiology or Medicine, RNAi received the 2006 Nobel Prize in Medicine, and many have suggested that it is just a matter of time until microarray technology garners the honour. In Fig. 1 a graph indicates the total number of publications per year from 1994 to 2005 for these four topics. Not surprisingly to anyone conducting biomedical research, there have been more publications every year of the past decade utilizing GFP than on any of the other three

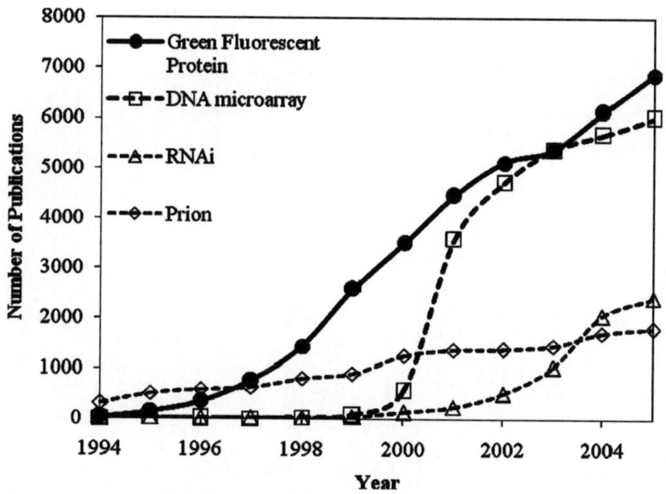

Fig. 1. The number of publications found in the SciFinder database on GFP, prions, DNA microarrays and RNAi.

topics. Does this mean that the development of GFP as a tool in biomedical research deserves the Nobel Prize? Unequivocally yes, and I am not alone in believing that it is just a matter of time until the committee comes to the same decision.

I have been using GFP in my courses at the US Air Force Academy for the past decade. We have done everything from simple bacterial transformation to computer comparisons of various crystal structures, to utilizing GFP-expressing transgenic mice and pigs to study bone marrow transplantation as a treatment for traumatic brain injury and to utilizing baculovirus expression of GFP in insect bioreactors. In that time, I have had the pleasure of collaborating with dozens of investigators from around the world who have done some amazing research enabled by GFP technology. My students and I are richer for that experience. I thank those students for challenging me to stay abreast of the rapidly advancing biomedical research front; it has been my pleasure to learn along with them. I also thank my wife and children for allowing me the extra hours away from home to do that work and to put this volume together and to go to the lab at strange hours of the night to look at glowing bacteria, mice brains or caterpillars. Without the motivation and support from those people, this volume would not have emerged.

Barry W. Hicks
USA

Contents

Preface v
Contributors ix

1. Using Fluorescent Proteins to Study Poxvirus Morphogenesis 1
   Brian M. Ward
2. Molecular Imaging of Human Embryonic Stem Cells 13
   Kazim H. Narsinh, Feng Cao, and Joseph C. Wu
3. Use of a GFP-PML-Expressing Cell Line as a Biosensor for Human Cytomegalovirus Infection 33
   Tomonori Ueno and Kiyoko Ogawa-Goto
4. Dual-Color Imaging of Tumor Angiogenesis 45
   Robert M. Hoffman
5. Use of GFP to Analyze Morphology, Connectivity, and Function of Cells in the Central Nervous System 63
   Alan R. Harvey, Erich Ehlert, Joris de Wit, Eleanor S. Drummond, Margaret A. Pollett, Marc J. Ruitenberg, Giles W. Plant, Joost Verhaagen, and Christiaan N. Levelt
6. Expression of EGFP by Adenovirus-Mediated Gene Transfer in the Central Nervous System 97
   Ryohei Tomioka
7. Adenovirus-Mediated Delivery of Short Hairpin RNA (shRNA) Mediates Efficient Gene Silencing in Terminally Differentiated Cardiac Myocytes 107
   Andreas Rinne, Christoph Littwitz, Kirsten Bender, Marie-Cécile Kienitz, and Lutz Pott
8. Microscopic Analysis of Adenoviral Decontamination Using GFP Adenovirus with Comparable Sensitivity to Flow Cytometry 125
   Michael Grusch, Annemarie Losert, Andreas Lackner, Alev Deli, Irene Herbacek, and Klaus Holzmann
9. Production of Multicistronic HIV-1-Based Lentiviral Vectors 137
   Nachimuthu Chinnasamy, James Shaffer, and Dhanalakshmi Chinnasamy
10. Using an EGFPmeter to Evaluate the Lentiviral Vector Production: Tricks and Traps 151
    Claude Bagnis , Pascal Bailly, and Sylvie Chapel-Fernandes
11. Use of a Macrophage-Tropic GFP-Tagged Human Immunodeficiency Virus Type 1 (HIV-1) to Study Viral Reservoirs 165
    Amanda M. Brown

12. GFP-Lentiviral Vectors Targeting for NeuroAIDS 177
Yuanan Lu

13. Construction of Cell-Type-Specific Promoter Lentiviruses for Optically Guiding Electrophysiological Recordings and for Targeted Gene Delivery 199
Aaron M. Jasnow, Donald G. Rainnie, Kimberly A. Maguschak, Jasmeer P. Chhatwal, and Kerry J. Ressler

14. Selective Protein Expression Within the CNS Using Hybrid Lentivirus 215
Fredrik Gussing, Ingrid van Marion, Igor Ralets, and Cecilia Lundberg

15. Construction and Analysis of Alphaherpesviruses Expressing Green Fluorescent Protein 227
Bruce W. Banfield and Gregory A. Bird

16. The Use of Green Fluorescent Fusion Proteins to Monitor Herpes Simplex Virus Replication 239
Travis J. Taylor and David M. Knipe

17. Modified Bovine Herpesvirus 1 for Protein Secretion 249
Günther M. Keil

18. Identification of Viral Peptide Fragments for Vaccine Development 261
Zhanglin Lin, Shuang Li, and Yong Chen

19. Use of GFP to Investigate Expression of Plant-Derived Vaccines 275
Hong-Ye Li and Mee-Len Chye

20. Studies of the Role and Function of Barley Stripe Mosaic Virus Encoded Proteins in Replication and Movement Using GFP Fusions 287
Sophie Haupt, Angelika Ziegler, Graham Cowan, and Lesley Torrance

21. Sendai Virus for Cancer Immunotherapy 299
Yasuji Ueda, Mamoru Hasegawa, and Yoshikazu Yonemitsu

22. Poxvirus Tropism for Primary Human Leukocytes and Hematopoietic Cells . . . 309
Qigui Yu, Ningjie Hu, and Mario Ostrowski

23. Reverse Genetics of Nipah Virus to Probe Viral Pathogenicity 329
Misako Yoneda, Kentaro Fujita, Hiroki Sato, and Chieko Kai

24. Baculovirus-Mediated Gene Transfer into Mesenchymal Stem Cells 339
Chien-Tai Tsai, Ching-Kuang Chuang, and Yu-Chen Hu

Index 353

Contributors

CLAUDE BAGNIS • *Laboratoire d'Hématologie Moléculaire, EFS Alpes Méditerranée, Marseille, France*
PASCAL BAILLY • *Laboratoire d'Hématologie Moléculaire, EFS Alpes Méditerranée, Marseille, France*
BRUCE W. BANFIELD • *Department of Pediatrics, Infectious Diseases Section, University of Colorado Health Sciences Center, Denver, CO*
KIRSTEN BENDER • *Institut fuer Physiologie, Ruhr-Universitaet, Bochum, Germany*
GREGORY A. BIRD • *Department of Pediatrics, Infectious Diseases Section, University of Colorado Health Sciences Center, Denver, CO*
AMANDA M. BROWN • *Johns Hopkins University School of Medicine, Baltimore, MD*
FENG CAO • *Molecular Imaging Program at Stanford, Stanford University School of Medicine, Stanford, CA*
SYLVIE CHAPEL-FERNANDES • *Laboratoire d'Hématologie Moléculaire, EFS Alpes Méditerranée, Marseille, France*
YONG CHEN • *Department of Chemical Engineering, Tsinghua University, Beijing, P.R. China*
JASMEER P. CHHATWAL • *Department of Psychiatry and Behavioral Sciences, Emory University, Atlanta, GA*
DHANALAKSHMI CHINNASAMY • *Vince Lombardi Gene Therapy Laboratory, Immunotherapy Program, Aurora St. Luke's Medical Center, Milwaukee, WI*
NACHIMUTHU CHINNASAMY • *Vince Lombardi Gene Therapy Laboratory, Immunotherapy Program, Aurora St. Luke's Medical Center, Milwaukee, WI*
CHING-KUANG CHUANG • *Department of Chemical Engineering, National Tsing Hua University, Hsinchu, Taiwan*
MEE-LEN CHYE • *Department of Botany, The University of Hong Kong, Pokfulam Road, Hong Kong, China*
GRAHAM COWAN • *Scottish Crop Research Institute, University of Dundee, Dundee, UK*
ELEANOR S. DRUMMOND • *School of Anatomy and Human Biology, The University of Western Australia, Crawley, Australia*
ERICH EHLERT • *Netherlands Institute for Neuroscience, Royal Netherlands Academy of Arts and Sciences, Amsterdam, the Netherlands*
KENTARO FUJITA • *Laboratory Animal Research Center, Institute of Medical Science, The University of Tokyo, Tokyo, Japan*
FREDRIK GUSSING • *CNS Gene Therapy Unit, Wallenberg Neuroscience Center, Department of Experimental Medical Science, Lund University, Lund, Sweden*
ALAN R. HARVEY • *School of Anatomy and Human Biology, The University of Western Australia, Crawley, Australia*
MAMORU HASEGAWA • *Department of Gene Therapy, Chiba University, Chiba, Japan*
SOPHIE HAUPT • *Scottish Crop Research Institute, University of Dundee, Dundee, UK*

Robert M. Hoffman • *AntiCancer, Inc., San Diego, CA*
Yu-Chen Hu • *Department of Chemical Engineering, National Tsing Hua University, Hsinchu, Taiwan*
Ningjie Hu • *Hawaii AIDS Clinical Research Program, Department of Medicine, University of Hawaii at Manoa, Honolulu, HI*
Aaron M. Jasnow • *Department of Psychiatry and Behavioral Sciences, Emory University, Atlanta, GA*
Chieko Kai • *Laboratory Animal Research Center, Institute of Medical Science, The University of Tokyo, Tokyo, Japan*
Günther M. Keil • *Friedrich-Loeffler-Institut, Federal Research Institute for Animal Health, Riems, Germany*
Marie-Cecile Kienitz • *Institut fuer Physiologie, Ruhr-Universitaet, Bochum, Germany*
David M. Knipe • *Department of Microbiology and Molecular Genetics, Harvard Medical School, Boston, MA*
Christiaan Levelt • *Research Group Molecular Visual Plasticity, Netherlands Institute for Neuroscience, Royal Netherlands Academy of Arts and Sciences, Amsterdam, the Netherlands*
Hong-Ye Li • *Department of Botany, The University of Hong Kong, Pokfulam Road, Hong Kong, China*
Shuang Li • *Department of Chemical Engineering, Tsinghua University, Beijing, P.R. China*
Zhanglin Lin • *Department of Chemical Engineering, Tsinghua University, Beijing, P.R. China*
Christoph Littwitz • *Institut fuer Physiologie, Ruhr-Universitaet, Bochum, Germany*
Yuanan Lu • *Department of Public Health, University of Hawaii, Honolulu, HI*
Cecilia Lundberg • *CNS Gene Therapy Unit, Wallenberg Neuroscience Center, Department of Experimental Medical Science, Lund University, Lund, Sweden*
Kimberly A. Maguschak • *Department of Psychiatry and Behavioral Sciences, Emory University, Atlanta, GA*
Ingrid van Marion • *CNS Gene Therapy Unit, Wallenberg Neuroscience Center, Department of Experimental Medical Science, Lund University, Lund, Sweden*
Kazim H. Narsinh • *Department of Medicine, University of California, San Diego School of Medicine, La Jolla, CA*
Kiyoko Ogawa-Goto • *Nippi Research Institute of Biomatrix, Toride, Ibaraki, Japan*
Mario Ostrowski • *Department of Immunology, University of Toronto, Toronto, ON, Canada*
Giles W. Plant • *School of Anatomy and Human Biology and Red's Spinal Cord Research Laboratory, The University of Western Australia, Crawley, Australia*
Margaret A. Pollett • *School of Anatomy and Human Biology, The University of Western Australia, Crawley, Australia*
Lutz Pott • *Institut fuer Physiologie, Ruhr-Universitaet, Bochum, Germany*
Donald G. Rainnie • *Department of Psychiatry and Behavioral Sciences, Emory University, Atlanta, GA*

Igor Ralets • *CNS Gene Therapy Unit, Wallenberg Neuroscience Center, Department of Experimental Medical Science, Lund University, Lund, Sweden*
Kerry J. Ressler • *Department of Psychiatry and Behavioral Sciences, Emory University, Atlanta, GA*
Andreas Rinne • *Institut fuer Physiologie, Ruhr-Universitaet, Bochum, Germany*
Marc J. Ruitenberg • *School of Anatomy and Human Biology and Red's Spinal Cord Research Laboratory, The University of Western Australia, Crawley, Australia*
Hiroki Sato • *Laboratory Animal Research Center, Institute of Medical Science, The University of Tokyo, Tokyo, Japan*
James Shaffer • *Vince Lombardi Gene Therapy Laboratory, Immunotherapy Program, Aurora St. Luke's Medical Center, Milwaukee, WI*
Travis J. Taylor • *Department of Microbiology and Molecular Genetics, Harvard Medical School, Boston, MA*
Ryohei Tomioka • *Lab for Cortical Organization and Systematics, RIKEN Brain Science Institute, Wako-shi, Saitama, Japan*
Lesley Torrance • *Scottish Crop Research Institute, University of Dundee, Dundee, UK*
Chien-Tai Tsai • *Department of Chemical Engineering, National Tsing Hua University, Hsinchu, Taiwan*
Yasuji Ueda • *Department of Gene Therapy, Chiba University, Chiba, Japan*
Tomonori Ueno • *Nippi Research Institute of Biomatrix, Toride, Ibaraki, Japan*
Joost Verhaagen • *Netherlands Institute for Neuroscience, Royal Netherlands Academy of Arts and Sciences, Amsterdam, the Netherlands*
Brian M. Ward • *Department of Microbiology and Immunology, University of Rochester Medical Center, Rochester, NY*
Joris De Wit • *Division of Biology, Neurobiology Section, University of California San Diego, La Jolla, CA*
Joseph C. Wu • *Molecular Imaging Program at Stanford (MIPS), Department of Medicine, Division of Cardiology, Stanford University School of Medicine, Stanford, CA*
Misako Yoneda • *Laboratory Animal Research Center, Institute of Medical Science, The University of Tokyo, Tokyo, Japan*
Yoshikazu Yonemitsu • *Department of Gene Therapy, Chiba University, Chiba, Japan*
Qigui Yu • *Hawaii AIDS Clinical Research Program, Department of Medicine, University of Hawaii at Manoa, Honolulu, HI*
Angelika Ziegler • *Scottish Crop Research Institute, University of Dundee, Dundee, UK*

Chapter 1

Using Fluorescent Proteins to Study Poxvirus Morphogenesis

Brian M. Ward

Summary

Fluorescent protein (FP) fusions not only allow for the convenient visualization of a protein of interest's subcellular localization but also permit the real-time monitoring of their subcellular trafficking. The subcellular fluorescent pattern of FP-fusions can also serve as a visual marker for various subcellular processes using either live or static microscopy. We have employed FP-fusions for the study of poxvirus morphogenesis. Fusion of FP with either a viral core protein or an extracellular virion-specific protein can serve as a visual read-out for normal poxvirus morphogenesis at the subcellular level. Recombinant viruses expressing a FP-fusion, in conjunction with the deletion of a gene involved in either morphogenesis or egress, usually display an aberrant FP pattern. Functional domains in the missing protein are then mapped by complementation in-trans followed by fluorescent microscopy for analysis of the FP pattern. The methods presented here describe how to infect and transfect cells for trans-complementation for the purpose of functional domain mapping. The imaging and analysis of these cells is described.

Key words: Vaccinia virus, B5R, A27L, Green fluorescent protein (GFP), HcRed, Intracellular enveloped virus, Intracellular mature virus.

1. Introduction

Although variola virus, the causative agent of smallpox, was eradicated over 20 years ago, poxviruses remain the subject of intense study due in large part to their use for protein production and potential use as a vaccine vector against a multitude of diseases. In conjunction with their potential use as a bioweapon, research on poxviruses has intensified over the past several years. Vaccinia virus, the live vaccine for smallpox, is the prototypical member of the Orthopoxvirus genus and the most widely studied poxvirus *(1)*. Remarkably, cells infected with vaccinia virus produce both

Barry W. Hicks (ed.), *Methods in Molecular Biology, Viral Applications of Green Fluorescent Protein, vol. 515*

DOI: 10.1007/978-1-59745-559-6_1

intracellular and extracellular forms of infectious virions (**Fig. 1**). These two forms are essentially identical with the exception that the extracellular form has an additional outer membrane that contains additional, unique, viral proteins. Unlike most DNA viruses, vaccinia replication occurs entirely in the cytoplasm in a defined area appropriately termed the viral factory (reviewed in *(1–3)*). Replication results in the production of the infectious intracellular form termed mature virions (MV). A subset of MVs is transported along microtubules to be enveloped by an extra double membrane derived from the trans-Golgi network or endosomal cisternae. After intracellular envelopment, they are called wrapped virions (WV). WVs are transported to the cell surface where the outermost membrane fuses with the plasma membrane releasing an infectious extracellular virion (EV). The majority of EVs remain attached to the infected cell's surface where actin polymerization occurs directly beneath the virion on the cytosolic side to produce actin tails that propel them toward neighboring cells. While MV make up the majority of progeny

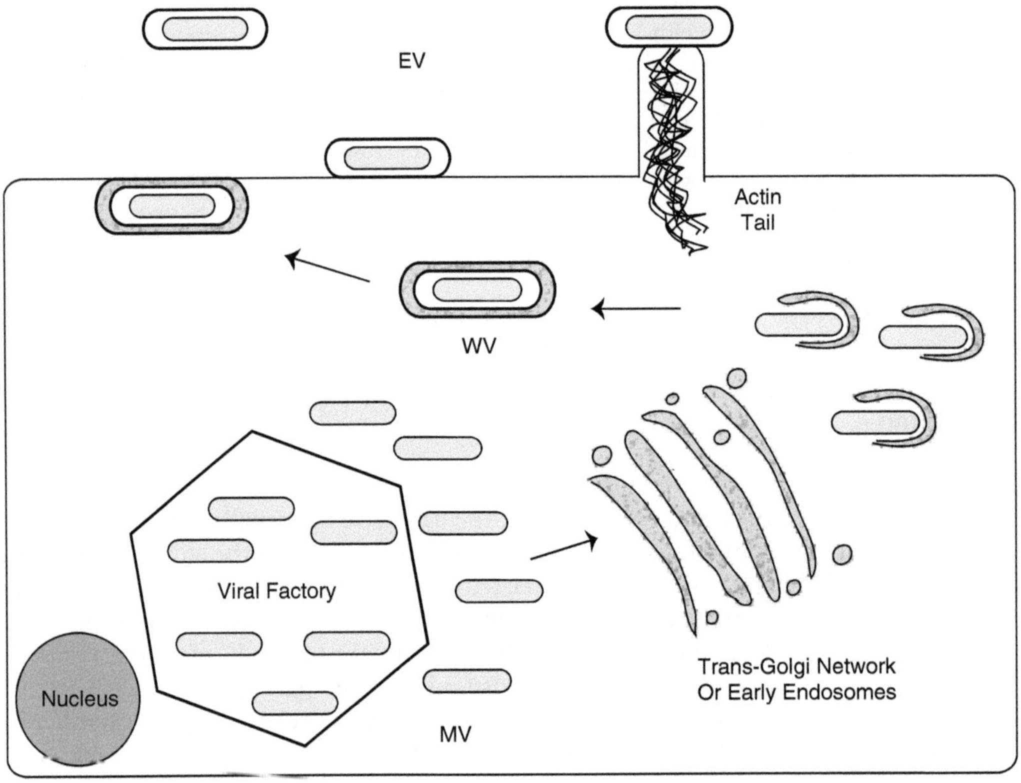

Fig. 1. Vaccinia virus morphogenesis. The first infectious progeny, mature virions (*MV*) are formed at the viral factory. Some MVs move along microtubules to the site of wrapping at the trans-Golgi network/early endosomes to form wrapped virions (*WV*). WVs are transported by kinesin along microtubules to the plasma membrane. The outer membrane of WV fuses with the plasma membrane and releases extracellular virions (*EV*). EV attached to the cell coordinate the polymerization of actin tails on the cytoplasmic side of the membrane.

virions, they remain in the infected cell and are only released if the cell is lysed. Therefore, EV production is required for active cell-to-cell spread and dissemination. Despite a long history of research and the identification of several viral proteins required for morphogenesis, many of the molecular functions carried out by these proteins remain to be discovered.

Two such viral proteins involved in morphogenesis and egress are encoded by the A36R and A27L open reading frames (ORFs) (*see* **Note 1**). The A36R ORF encodes a type 1b transmembrane protein (A36) that has a single transmembrane domain at the beginning of the protein and a long cytoplasmic domain **(Fig. 2)** *(4)*. The protein is required for actin tail formation. Subsequently, it has been shown that phosphorylation of tyrosine residues 112 and 124 is required for interaction with the cellular protein Nck, which leads to the recruitment of N-Wasp and the actin polymerization machinery for the formation of actin tails *(5–9)*. In addition, it was shown that residues 91–111 interact with the WV-specific protein A33 for incorporation into

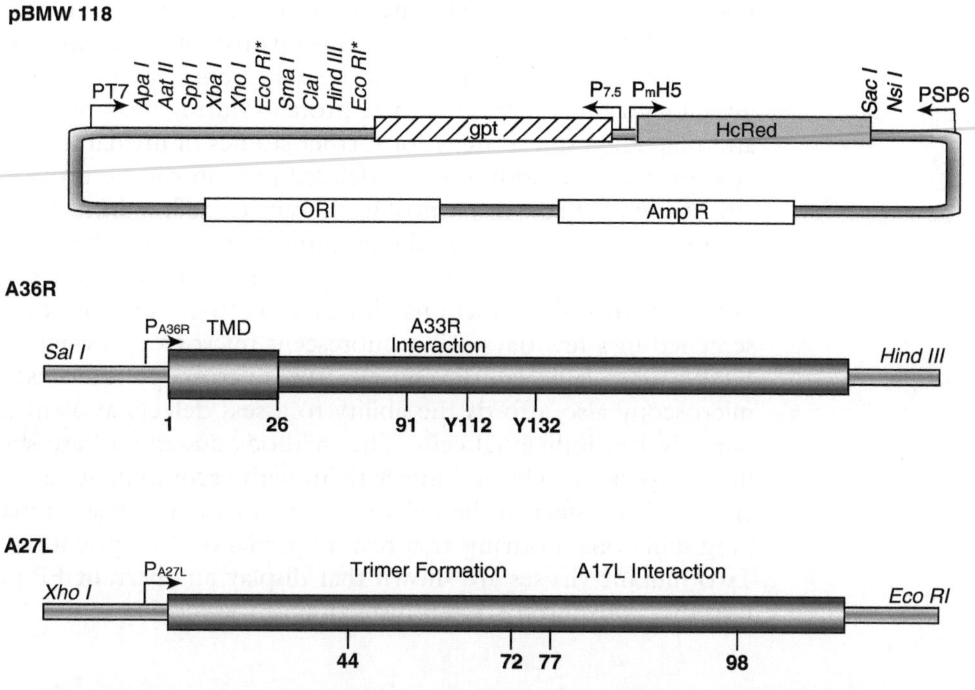

Fig. 2. Diagrammatic representation of pBMW-118, A36R, and A27L used for trans-complementation. For pBMW-118, the position of unique endonuclease sites is shown with the exception of *Eco*RI, which has two sites. The fluorescent protein (*FP*) HcRed is expressed from the modified viral H5 promoter (PmH5). The *E. coli* guanine phosphoribosyltransferase (*gpt*) gene is expressed from the viral 7.5 promoter (P7.5) and allows for transient dominant selection with MPA. PT7, T7 promoter; PSP6, SP6 promoter; ORI, f1 origin of replication; AmpR, ampicillin resistance. For A36R, the predicted transmembrane domain (TMD), and A33 interaction domain and phospho-tyrosines critical for actin tail formation are shown. For A27L, the regions for trimer formation and A17L interaction are shown. For both, the locations of endonuclease sites engineered at the ends of the flanks are designated.

WV membranes *(10)*. The protein encoded by the A27L ORF (A27) is found as a trimer that associates with the MV membrane through interaction with the viral A17 transmembrane protein. Trimerization and A17 interaction domains have been mapped to residues 44–72 and 77–98, respectively (**Fig. 2**) *(11–15)*. It was previously believed that A27 was required for virus entry. Subsequently, deletion of A27L showed that it is dispensable for infectious MV production but plays a critical role in WV formation as the deletion virus formed miniscule plaques on cell monolayers indicating a severe defect in EV formation *(16)*.

Several laboratories have described recombinant vaccinia viruses that express fluorescent proteins (FPs) appended to either the viral core protein A4 *(16, 17)*, which labels all forms of progeny virions (MV, WV, and EV), or various EV-specific proteins, which label only EV *(18–20)*. These FP-fusions have proven indispensable for studying both virion transport out of the cell and the transport of viral cores into the cell during the initial stages of infection. These recombinants also provide useful tools for assessing the function of other viral proteins during morphogenesis. The viral FP-protein chimera results in a characteristic fluorescence pattern in cells infected with the recombinant. Mutations and deletions of other proteins involved in morphogenesis normally alter this pattern. The resulting aberrant pattern can give clues to the function of these proteins during morphogenesis and may suggest a strategy for further studies of uncharacterized viral proteins. In addition, the deleted protein can be provided, either in its entirety or parts of it, in trans to define domains that are required to either partially or completely restore the normal fluorescent phenotype. Trans-complementation accelerates the process of mapping functional domains because mutations can be screened in a few days using fluorescent microscopy as opposed to the more time consuming electron microscopy. Fluorescent microscopy also affords the ability to assess defects at a subcellular level in individual cells. The methods described here show how to prepare cells and infect them with recombinant vaccinia virus and transfect in the deleted genes for the purpose of mapping functional domains that restore partial or total phenotypes. Two mutant viruses are shown that display an aberrant FP pattern that is restored upon transfection of the missing gene.

2. Materials

2.1. Virus and Cells

1. Crude cell lysate stocks of recombinant vaccinia virus vA4-YFP, vB5R-GFP, vA4-YFP/ΔA27L, and vB5R-GFP/ΔA36R (*see* **Note 2**). Titer as described in *(21)*.

2. DMEM-10: Dulbecco's modified Eagle's medium (DMEM; Mediatech, Inc.) supplemented with 0.03% glutamine, 100 U/mL penicillin, 100 μg/mL streptomycin, and 10% serum (DMEM-10).
3. DMEM-2.5 and DMEM-0: DMEM-10 containing either 2.5% or 0% serum, respectively.
4. HeLa cell monolayer (ATCC #CCL-2) maintained in DMEM-10. For infection use DMEM supplemented with 2.5% serum (DMEM-2.5).
5. 2× DMEM-2.5: DMEM containing 0.03% glutamine, 200 U/mL penicillin, 200 μg/mL streptomycin, and 5% serum.
6. Lipofectamine reagent (Invitrogen).
7. 2× Trypsin. Salt-free trypsin (Worthington). Dilute to 0.25 mg/mL in water and filter sterilize. Store indefinitely at –20 °C.
8. Materials and equipment for routine tissue culture, including PBS, 0.25% trypsin/0.02% EDTA, plastic ware, laminar-flow hood, CO_2 incubator, and centrifuge.
9. Cup sonicator.
10. Wizard Plus midiprep DNA purification system (Promega).

2.2. Microscopy

1. No. 1 cover glass, 18 mm in diameter. Acid treated and stored in absolute ethanol.
2. Mowiol 4.88 (Calbiochem)-based mounting media *(22)*.
3. Fluorescent microscope with the means to capture images, a high quality, oil emersion objective, and filter sets for visualizing Hoechst, GFP, and Texas red fluorescence.
4. Sterile PBS.
5. 16% (w/v) aqueous paraformaldahyde, methanol free (Alfa Aesar).
6. 5 mg/mL (100×) Bisbenzimide H (Hoechst) in water. 1×Hoechst: dilute Bisbenzimide H stock 1:100 into sterile PBS.

3. Methods

3.1. Plasmid Description

Figure 2 shows the plasmid we routinely use for trans-complementation. It encodes for the fluorescent protein HcRed (Clonetech) under control of the modified vaccinia virus promoter H5 (P_{mH5}) for strong early expression of red fluorescence in cells that are infected and transfected. This serves as a visual marker for identifying cells that are infected/transfected and should be expressing the trans-protein of interest. In addition, there is a multiple cloning site for the insertion of viral genes to be studied. These should contain their natural viral promoter to assure expression of the protein at the proper time during infection (*see* **Note 3**). For the purpose of this chapter, we have inserted either the A36R ORF or the A27L ORF along with ~500 bp of flanking DNA on either side of each gene into pBMW-118 (pBMW-118/A36R or pBMW-118/A27L, respectively) using the restriction sites shown (**Fig. 2**). The sites were added by PCR and the amplified fragments were inserted using standard cloning techniques. Inserts were verified by sequencing, and the constructs were purified using the Wizard Plus midiprep DNA purification kit.

3.2. Transfection of Infected Cells

HeLa cells are seeded into a 35-mm plate that contains sterile coverslips. The following day, cells are infected with virus and transfected with the trans-complementation plasmid. Infected/transfected cells are incubated overnight and fixed for microscopy the next day.

1. Remove a single coverslip from the ethanol with metal forceps and place briefly into a flame to sterilize (*see* **Note 4**).
2. Once all the ethanol has burned off, place the coverslip into a well of a sterile 35-mm plate. Repeat with up to four coverslips per plate.
3. Remove the media from a HeLa cell monolayer and trypsinize for passage as normal.
4. Seed about 1×10^6 cells/coverslip-containing plate, cover the cells with 2 mL of DMEM-10, and incubate overnight at 37 °C.
5. The next day thaw a stock of vaccinia virus and mix with an equal volume of 2× trypsin. Vortex, and incubate in a 37 °C water bath for 30 min, mixing by vortex every 10 min (*see* **Note 5**).
6. Sonicate the trypsinized virus in a cup sonicator for 30 s.
7. Dilute the virus in DMEM-2.5 to a titer of 1×10^5 plaque forming units/mL.
8. Remove media from plate and reposition coverslips with a sterile pipette tip to be sure they are not overlapping each other.
9. Add 0.5 mL of DMEM-2.5 and 100 μL of diluted virus per plate.
10. Place back in CO_2 incubator for 2 h, rocking gently every 20 min to spread the inoculum and keep the cells moist.
11. One hour postinfection, start setting up the transfection by placing 100 μL of DMEM-0 in a sterile screw-cap tube.

12. Add 5 µL of lipofectamine to the DMEM-0.
13. Add 100 µL of DMEM-0 to a sterile 5 mL polystyrene round bottom tube.
14. Add 1 µg of purified plasmid DNA to the tube.
15. Add the liposome/DMEM-0 mixture to the DNA/DMEM-0 mixture and mix gently. Let sit at room temperature for at least 30 min.
16. Two hours postinfection, remove the media from the cells and add 2 mL DMEM-0.
17. Add 800 µL of DMEM-0 to the lipsome/DNA mixture.
18. Aspirate the DMEM-0 off of the cells and add the DNA/liposomes to the cells.
19. Place back into the 37 °C CO_2 incubator for 4 h.
20. Add 1 mL pre-warmed 2× DMEM-2.5 to the cells and incubate overnight.

3.3. Microscopy of Fixed Cells

The next day, coverslips with infected/transfected cells on them are washed, fixed, and mounted for observation (*see* **Note 6**).

1. Dilute paraformaldehyde to 4% in PBS and add 500 mL/well of a 24-well plate for each coverslip that needs to be fixed.
2. Remove the media from the infected/transfected cells and replace with PBS.
3. Working quickly (*see* **Note 7**), remove a coverslip and place it into a well containing fixative. Repeat for each coverslip.
4. Incubate coverslips in fixative for 10 min.
5. Aspirate off fixative and replace with sterile PBS (*see* **Notes 8** and **9**).
6. If desired, stain DNA with 1× Hoechst diluted in PBS for 10 min.
7. Wash twice with PBS.
8. Mount in Mowiol (*see* **Note 10**).
9. Visualize by fluorescent microscopy (*see* **Note 11**).

3.3.1. Interpreting Microscopy Results

Cells infected with vB5R-GFP show a characteristic pattern of fluorescence with bright fluorescence at the site of wrapping in the juxtanuclear region, wrapped virions throughout the cytoplasm, and a collection of signal at the vertices (**Fig. 3**, top row). In contrast, the deletion of the viral A36R gene results in an altered fluorescent pattern with only a collection of signal at the site of wrapping (**Fig. 3**). Cells infected with vB5R-GFP/ΔA36R and transfected with pBMW-118/A36R, which expresses the full length A36R protein, restore the normal fluorescence pattern. Using truncation mutagenesis, it was shown that expression of

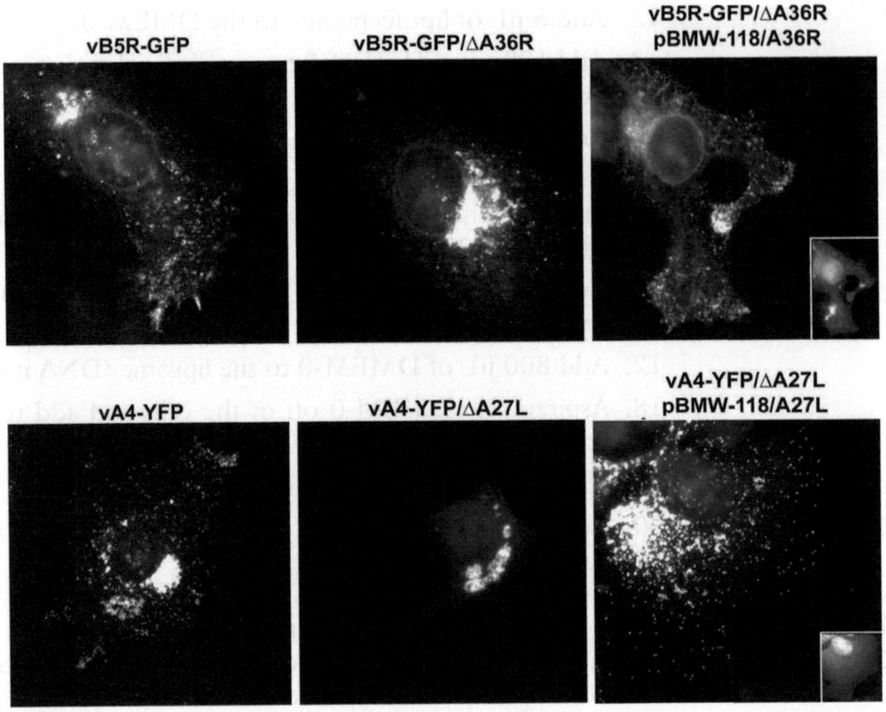

Fig. 3. Subcellular localization of FP-proteins. HeLa cells were infected with the indicated viruses and transfected with the indicated plasmids. Images show the subcellular localization of the FP fusion proteins. Insets show the expression of HcRed indicating that the cells were infected and transfected.

residues 1–111 of A36 is capable of restoring the normal FP pattern while expression of residues 1–91 is not (supplemental material for *(10)*). These results led to the mapping of a region of A36 (91–111) that interacts with A33 and the light chain of kinesin *(10, 23)*. These two interactions are required for the incorporation of A36 into WV and intracellular transport of WV on microtubules and the normal FP pattern.

Cells infected with vA4-YFP show a collection of signal at the viral factory in addition to the dissemination of virus-sized particles throughout the cytoplasm (**Fig. 3**, bottom row). In stark contrast, deletion of the viral A27L gene (an MV protein) results in a cluster of particle around the factory. Transfection of the plasmid expressing A27L in cells infected with vA4-YFP/ΔA27L restores the normal FP pattern. A subsequent study showed that in the absence of A27L, MVs are not wrapped to form WV. The defect in wrapping was not due to transport as live imaging of cells infected with vA4-YFP/ΔA27L revealed that MVs were capable of intracellular transport *(16)*. The use of this system should allow the mapping of residues on A27L that are required for restoration of the phenotype and intracellular wrapping.

Notes

1. In accordance with recent manuscripts about vaccinia virus, open reading frames (ORFs) are designated by a capital letter indicating a *Hin*dIII restriction endonuclease fragment, a number indicating the position in the *Hin*dIII fragment, and a letter (L or R) indicating the direction of transcription, for example, A36R. The corresponding protein is designated by a capital letter and number, for example, A36.
2. Vaccinia virus is a BSL-2 pathogen and should be handled as such. The focus of this chapter is not in production of the recombinant viruses, but on their use in determining the function of viral proteins. Additional detail on their construction can be obtained in the literature *(10, 16, 20, 24)*.
3. While expression from the natural promoter would best emulate normal expression during infection, expression from the plasmid's T7 promoter is possible if a virus such as vTF7-3 *(25)* is used but this may not result in proper temporal expression of the gene. It should be noted though that several late proteins involved in morphogenesis have been studied this way *(26–29)*.
4. Extreme caution should be exercised when flaming with alcohol. Always be sure the coverslip is below the hand so that burning ethanol cannot drip or run down onto the skin. In addition, cover the stock of coverslips so they do not catch on fire.
5. Several reports have indicated that protease treatment of vaccinia virus enhanced infectivity *(30–32)*.
6. While details are given for observation of fixed cells, the detailed protocols for the live visualization of infected cells is provided in *(33)*.
7. PBS will eventually cause the cells to detach from the coverslip before fixation.
8. If good sterile technique is used, the coverslips can be stored in PBS in a sealed 24-well plate for several weeks at 4 °C.
9. Cells can also be permeabilized and fluorescently stained for other cell or viral proteins.
10. Mounted coverslips should be stored at 4 °C overnight before prolonged observation to allow any antifading agent in the mounting media time to diffuse into the samples. In addition, mounted coverslips can be stored indefinitely at 4 °C.
11. In addition, a laser scanning confocal microscope can be used. Most of these are able to image GFP/YFP and HcRed using settings to visualize FITC and Texas red, respectively.

Acknowledgments

The author thanks the members of his laboratory. Research in the author's laboratory was supported by NIH grants AI067391, AI54392, and AI50020.

References

1. Moss, B. (2001) *Poxviridae*: The viruses and their replication in "Fields Virology" (Fields, B. N., Knipe, D. M., and Howley, P. M., Eds.), Vol. 2, pp. 2849-83, Lippincott-Raven Publishers, Philadelphia.
2. Moss, B. (2006) Poxvirus entry and membrane fusion. Virology 344, 48-54.
3. Smith, G. L., Vanderplasschen, A., and Law, M. (2002) The formation and function of extracellular enveloped vaccinia virus. J. Gen. Virol. 83, 2915-31.
4. van Eijl, H., Hollinshead, M., and Smith, G. L. (2000) The vaccinia virus A36R protein is a type Ib membrane protein present on intracellular but not extracellular enveloped virus particles. Virology 271, 26-36.
5. Frischknecht, F., Cudmore, S., Moreau, V., Reckmann, I., Rottger, S., and Way, M. (1999) Tyrosine phosphorylation is required for actin-based motility of vaccinia but not Listeria or Shigella. Curr. Biol. 9, 89-92.
6. Frischknecht, F., Moreau, V., Rottger, S., Gonfloni, S., Reckmann, I., Superti-Furga, G., and Way, M. (1999) Actin-based motility of vaccinia virus mimics receptor tyrosine kinase signalling. Nature 401, 926-29.
7. Moreau, V., Frischknecht, F., Reckmann, I., Vincentelli, R., Rabut, G., Stewart, D., and Way, M. (2000) A complex of N-WASP and WIP integrates signalling cascades that lead to actin polymerization. Nat. Cell Biol. 2, 441-48.
8. Scaplehorn, N., Holmstrom, A., Moreau, V., Frischknecht, F., Reckmann, I., and Way, M. (2002) Grb2 and Nck act cooperatively to promote actin-based motility of vaccinia virus. Curr. Biol. 12, 740-5.
9. Zettl, M., and Way, M. (2001) New tricks for an old dog? Nat. Cell Biol. 3, E74-75.
10. Ward, B. M., Weisberg, A. S., and Moss, B. (2003) Mapping and functional analysis of interaction sites within the cytoplasmic domains of the vaccinia virus A33R and A36R envelope proteins. J. Virol. 77, 4113-26.
11. Lai, C. F., Gong, S. C., and Esteban, M. (1990) Structural and functional properties of the 14-kDa envelope protein of vaccinia virus synthesized in Escherichia coli. J. Biol. Chem. 265, 22174-80.
12. Lai, C. F., Gong, S. C., and Esteban, M. (1991) The purified 14-kilodalton envelope protein of vaccinia virus produced in Escherichia coli Induces virus immunity in animals. J. Virol. 65, 5631-35.
13. Rodriguez, D., Rodriguez, J. R., and Esteban, M. (1993) The vaccinia virus 14-kilodalton fusion protein forms a stable complex with the processed protein encoded by the vaccinia virus A17L gene. J. Virol. 67, 3435-40.
14. Rodriguez, J. F., Janeczko, R., and Esteban, M. (1985) Isolation and characterization of neutralizing monoclonal antibodies to vaccinia virus. J. Virol. 56, 482-88.
15. Rodriguez, J. F., Paez, E., and Esteban, M. (1987) A 14,000-Mr envelope protein of vaccinia virus is involved in cell fusion and forms covalently linked trimers. J. Virol. 61, 395-404.
16. Ward, B. M. (2005) Visualization and characterization of the intracellular movement of vaccinia virus intracellular mature virions. J. Virol. 79, 4755-63.
17. Carter, G. C., Rodger, G., Murphy, B. J., Law, M., Krauss, O., Hollinshead, M., and Smith, G. L. (2003) Vaccinia virus cores are transported on microtubules. J. Gen. Virol. 84, 2443-58.
18. Geada, M. M., Galindo, I., Lorenzo, M. M., Perdiguero, B., and Blasco, R. (2001) Movements of vaccinia virus intracellular enveloped virions with GFP tagged to the F13L envelope protein. J. Gen. Virol. 82, 2747-60.
19. Hollinshead, M., Rodger, G., Van Eijl, H., Law, M., Hollinshead, R., Vaux, D. J., and Smith, G. L. (2001) Vaccinia virus utilizes microtubules for movement to the cell surface. J. Cell Biol. 154, 389-402.
20. Ward, B. M., and Moss, B. (2001) Visualization of intracellular movement of vaccinia virus virions containing a green fluorescent protein-B5R membrane protein chimera. J. Virol. 75, 4802-13.
21. Earl, P. L., and Moss, B. (1991) Generation of recombinant vaccinia viruses in "Current Protocols

in Molecular Biology" (Ausubel, F. M., Brent, R., Kingston, R. E., Moore, D. D., Seidman, J. G., Smith, J. A., and Struhl, K., Eds.), Vol. 2, pp. 16.17.1-16.17.16, Greene Publishing Associates & Wiley Interscience, New York.

22. Pagh, K. I., and Adelman, M. R. (1988) Assembly, disassembly, and movements of the microfilament-rich ridge during the amoeboflagellate transformation in Physarum polycephalum. Cell Motil. Cytoskeleton 11, 223-34.
23. Ward, B. M., and Moss, B. (2004) Vaccinia virus A36R membrane protein provides a direct link between intracellular enveloped virions and the microtubule motor kinesin. J. Virol. 78, 2486-93.
24. Ward, B. M., and Moss, B. (2001) Vaccinia virus intracellular movement is associated with microtubules and independent of actin tails. J. Virol. 75, 11651-63.
25. Mohamed, M. R., and Niles, E. G. (2004) Transient and inducible expression of vaccinia/T7 recombinant viruses. Methods Mol. Biol. 269, 41-50.
26. Senkevich, T. G., Ward, B. M., and Moss, B. (2004) Vaccinia virus entry into cells is vdependent on a virion surface protein encoded by the A28L gene. J. Virol. 78, 2357-66.
27. Senkevich, T. G., Ward, B. M., and Moss, B. (2004) Vaccinia virus A28L gene encodes an essential protein component of the virion membrane with intramolecular disulfide bonds formed by the viral cytoplasmic redox pathway. J. Virol. 78, 2348-56.
28. Szajner, P., Jaffe, H., Weisberg, A. S., and Moss, B. (2003) Vaccinia virus G7L protein Interacts with the A30L protein and is required for association of viral membranes with dense viroplasm to form immature virions. J. Virol. 77, 3418-29.
29. Townsley, A. C., Senkevich, T. G., and Moss, B. (2005) The product of the vaccinia virus L5R gene is a fourth membrane protein encoded by all poxviruses that is required for cell entry and cell-cell fusion. J. Virol. 79, 10988-98.
30. Ichihashi, Y., and Oie, M. (1980) Adsorption and penetration of the trypsinized vaccinia virion. Virology 101, 50-60.
31. Ichihashi, Y., Tsuruhara, T., and Oie, M. (1982) The effect of proteolytic enzymes on the infectivity of vaccinia virus. Virology 122, 279-89.
32. Townsley, A. C., and Moss, B. (2007) Two distinct low-pH steps promote entry of vaccinia virus. J. Virol. 81, 8613-20.
33. Ward, B. M. (2004) Pox, dyes, and videotape: making movies of GFP-labeled vaccinia virus. Methods Mol. Biol. 269, 205-18.

Chapter 2

Molecular Imaging of Human Embryonic Stem Cells

Kazim H. Narsinh, Feng Cao, and Joseph C. Wu

Summary

Human embryonic stem cells (hESCs) are a renewable source of differentiated cell types that may be employed in various tissue regeneration strategies. However, clinical implementation of cell transplantation therapy is hindered by legitimate concerns regarding the in vivo teratoma formation of undifferentiated hESCs and host immune reactions to allogeneic cells. Investigating in vivo hESC behavior and the ultimate feasibility of cell transplantation therapy necessitates the development of novel molecular imaging techniques to longitudinally monitor hESC localization, proliferation, and viability in living subjects. An innovative approach to harness the respective strengths of various imaging platforms is the creation and use of a fusion reporter construct composed of red fluorescent protein (RFP), firefly luciferase (fluc), and herpes simplex virus thymidine kinase (HSV-tk). The imaging modalities made available by use of this construct, including optical fluorescence, bioluminescence, and positron emission tomography (PET), may be adapted to investigate a variety of physiological phenomena, including the spatio-temporal kinetics of hESC engraftment and proliferation in living subjects. This chapter describes the applications of reporter gene imaging to accelerate basic science research and clinical studies involving hESCs through (1) isolation of a homogeneous hESC population, (2) noninvasive, longitudinal tracking of the location and proliferation of hESCs administered to a living subject, and (3) ablation of the hESC graft in the event of cellular misbehavior.

Key words: Human embryonic stem cells, Red fluorescent protein, Firefly luciferase, Herpes simplex virus thymidine kinase, Fluorescence, Bioluminescence, Positron emission tomography.

1. Introduction

The pluripotency of human embryonic stem cells (hESCs) offers investigators the prospect of being able to create virtually any cell type in the body in a renewable fashion. Investigators have already successfully demonstrated the ability of hESCs to

Barry W. Hicks (ed.), *Methods in Molecular Biology, Viral Applications of Green Fluorescent Protein, vol. 515*

DOI: 10.1007/978-1-59745-559-6_2

differentiate into cardiac *(1)*, neuronal *(2, 3)*, and pancreatic islet *(4)* cells. Theoretically, the newly synthesized cells derived from undifferentiated hESCs can be used to replace the diseased cells of an organism. The successful implementation of such cell transplantation therapy would affect the treatment of a wide range of diseases with high rates of morbidity and mortality, including heart disease, neurological injury, and diabetes. Understandably, the prospect of tissue regeneration using stem cells has generated significant enthusiasm amongst the scientific and clinical communities, as well as amongst the general public.

However, much remains to be explained about the cellular underpinnings of stem cell biology prior to their therapeutic application. Although the pluripotency and self renewal of hESCs is precisely what makes them attractive candidates for tissue regeneration, these properties also create a high risk of teratoma formation after cell transplantation. Accordingly, one must ensure that the cell population delivered to the organism is homogenous in its differentiation stage and gene expression profile. An additional hurdle to the implementation of cell transplantation therapy is the potential for host immune reactions causing rejection of an allogeneic stem cell graft.

These and other challenges highlight the need for an ability to assess and control hESC behavior in vivo over an extended period of time. The development of molecular imaging techniques allows for the noninvasive, repetitive visualization of hESC-derived cell location, migration, proliferation, and ablation in vivo (5). Here, we seek to highlight the varied applications of several molecular imaging techniques while using a single fusion reporter gene.

1.1. Fluorescent Reporter Proteins

One of the many basic questions to be addressed regarding stem cell therapy is which progenitor cell types are responsible for the observed effects of cell transplantation, whether they be positive effects such as the enhancement of left ventricular ejection fraction *(6)*, or negative effects such as teratoma formation *(7, 8)*. Efforts to identify the particular cell types responsible for certain observed effects have been hindered by difficulty obtaining homogeneous progenitor cell populations prior to administration. Any clinical implementation of cell transplantation therapy would likely require purification of the specific cell population that produces the desired therapeutic effect. This challenge has been partially overcome by the development of hESC lines stably transfected or transduced with green fluorescent protein (GFP) *(9, 10)*. In addition to facilitating visualization of the hESC-derived portion of tissue grafts by fluorescence microscopy or immunostaining, GFP expression can be used to select a nearly homogeneous population of cells with fluorescence activated cell sorting (FACS). For instance, transfection of a reporter construct containing GFP driven by a lineage-specific promoter would allow isolation of the cell population of that particular lineage out of a heterogeneous progenitor cell population *(11, 12)*. Alternatively, one could isolate only stably transfected cells containing a GFP-linked gene of interest driven by an inducible or constitutive promoter. Hence, real-time visualization of single cells and the selective isolation of a particular cell type can both be achieved with fluorescent reporter genes such as GFP, red fluorescent protein (RFP), and yellow fluorescent protein (YFP).

1.2. Indirect Molecular Imaging

However, fluorescent reporter genes cannot be used reliably to assess the characteristics of transplanted cells in vivo due to poor tissue penetration and the need for extrinsic excitation light, which produces an unacceptable level of background signal. Instead, GFP-expressing cells are usually identified histologically. Such an assessment of hESC-derived cell fate using histology requires postmortem tissue analysis and the sacrifice of a large number of animals in order to amass enough tissue "snapshots" to overcome sampling error and obtain a realistic picture of cell survival and viability over time. The use of molecular imaging reporter constructs offers several advantages over the conventional histological approaches to stem cell visualization.

Molecular imaging can be broadly defined as the visualization of molecular and cellular processes in the living subject. For in vivo molecular imaging to work, two basic elements are required: a molecular probe that detects a quantifiable signal based on the presence of a gene, RNA, or protein, and a method to monitor these probes in vivo. In general, molecular imaging can be divided into two categories: direct imaging of probe-target interactions or indirect imaging based on a reporter gene and reporter probe.

The concept behind indirect molecular imaging is an expansion upon basic reporter gene technology, whereby a promoter or enhancer region of interest is linked to the imaging reporter gene. Once introduced into the target cell, the reporter gene produces the reporter protein, which then interacts with the introduced reporter probe, producing an analytic signal that can be detected by the detector system (**Fig. 1**). Depending on the reporter gene used, available imaging modalities include positron emission tomography (PET), single photon emission computed tomography (SPECT), magnetic resonance imaging (MRI), or a charged-coupled device (CCD) camera. The two most widely used reporter gene imaging systems are firefly luciferase (Fluc)-based optical bioluminescence imaging and herpes simplex virus thymidine kinase (HSV-tk)-based PET imaging. These reporter constructs rely on the delivery of the exogenous reporter probes D-luciferin and 9-(4-[^{18}F]fluoro-3-hydroxymethyl)butylguanine ([^{18}F]FHBG), respectively, in order to generate signal in vivo.

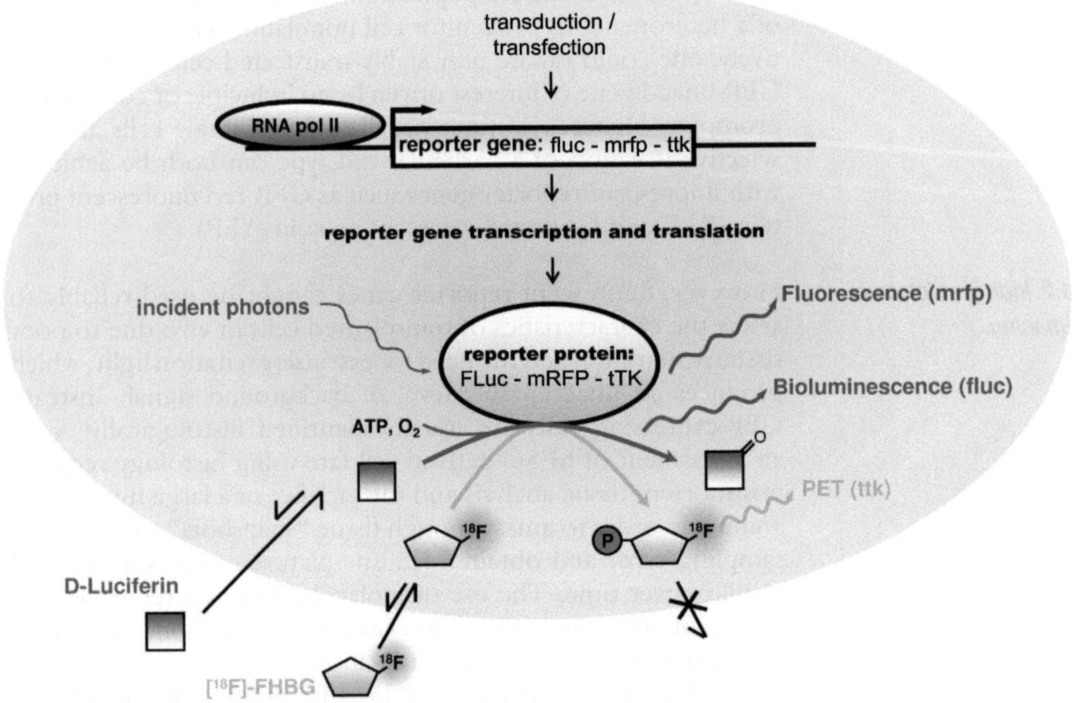

Fig. 1. The conceptual basis of reporter gene imaging. Transfected or transduced cells expressing the reporter genes will produce reporter proteins. Exogenously administered reporter probe can then interact with the reporter protein to produce a detectable signal.

1.2.1. Luciferase

Fluc encodes an enzyme responsible for conversion of exogenously administered D-luciferin to an optically active metabolite in an ATP-dependent, O_2-dependent fashion. Optical activity can be detected by the use of an ultrasensitive CCD camera. Although bioluminescence imaging is attractive because of its relatively low cost and high-throughput, it is difficult to use for single cell imaging and lacks fine spatial resolution.

1.2.2. Thymidine Kinase

HSV-tk and its variants, such as truncated thymidine kinase (HSV-ttk), serve as reporter constructs for PET imaging. (Note: tk refers to the gene while TK refers to the protein). HSV-TK phosphorylates nucleoside analog reporter probes, such as [^{18}F] FHBG, thereby trapping them within cells expressing HSV-tk *(13)*. [^{18}F]FHBG is a relatively poor substrate for mammalian thymidine kinase, thus producing a desirable low background signal from host cells not expressing HSV-tk. The [^{18}F] radioisotope emits high energy photons (511 keV) which are not attenuated within deep tissues, in contrast to the low energy photons (2–3 eV) of fluorescence or bioluminescence imaging. PET is also able to detect picomolar to nanomolar (10^{-11} to 10^{-12} mol/L) concentrations of radiolabeled reporter probes, making it several orders of magnitude more sensitive than the 10^{-3} to 10^{-5} mol/L scanning sensitivity of MRI. [^{18}F]FHBG has already been shown to be safe and pharmacokinetically stable in human volunteer subjects *(14, 15)*, making PET scanning in conjunction with HSV-tk and [^{18}F]FHBG most suitable for translation into clinical trials involving human subjects. The Food and Drug Administration (FDA) has already approved [^{18}F]FHBG as an investigational new drug by the Stanford group.

HSV-tk has the additional property of serving as a suicide gene upon administration of ganciclovir (GCV) at pharmacological doses *(16)*. The HSV-TK enzyme phosphorylates the GCV prodrug to produce a cytotoxic product that accumulates selectively within cells expressing the HSV-tk gene. This enzyme-prodrug strategy has already been utilized in cancer chemotherapy to increase target specificity. Transfer of the wild-type herpes simplex virus thymidine kinase (HSV-tk) gene to tumor cells followed by GCV treatment has been validated with both in vitro and in vivo applications *(17–19)*.

In the field of stem cell therapy, GCV treatment of animals injected with HSV-tk(+) stem cells is an effective strategy for teratoma ablation *(20, 21)* (**Fig. 2**). Since hESC transplantation involves complex biological processes that can only be studied in the context of the entire living subject, noninvasive longitudinal tracking and selective ablation of stem cell misbehavior (e.g., teratoma formation) will be an essential component of any future cell transplantation therapy.

1.3. Fusion Constructs

A strategy to combine the different imaging modalities in order to harness their respective strengths is the construction of a fusion gene composed of different domains containing the individual reporter genes. Provided that each individual reporter gene retains a moderate level of activity within the context of the fusion construct, it is possible to combine PET and optical imaging for applications in the same living subject.

With use of a multigene fusion constructs, the signal obtained may be decreased as compared to signal obtained with single gene constructs. However, a variety of improvements in the stability and activity of reporter gene enzymes and substrates has enabled the generation of fusion genes for multimodality imaging despite some modest decreases in activity compared to single gene constructs. For instance, a mutant version of HSV-tk (obtained via site-directed mutagenesis) that is more effective at reporter probe phosphorylation and results in increased PET signal has been reported *(22)*. Truncation of the first 135 bp (44 aa) of the already mutated gene removes the nuclear localization signal, causing increased cytoplasmic concentrations of enzyme and even more enhancement of PET signal. Similarly, Tisi et al. developed a thermostable mutant luciferase with a longer half-life at 37 °C as compared to wild-type luciferase *(23)*. Deletion of the peroxisome localization signal of the already mutated gene results in increased

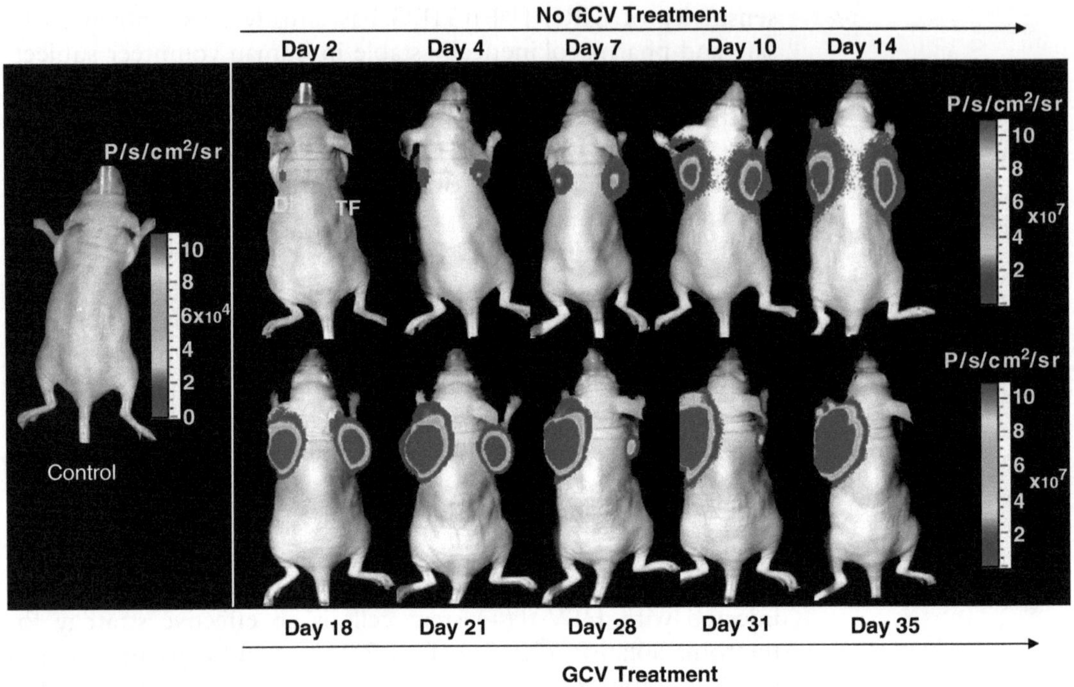

Fig. 2. Suicide gene ablation by ganciclovir (*GCV*) administration. Adult athymic nude mice were injected subcutaneously in each shoulder with 10^6 mouse embryonic stem cells (*ESCs*). Mouse ESCs injected in the right shoulder were stably transduced with the triple fusion *ttk*-containing reporter construct while mouse ESCs injected in the left shoulder were stably transduced with a *fluc-egfp* double fusion construct lacking *ttk*. After initiating 50 mg/kg of GCV treatment for 2 weeks, tumor size on the right gradually decreased while the tumor on the left did not, as seen by optical bioluminescence imaging using a cooled charged-coupled device (*CCD*) camera. Reproduced with permission from **ref.** *(21)*.

cytoplasmic enzyme concentrations and further improvement in bioluminescence signal *(24)*. These enhancements to molecular imaging techniques have allowed for the creation of libraries of fusion constructs that can be screened for activity by the appropriate biological assays, as shown in **Table 1**. We can see from **Table 1** that there are modest decreases in luciferase activity in multigene fusion constructs when compared to single gene constructs. Thus, screening for the least affected constructs may be necessary to avoid an unacceptable reduction in signal detection.

Recently, our group has found the triple fusion construct bearing firefly luciferase (fluc), monomeric red fluorescent protein (mrfp), and herpes simplex virus truncated thymidine kinase (ttk) to be a versatile construct for multimodality tracking of stem cell behavior and suicide gene ablation in small animals *(21, 25, 26)*. This triple fusion construct causes no significant adverse effects on mouse ESC viability, proliferation, differentiation, or proteomic expression *(27)*. Several conclusions regarding the design of fusion reporter constructs can be drawn from previous studies *(24, 26)*. For instance, thymidine kinase activity is greatest when the gene is

Table 1
Reporter enzyme activities in multigene fusion constructs compared to enzyme activity in single gene constructs

Constructs	% tTK activity	% wTK activity	% hRL activity	% FLuc activity	RFP/eGFP/mRFP (fluorescence activity by microscopy)
hrl–rfp–ttk	40		28		Medium
fluc–rfp–ttk	30			22	Low
hrl–egfp–ttk	50		33		High
fluc–egfp–ttk	43			20	High
hrl–mrfp–ttk	149		54		High
fluc–mrfp–ttk	100			53	Medium
hrl–rfp–wtk		76	45		Medium
fluc–rfp–wtk		62		63	Low
ttk	100				
wtk		100			
hrl			100		
fluc				100	
mrfp1					Very high

The *fluc-mrfp-ttk* fusion construct retains acceptable activity levels. *hrl* = Renilla luciferase, *fluc* = firefly luciferase, *rfp* = red fluorescent protein, *mrfp* = monomeric red fluorescent protein, *egfp* = enhanced green fluorescent protein, *wtk* = wild-type thymidine kinase, *ttk* = truncated thymidine kinase. Capital letters indicate proteins. Reproduced with permission from ref. 24.

placed at the 3′ end of a fusion construct, suggesting that the carboxy-terminal end of the enzyme is crucial for activity *(26, 28)*. In addition to using reporter gene mutations and orientations optimized for stability, cytoplasmic localization, and high enzymatic activity, one must ensure that the linker sequences joining each individual domain are stable and resistant to cleavage in vivo. Also, RFPs are preferable over GFPs for imaging experiments because red and near-infrared light penetrates tissues better *(29)*. Additionally, we advise against the usage of multimeric fluorescent proteins such as tetrameric RFP (DsRed2) in fusion constructs. Presumably, the obligate formation of the RFP tetramer imposes structural and functional limitations on the functioning of the luciferase and thymidine kinase components of the fusion protein.

1.4. Stable Transfection

Reporter genes can be incorporated into hESCs using either viral or nonviral vectors. Among nonviral methods, lipofection *(30)* and electroporation *(31)* have resulted in relatively low transfection efficiency, although nucleofection *(33)* has recently been shown to transfect hESCs well. Others have shown greater success using lentiviral vectors to obtain stable gene expression with minimal cytotoxicity in human and murine ESCs *(33)*. Subsequently, stably transduced cells can be isolated using FACS as mentioned above. Stably transduced cells have the advantage of containing the reporter constructs integrated within chromosomal DNA, and thereby passing reporter gene DNA to daughter cells. Because expression of the reporter gene product is required for signal generation, only viable parent and daughter cells will contribute detectable signal. In alternative visualization approaches such as radionuclide- or paramagnetic-labeling, signal intensity is diluted with each cell division. However, all cells derived from stably transduced hESCs will express molecular imaging constructs equally, thus allowing one to longitudinally monitor hESC survival and proliferation in vivo *(34)*.

Over time, stably integrated reporter genes may be subject to gene silencing by the endogenous chromosomal machinery. A reporter gene's susceptibility to gene silencing is closely related to the choice of promoter driving its expression. For instance, the cytomegalovirus promoter (pCMV) has been shown to produce particularly poor stable transgene expression in hESCs *(35)*. We have successfully used the human ubiquitin-C promoter (pUbiC) to drive expression of the triple fusion construct in multiple hESC cell lines and have observed no significant signal depreciation.

Lentivirally mediated transduction has the disadvantage of "random" integration specificity, which incurs a risk of insertional mutagenesis. In the future, site-specific integration of reporter genes into loci distanced from proto-oncogenes and less prone to gene silencing may be required *(36)*.

1.5. Multimodality Imaging

hESCs stably transfected or transduced with triple fusion constructs can be purified using FACS. Subsequently, bioluminescence can be used for small animal studies involving even a small number of hESCs. Ultimately, PET imaging can be used for tomographic imaging of living subjects including larger animals and humans to assist in translational studies and clinical implementation.

2. Materials

2.1. Viral Particle Development

1. 293T cell growth medium: minimal essential media (MEM) supplemented with 10% fetal bovine serum (FBS) and 1% penicillin (100 µg/mL)/streptomycin (292 µg/mL).
2. 2.5M $CaCl_2$ stock solution (sterilize through a 0.45-µm filter and store at –20 °C).

3. 2× BES-buffered saline: 2× BBS containing 50mM BES (pH 6.95), 280 mM NaCl, and 1.5 mM Na_2HPO_4. Sterile filter and store at −20 °C. The pH can be adjusted with HCl at room temperature.
4. HIV-1 packaging vector (pCMVΔR8.2) and vesicular stomatitis virus G glycoprotein-pseudotyped envelop vector (pMD.G) (gift from Dr. Sanjiv Gambhir, Stanford University, CA, *see* **Note 1**).
5. Ultracentrifuge.

2.2. hESC Maintenance and Culture

1. 0.1% gelatin solution: Add 0.5 g gelatin to 500 mL endotoxin-free water and autoclave before using. Do not use glass bottles that have seen detergent. Glass bottles should be cleaned with NaOH when first obtained and dedicated to sterile gelatin solution storage.
2. bFGF solution: Add 10 μg human recombinant bFGF (Invitrogen) to 5mL of 0.1% BSA in PBS.
3. 200mM L-glutamine+2-mercaptoethanol solution: Add 7 μL of 2-mercaptoethanol to 5 mL of 200 mM l-glutamine in fume hood. L-Glutamine aliquots should be kept frozen and thawed immediately before use.
4. Frozen mouse embryonic fibroblasts (MEFs).
5. MEF culture medium: Mix 450 mL of high glucose Dulbecco's Modified Eagle's Medium (DMEM), 50 mL of FBS that has been heat inactivated for 30min at 56 °C, and 5 mL of 100× Non-Essential Amino Acids Solution (Gibco® cat. no. 1140-050). Sterile filter and store at 4 °C.
6. hESC culture media: Mix 200 mL DMEM-F12 media with 50mL of Knockout Serum Replacer, 1.25 mL of 200 mM L-glutamine+2-mercaptoethanol solution, 2.5 mL of 100× Non-Essential Amino Acids Solution, and 1mL of bFGF solution (2 μg/mL).
7. 0.05% TrypsinEDTA solution (Gibco® cat. no. 25300).
8. Collagenase IV solution: Dissolve 30 mg Collagenase Type IV in 30 mL DMEM-F12 media. Sterile filter and store at 4 °C (to be used within one week).
9. Phosphate buffered saline (PBS).
10. Swinging bucket centrifuge (able to hold 15 mL conical tubes).
11. Sterile tissue culture hood.
12. Humidified incubator set at 37 °C and 5% CO_2.
13. Aspirator.
14. Baked Pasteur pipets.
15. 150 × 20 mm tissue culture dish (TPP® ca. no. 93150).

16. Frozen stock of MEF (can be derived from CF-1 strain timely pregnant mice or bought commercially).
17. Microscope.
18. Radiation source to provide 6,000 rads of exposure (such as the Mark I 137-Cesium irradiator).

2.3. Methods

2.3.1. Construction of Triple Fusion Reporter Gene

Construction of the triple reporter gene has been published. The focus of this chapter is on its applications, so detailed procedures are not provided here. However, the general strategy pursued in creating that construct is provided below to assist anyone wishing to generate similar materials.

Mutations altering the substrate specificity and enzymatic activity of thymidine kinase have been documented elsewhere *(37, 38)*. The creation of the HSV-sr39tk mutant which has been shown to be a more effective PET signaling construct involves the mutation of base pairs 784–816 from CTC ATC TTC GCC CTC to ATC TTC CTC TTC ATG (amino acid mutation from LIFAL to IFLFM) as documented *(22)*. The first 135 bp of the gene (44 aa of the protein) may be deleted by PCR to remove the nuclear localization signal and increase cytoplasmic localization (the resulting pCDNA 3.1-*sr39-truncated tk* was originally developed by Dr. D. Kaufman, University of California, Los Angeles, CA).

Mutations to enhance thermostability *(23)* and remove the peroxisome localization sequence *(24)* can be performed on commercially available luciferases such as pCDNA 3.1-*CMV-hrl* or *CMV-fluc* (Promega, Madison, WI). A PCR-amplified luciferase gene fragment can be inserted in frame with *ttk* into the pCDNA 3.1-*sr39-truncated tk* mentioned above. A 22-aa long spacer (LENSHASAGYQACGTAGPGSTG) separates these two reporter genes in the initial double fusion construct.

The development of *mrfp1* has been documented elsewhere *(39)*. The PCR-amplified *mrfp1* gene fragment (without stop codon) can be inserted in the middle of the spacer listed above (at the position of Cys-Gly) to generate a *flucmrfpttk* triple fusion construct. Specifically, the double fusion construct vector, which contains *fluc* and *ttk* separated by a 66-bp spacer within pCDNA 3.1+, is digested with *Hin*dIII and *Sac*II and ligated in frame to PCR-amplified, *Hin*dIII/*Sac*II-digested mrfp1 gene fragment (without stop codon) from pRSETB vector (Invitrogen) to generate the triple fusion *flucmrfpttk* construct *(26)*.

3.2. Production of Virus Carrying the Triple Fusion Report Gene

3.2.1. Construction of Lentiviral Vector

The triple fusion gene was originally located downstream of the cytomegalovirus promoter in pCDNA 3.1(+) *(26)*. This 3.3 kbp fragment was excised using *Bam*HI and *Not*I digestion and subsequently blunt-end ligated into the multiple cloning site of the lentiviral transfer vector FUG. FUG is a self-inactivating (SIN) lentiviral vector lacking viral promoter and enhancer sequences of the U3 region of the 3′ LTR, including the TATA box and binding sites for transcription factors, Sp1 and NF-κB *(40)*. These deletions prevent mobilization of replication-competent virus and allow Tat-independent reporter gene expression driven by the human ubiquitin-C promoter, without a concomitant loss in transduction efficiency *(41)*. A schema of the lentiviral vector is depicted in **Fig. 3a**.

3.2.2. 293T Cell Culture

1. Culture 293T (human embryonic kidney fibroblast) cells in growth medium.
2. Trypsinize and seed 293T cells at 5 × 10^5 cells per 10-cm plate.
3. Incubate overnight in 10 mL of growth medium.

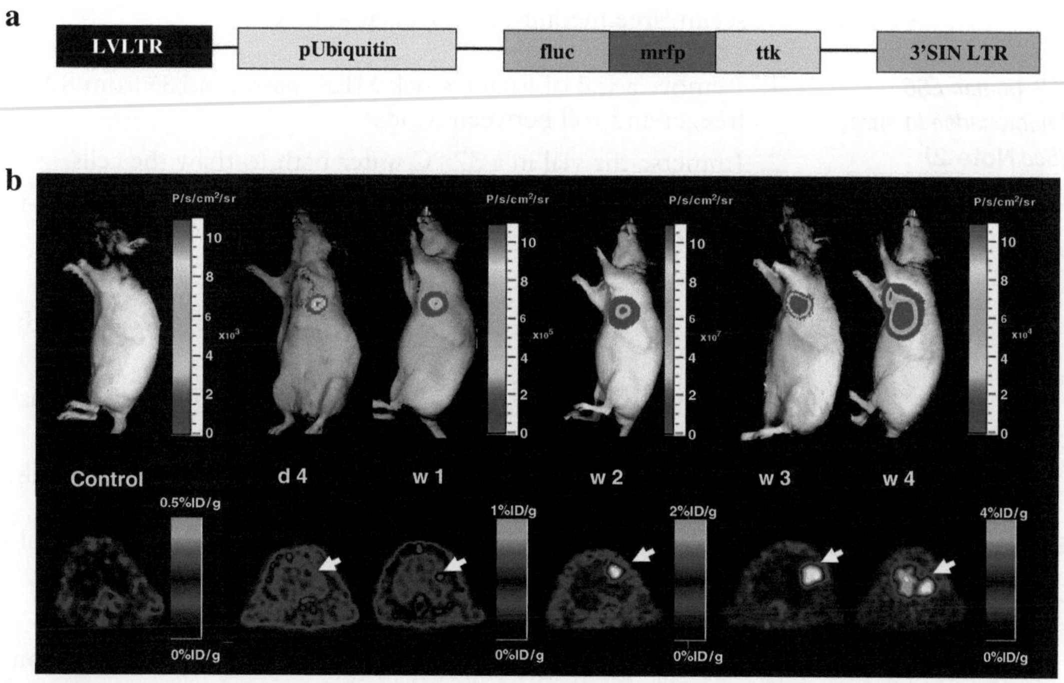

Fig. 3. a The triple fusion construct cassette containing firefly luciferase (*fluc*), monomeric red fluorescent protein (*mrfp*), and herpes simplex virus truncated thymidine kinase (*ttk*), as part of a lentiviral vector used for stable transduction of mouse and human embryonic stem cells (*ESCs*). b Tracking of mouse ESCs stably transduced with the triple fusion construct and injected into rat myocardium by bioluminescence imaging (*above*) and positron emission tomography (PET) imaging (*below*) over the course of 4 weeks. Reproduced with permission from **ref.** *(25)*.

3.2.3. 293T Cell Transfection

293T cells are transiently transfected using the standard calcium phosphate method *(42)*.

1. Mix 15μg pFUG-TF (containing the triple fusion construct), 10μg HIV-1 packaging vector (pCMVΔR8.2) and 5μg vesicular stomatitis virus G glycoprotein-pseudotyped envelop vector (pMD.G) *(41, 43)* with 0.5mL of 0.25M $CaCl_2$.
2. Add 0.5 mL 2× BBS.
3. Incubate at room temperature for 20–30 min.
4. Add the calcium phosphate–DNA solution to a plate of 293T cells dropwise and swirl gently to mix.
5. Incubate the cells for 15–24h in tissue culture incubator before aspirating media and refeeding cells with fresh medium.
6. About 48–72h after transfecting the cells, harvest the lentivirus-containing supernatant.
7. Centrifuge the supernatant at low speed (3,000 rpm for 5 min), and purify by passing it through a 0.45-μm filter.
8. Concentrate the lentivirus by sediment centrifugation of the medium with an SW29 rotor at 50,000 × *g* for 2h.
9. After centrifugation, dissolve the viral pellets in 100 μL of serum-free medium and store at –70 °C.

3.3. Human ESC Maintenance In Vitro (See Note 2*)*

3.3.1. MEF Feeder Layer

Culture of MEFs

1. Remove a vial of frozen stock MEF (passages 13) from 80 °C freezer and roll between hands.
2. Immerse the vial in a 37 °C water bath to thaw the cells.
3. Spray with 95% ethanol to disinfect and allow vial to air dry in sterile tissue culture hood.
4. Transfer the cells to a 15-mL conical tube.
5. Add 4mL of MEF culture media to the 15-mL conical tube and pipet up and down gently to mix.
6. Transfer the cell suspension to a 150 × 20-mm Petri dish (TPP® ca. no. 93150) that is not coated with gelatin.
7. Place the dish in an incubator and monitor the cell density daily.

Split MEF Cells

1. Aspirate the MEF culture media. The cell monolayer should remain attached to the plastic surface.
2. Wash the dish with 5mL of PBS. Aspirate the PBS.
3. Add 3mL 0.05% TrypsinEDTA solution. Let it sit at room temperature for 3–5min.
4. Dislodge the cells from the bottom of the dish by quickly moving the dish back and forth and washing the dish bottom by pipeting. Continue until you can see that the cell layer is dislodged.
5. Add 5mL of MEF culture media. The media contains serum and will inhibit further trypsin action.

6. Mix well to form a cell suspension. Pool all the cells into one 50-mL conical tube.
7. Add MEF media as needed (40-mL for one plate) for a total of ~50 mL and mix thoroughly.
8. Add 10mL of the resulting cell suspension to each of the five new 150 × 20-mm cell culture dishes.

Gelatin-Coated Plates

1. Autoclave a 0.1% gelatin solution for 30min. The gelatin will solubilize. Store this solution at room temperature.
2. Place 1mL of the gelatin solution into each well of a 6-well plate, or use 6mL gelatin solution for each 100 × 20-mm cell culture dish.
3. Place the plates in 37 °C incubator at least 2h, or overnight until needed.
4. Aspirate the remaining gelatin solution from each plate prior to plating MEF.

Trypsinize, Count, and Irradiate MEF Cells (*See* **Note 3**)

1. Aspirate the MEF media from the cell culture dishes (150 × 20mm).
2. Wash the cells with 5 mL PBS.
3. Add 3mL 0.05% TrypsinEDTA solution for 3-5min.
4. Dislodge cells from cell culture dish by shaking as before.
5. Add 5mL of MEF culture media to the dish.
6. Mix to form cell suspension. Pool all cells into one cell culture dish.
7. Remove 0.5–1 mL of the suspension and transfer it to a 15-mL conical tube.
8. Pipet cell suspension vigorously to break up cell aggregates.
9. Remove 20 μL cell suspension and count cells with a hemacytometer to determine the concentration of cells/mL in your sample.
10. Determine how many cells you will need to make new feeder plates. Normally, feeder plate density is 2×10^5 MEF cells/mL.
11. Remove the appropriate volume of cell suspension containing the above-calculated number of cells and transfer it into a 15-mL conical tube. Add PBS as necessary to obtain desired concentration of cells. If the MEFs are under passage 4, MEFs that are not to be irradiated can be plated into sterile tissue culture dishes.
12. Irradiate the cells. The rads of exposure needed to inactivate MEF cell batches may vary, but is usually between 5,000 and 8,000 rads. We typically irradiate MEF cells at 6,000 rads.
13. Centrifuge the irradiated cells for 5 min at 200 × *g*.

14. Remove the supernatant.
15. Resuspend the cells at 10^6 cells/mL in MEF culture media. Use the cell count calculations made prior to irradiation.

Plate MEF Feeder Layer onto Gelatin-Coated Plates

1. Further dilute the cell suspension of irradiated MEF to the required cell concentration.
2. Remove the gelatin-coated plates prepared in the section "Gelatin-Coated Plates" from the incubator and aspirate any excess gelatin.
3. Add the cell suspension dropwise to each well of the plate.
4. Place the plates into an incubator and allow cells to attach overnight.

3.3.2. Human ESCs

hESC Cultures

1. Remove a vial of hESCs from liquid nitrogen storage. Roll the vial between fingers to remove frost (*see* ***Note* 4**).
2. Immerse the vial in a 37 °C water bath to thaw cells.
3. Spray the vial with 95% ethanol to disinfect the surface.
4. Gently pipet the cells into a 15-mL conical tube.
5. Slowly add 4 mL of hESC culture media dropwise to the hESCs to avoid osmotic shock. (The suspension is diluted to reduce the DMSO concentration.) Gently shake the tube to mix cells.
6. Centrifuge the cells at 200 × g for 5min.
7. Remove and discard the DMSO-containing supernatant.
8. Gently resuspend the cell pellet in 2.5mL of hESC culture media.
9. Add 2.5 mL cell suspension dropwise into one well of a 6-well tissue culture plate containing a MEF feeder layer (prepared in the section "Plate MEF Feeder Layer onto Gelatin-Coated Plates," *see* **Note 5**).
10. Place the plate in a 37 °C incubator.
11. Refresh hESC culture media daily.

Split hESCs

1. Cells should be split when either the MEF feeder layer is 1week old, or hESC colonies become too large or too dense.
2. Aspirate the spent media from well.
3. Add 1 mL freshly made collagenase IV solution to each well of the 6-well plate and incubate at 37 °C for at least 5min. Confirm that colonies have partly separated from the plate by observing that the colony edges appear folded back under the microscope.
4. Remove the collagenase solution and replace with 1mL fresh hESC culture medium.

5. Use a cell scraper to scrape cells off the surface of the plate. Pipet the cell suspension up and down to wash the cells off of the surface of the plate.
6. Pool the suspension in a 15-mL conical tube.
7. Wash each well with 1 mL hESC media and transfer wash to 15-mL conical tube. Gently pipet to mix. Discard the used plate. Wait at room temperature for 5min and allow hESC clumps to settle to bottom of tube before gently aspirating supernatant. Old MEF cell contamination is contained in the supernatant.
8. Aspirate the supernatant.
9. Wash the cell pellet with 2–3 mL hESC media and gently reconstitute pellet.
10. Centrifuge again at 200 for 5 min.
11. While hESCs are centrifuging, aspirate MEF media from fresh feeder plates.
12. Add about 1mL of PBS to each well of the fresh 6-well feeder plate to wash away serum. Do not keep PBS on MEF for more than 5–10 min.
13. Aspirate the supernatant from the pelleted hESCs and again gently resuspend the pellet in 2–3 mL hESC media (*see* **Note 6**).
14. Dilute the cell suspension with a sufficient volume of hESC media in order to ensure 2.5 mL cell suspension per well (15mL cell suspension per 6-well plate). Mix well with a pipet.
15. Aspirate the PBS from wells of MEF feeder plate.
16. Add the cell suspension dropwise to each well of the plate.
17. Place the 6-well plate into the incubator. Move the plate in several quick, short back-and-forth and side-to-side motions to ensure an even distribution of cell colonies across each well.
18. Incubate the cells overnight to allow colonies to attach. Refresh hESC media daily.

3.4. Lentiviral Transduction

hESCs can be transduced 3-5 days after passage at a multiplicity of infection (MOI) of 10 (viral titer of ~10^7 incubated with 10^6 cells).

1. The thawed viral stock can be added directly to fresh hESC media.
2. Refresh the media 12 and 24 h later.
3. After 48 h, transduction efficiency can by qualitatively assessed using fluorescence microscopy. Subsequently, FACS can be used to isolate infected cells (*see* **Note 7**).

3.5. Imaging

3.5.1. Bioluminescence

1. For cell imaging, aspirate media-containing supernatant and add 1 mL PBS per well (of 6-well plates).
2. Add 10 μL of reporter probe D-luciferin (45 mg/mL) to the cells and visualize using an exposure time of 1 min (*see* **Note 8**).
3. For whole animal bioluminescence imaging, first perform an intraperitoneal injection of the reporter probe D-luciferin (375mg/kg body weight for heart or other deep organs penetration; 125 mg/kg body weight for subcutaneous penetration).
4. Image animals for 30 min using 1 min acquisition intervals.
5. Bioluminescence is quantified in units of photons per second per centimeter square per steridian ($p/s/cm^2/sr$). The reporter probe can be administered before each imaging session, allowing for multiple imaging acquisitions of the same animal over time (**Fig. 3; ref.** *(25)*).

3.5.2. PET Imaging (See **Note 9**)

1. Inject animals intravenously with 200 μCi of reporter probe [^{18}F]-FHBG (10μCi/g body weight). PET images can be acquired using the P4 Concorde MicroPET system (**Fig. 4**).
2. Reconstruct images from 60 to 75 min postinjection by filtered back projection algorithm as described *(44)*.
3. Regions of interest (ROI) may be drawn over the anterolateral wall of the heart in the horizontal view.
4. The counts/pixel/min are converted to counts/mL/min using a calibration constant obtained from scanning a cylindric phantom using ASIPro VM software.
5. The ROI counts/mL/min are converted to counts/g/min (assuming a tissue density of 1g/mL) and divided by the injected dose to obtain an image ROI-derived [^{18}F]FHBG percentage injected dose per gram of heart (% ID/g).

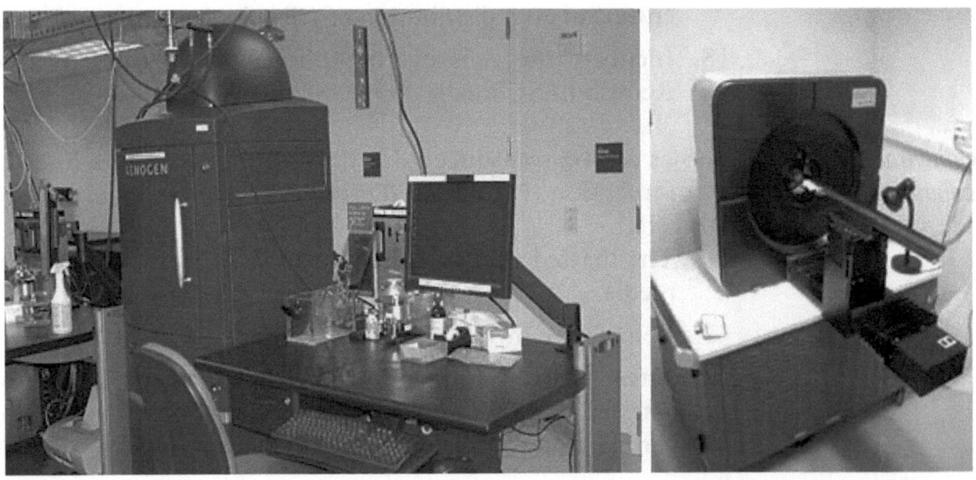

Fig. 4. IVIS 200 Imaging System (Xenogen Corporation, Alameda, CA, *left*) and P4 microPET scanner (Concorde Microsystems, Knoxville, TN, *right*) at Stanford's Center for In Vivo Imaging (SCI^3).

2.3.5.3. Fluorescence Microscopy

In addition to initially utilizing RFP/eGFP expression for cell sorting, fluorescent proteins may be visualized in frozen sections prepared from excised tissue postmortem (*see* **Note 10**).

Notes

1. Plasmids containing genes that produce the lentiviral packaging proteins (*gag/pol* and *rev*) and the VSV-G envelope protein are also commercially available as a ViraPower™ Packaging Mix (Invitrogen).
2. In our group, hESCs are mainly cultured on a MEF feeder layer according to protocols obtained from the WiCell Research Institute. To eliminate MEFs, hESCs can be cultured in a feeder-free in vitro system with the use of Matrigel® and MEF-conditioned medium *(45)*.
3. Here, we describe the inactivation of the MEF cells by irradiation. Alternatively, one may use mitomycin-C to inactivate MEFs *(46)*, although we find irradiation to be more reliable.
4. Be sure to wear eye protection as vials may explode after being kept in liquid nitrogen.
5. Confirm under the microscope that the MEF feeder layer has been plated at the appropriate confluency prior to plating hESCs. If by visual inspection it is apparent that MEF cells have been inadequately irradiated and are still proliferating, MEF cells may be re-irradiated while in the tissue culture dish (before hESCs have been plated), although this is not generally advised.
6. *Do not* attempt to break up cell clumps to a single cell suspension at this time. Ideally hESCs should remain as clumps of ~100 cells that are easily visualized by the unaided eye by holding the 15-mL conical tube against a light source.
7. To determine the infectivity after 48h, we detect eGFP/RFP expression as analyzed on FACScan (BD Bioscience, San Jose, CA). After dissociating hESCs into a single-cell suspension, the RFP-positive cell populations can be isolated by FACS (Becton Dickinson Immunocytometry Systems, San Jose, CA) and subsequently plated on feeder layer cells for long-term culturing (to select stably transduced cells, for example). Flow cytometry data can be analyzed with FlowJo analysis software (Treestar, San Carlos, CA).
8. For cell and animal biolumineńsence imaging, we use the Xenogen In Vivo Imaging System (IVIS) (**Fig. 4**) as described *(47)*.

9. The PET reporter gene (HSV-ttk) can also serve as a suicide gene when GCV is administered in pharmacologic dosages *(16)*. In the field of stem cell therapy, we have previously shown that mouse ESCs (expressing HSV-ttk reporter gene) injected into the heart lead to both intracardiac and extracardiac teratomas *(25)*. In the event of such cellular misbehavior, the administered HSV-ttk-expressing cells can be ablated by intraperitoneal injection of GCV twice daily at 50mg/kg for 2 weeks (**Fig. 2; ref.** *(21)*).
10. A variety of tissue preparation techniques may be used. Excised organs may be immersion fixed in 4% paraformaldehyde (Sigma) in PBS at pH 7.4 for 1-2h and then immersed in 30% sucrose overnight, embedded in OCT, frozen, and prepared into 10-μm thick frozen sections with a 5030 series Microtome (Bright Instruments, Huntingdon, England). Endogenous GFP signals may be amplified for immunohistochemical analysis with Alexa Fluor 488-conjugated anti-GFP polyclonal antibody (Molecular Probes).

References

1. Cao, F. et al. Transcriptional and functional profiling of human embryonic stem cell-derived cardiomyocytes. *PLoS ONE* 3(10), e3474 (2008).
2. Baharvand, H. et al. Neural differentiation from human embryonic stem cells in a defined adherent culture condition. *Int J Dev Biol* 51, 371–8 (2007).
3. Schulz, T.C. et al. Directed neuronal differentiation of human embryonic stem cells. *BMC Neurosci* 4, 27 (2003).
4. Jiang, J. et al. Generation of insulin-producing Islet-like clusters from human embryonic stem cells. *Stem Cells* 17, 17 (2007).
5. Swijnenburg, R.J. van der Bogt, K.E.A., Sheikh, A.Y., Cao, F. & Wu, J.C. Clinical hurdles for the transplantation of cardiomyocytes derived from human embryonic stem cells: role of molecular imaging. *Curr Opin Biotechnol* 18, 38–45 (2007).
6. Wollert, K.C. et al. Intracoronary autologous bone-marrow cell transfer after myocardial infarction: the BOOST randomised controlled clinical trial. *Lancet* 364, 141–8 (2004).
7. Reubinoff, B.E., Pera, M.F., Fong, C.Y., Trounson, A. & Bongso, A. Embryonic stem cell lines from human blastocysts: somatic differentiation in vitro. *Nat Biotechnol* 18, 399–404 (2000).
8. Thomson, J.A. et al. Embryonic stem cell lines derived from human blastocysts. *Science* 282, 1145–7 (1998).
9. Sidhu, K.S. & Tuch, B.E. Derivation of three clones from human embryonic stem cell lines by FACS sorting and their characterization. *Stem Cells Dev* 15, 61–9 (2006).
10. Nicholas, C.R., Gaur, M., Wang, S., Pera, R.A. & Leavitt, A.D. A method for single-cell sorting and expansion of genetically modified human embryonic stem cells. *Stem Cells Dev* 16, 109–17 (2007).
11. Franz, W.M., Rothmann, T., Frey, N. & Katus, H.A. Analysis of tissue-specific gene delivery by recombinant adenoviruses containing cardiac-specific promoters. *Cardiovasc Res* 35, 560–6 (1997).
12. Huber, I. et al. Identification and selection of cardiomyocytes during human embryonic stem cell differentiation. *FASEB J* 13, 13 (2007).
13. Yaghoubi, S.S. & Gambhir, S.S. PET imaging of herpes simplex virus type 1 thymidine kinase (HSV1-tk) or mutant HSV1-sr39tk reporter gene expression in mice and humans using [18F]FHBG. *Nat Protoc* 1, 3069–75 (2006).
14. Yaghoubi, S. et al. Human pharmacokinetic and dosimetry studies of [(18)F]FHBG: a reporter probe for imaging herpes simplex virus type 1 thymidine kinase reporter gene expression. *J Nucl Med* 42, 1225–34 (2001).
15. Yaghoubi, S.S. et al. Preclinical safety evaluation of 18F-FHBG: a PET reporter probe for imaging herpes simplex virus type 1 thymidine kinase (HSV1-tk) or mutant HSV1-sr39tk's expression. *J Nucl Med* 47, 706–15 (2006).

16. Phelps, M.E. Inaugural article: positron emission tomography provides molecular imaging of biological processes. *Proc Natl Acad Sci USA* 97, 9226–33 (2000).
17. Moolten, F.L. Tumor chemosensitivity conferred by inserted herpes thymidine kinase genes: paradigm for a prospective cancer control strategy. *Cancer Res* 46, 5276–81 (1986).
18. Moolten, F.L., Wells, J.M., Heyman, R.A. & Evans, R.M. Lymphoma regression induced by ganciclovir in mice bearing a herpes thymidine kinase transgene. *Hum Gene Ther* 1, 125–34 (1990).
19. Vile, R.G. & Hart, I.R. Use of tissue-specific expression of the herpes simplex virus thymidine kinase gene to inhibit growth of established murine melanomas following direct intratumoral injection of DNA. *Cancer Res* 53, 3860–4 (1993).
20. Schuldiner, M., Itskovitz-Eldor, J. & Benvenisty, N. Selective ablation of human embryonic stem cells expressing a "suicide" gene. *Stem Cells* 21, 257–65 (2003).
21. Cao, F. et al. Molecular imaging of embryonic stem cell misbehavior and suicide gene ablation. *Cloning Stem Cells* 9, 107–17 (2007).
22. Gambhir, S.S. et al. A mutant herpes simplex virus type 1 thymidine kinase reporter gene shows improved sensitivity for imaging reporter gene expression with positron emission tomography. *Proc Natl Acad Sci USA* 97, 2785–90 (2000).
23. Tisi, L.C. et al. Development of a thermostable firefly luciferase. *Anal Chim Acta* 457, 115–23 (2002).
24. Ray, P., Tsien, R. & Gambhir, S.S. Construction and validation of improved triple fusion reporter gene vectors for molecular imaging of living subjects. *Cancer Res* 67, 3085–93 (2007).
25. Cao, F. et al. In vivo visualization of embryonic stem cell survival, proliferation, and migration after cardiac delivery. *Circulation* 113, 1005–1014 (2006).
26. Ray, P., De, A., Min, J.J., Tsien, R.Y. & Gambhir, S.S. Imaging tri-fusion multimodality reporter gene expression in living subjects. *Cancer Res* 64, 1323–30 (2004).
27. Wu, J.C. et al. Proteomic analysis of reporter genes for molecular imaging of transplanted embryonic stem cells. *Proteomics* 6, 6234–49 (2006).
28. Ray, P., Wu, A.M. & Gambhir, S.S. Optical bioluminescence and positron emission tomography imaging of a novel fusion reporter gene in tumor xenografts of living mice. *Cancer Res* 63, 1160–5 (2003).
29. Massoud, T.F. & Gambhir, S.S. Molecular imaging in living subjects: seeing fundamental biological processes in a new light. *Genes Dev* 17, 545–80 (2003).
30. Eiges, R. et al. Establishment of human embryonic stem cell-transfected clones carrying a marker for undifferentiated cells. *Curr Biol* 11, 514–8 (2001).
31. Mohr, J.C., de Pablo, J.J. & Palecek, S.P. Electroporation of human embryonic stem cells: Small and macromolecule loading and DNA transfection. *Biotechnol Prog* 22, 825–34 (2006).
32. Siemen, H. et al. Nucleofection of human embryonic stem cells. *Stem Cells Dev* 14, 378–83 (2005).
33. Ma, Y., Ramezani, A., Lewis, R., Hawley, R.G. & Thomson, J.A. High-level sustained transgene expression in human embryonic stem cells using lentiviral vectors. *Stem Cells* 21, 111–7 (2003).
34. Li, Z. et al. Comparison of reporter gene and iron particle labeling for tracking fate of human embryonic stem cells and differentiated endothelial cells in living subjects. Stem Cells 26(4), 864–73 (2008).
35. Liew, C.G., Draper, J.S., Walsh, J., Moore, H. & Andrews, P.W. Transient and stable transgene expression in human embryonic stem cells. *Stem Cells* 25, 1521–8 (2007).
36. Groth, A.C., Olivares, E.C., Thyagarajan, B. & Calos, M.P. A phage integrase directs efficient site-specific integration in human cells. *Proc Natl Acad Sci USA* 97, 5995–6000 (2000).
37. Black, M.E., Newcomb, T.G., Wilson, H.M. & Loeb, L.A. Creation of drug-specific herpes simplex virus type 1 thymidine kinase mutants for gene therapy. *Proc Natl Acad Sci USA* 93, 3525–9 (1996).
38. Kokoris, M.S., Sabo, P. & Black, M.E. In vitro evaluation of mutant HSV-1 thymidine kinases for suicide gene therapy. *Anticancer Res* 20, 959–63 (2000).
39. Campbell, R.E. et al. A monomeric red fluorescent protein. *Proc Natl Acad Sci USA* 99, 7877–82 (2002).
40. Miyoshi, H., Blomer, U., Takahashi, M., Gage, F.H. & Verma, I.M. Development of a self-inactivating lentivirus vector. *J Virol* 72, 8150–7 (1998).
41. De, A., Lewis, X.Z. & Gambhir, S.S. Noninvasive imaging of lentiviral-mediated reporter gene expression in living mice. *Mol Ther* 7, 681–91 (2003).

42. Chen, C. & Okayama, H. High-efficiency transformation of mammalian cells by plasmid DNA. *Mol Cell Biol* 7, 2745–52 (1987).
43. Naldini, L., Blomer, U., Gage, F.H., Trono, D. & Verma, I.M. Efficient transfer, integration, and sustained long-term expression of the transgene in adult rat brains injected with a lentiviral vector. *Proc Natl Acad Sci USA* 93, 11382–8 (1996).
44. Wu, J.C. et al. Molecular imaging of the kinetics of vascular endothelial growth factor gene expression in ischemic myocardium. *Circulation* 110, 685–91 (2004).
45. Xu, C., Inokuma, M.S., Denham, J., Golds, K., Kundu, P., Gold, J.D., Carpenter, M.K. Feeder-free growth of undifferentiated human embryonic stem cells. *Nat Biotechnol* 19, 971–4 (2001).
46. Eiges, R. Genetic manipulation of human embryonic stem cells by transfection. *Methods Mol Biol* 331, 221–39 (2006).
47. Wu, J.C., Inubushi, M., Sundaresan, G., Schelbert, H.R. & Gambhir, S.S. Optical imaging of cardiac reporter gene expression in living rats. *Circulation* 105, 1631–4 (2002).

Chapter 3

Use of a GFP-PML-Expressing Cell Line as a Biosensor for Human Cytomegalovirus Infection

Tomonori Ueno and Kiyoko Ogawa-Goto

Summary

Human cytomegalovirus (HCMV) infection has a marked effect on promyelocytic leukemia (PML) bodies. Here, we describe a novel real-time monitoring system for HCMV-infected cells *in vitro* using a newly established cell line that stably expresses GFP-PML protein. Upon infection, HCMV causes specific dispersion of GFP-PML bodies, thereby allowing the infected cells to be monitored by fluorescence microscopy without immunostaining. Quantitative protocols using either an NPB fluorescence assay or a GFP-PML imaging assay are also described. The NPB fluorescence assay is rapid, sensitive, and sufficiently simple for screening of inhibitory reagents, while the GFP-PML imaging assay is highly sensitive and applicable to drug susceptibility testing of low-titer clinical isolates.

Key words: PML body, GFP-PML, SE/15 cell, HCMV, Clinical isolate, Drug susceptibility testing.

1. Introduction

Promyelocytic leukemia (PML) bodies, also known as ND10 (for "nuclear domain 10"), PODs (for "PML oncogenic domains"), and Kr bodies, are present in the nuclei of most mammalian cells as 10–20 spherical structures ranging in size from 0.3 to 0.5 μm *(1)*. They are composed of multiprotein complexes, including the PML, Sp100, NDP55, and Daxx proteins *(2)*. Infection with DNA viruses, such as herpes simplex virus type 1 (HSV-1), human cytomegalovirus (HCMV), and adenoviruses, has a marked effect on PML bodies. HCMV deposits its genome adjacent to PML bodies and starts transcription of immediate early (IE) gene, thereby causing dispersion of the PML bodies in the

Barry W. Hicks (ed.), *Methods in Molecular Biology, Viral Applications of Green Fluorescent Protein, vol. 515*

DOI: 10.1007/978-1-59745-559-6_3

nucleus through de-SUMOylation of the PML protein *(3)*. We exploited a unique event observed in HCMV-infected cells, that is, the dispersion of PML bodies upon IE1 expression. For this purpose, we established a new cell line (SE/15) that stably expresses green fluorescent protein (GFP)-PML protein *(4)*. SE/15 cells have functional GFP-PML bodies, whose morphology is specifically changed to a diffuse GFP pattern throughout the nucleus upon HCMV infection (**Fig.1**). Thus, visual assessment by fluorescence microscopy is sufficient for sensitive detection of infected cells without the necessity for any further procedures, such as fixation and staining. Subsequently, we developed a new quantitative system for HCMV titration using this cell line, which is applicable to the screening of antiviral drugs. In addition, we established a simple, sensitive, and quantitative protocol for drug susceptibility testing of low-titer clinical isolates, most of which are cell associated.

2. Materials

2.1. Cultivation of SE/15 Cells (see Note 1*)*

1. SE/15 cells are available from Nippi Inc. (*see* **Notes 2** and **3**).
2. G418 (GIBCO/BRL) stock solution.
3. Zeocin (Invitrogen): 100 mg/mL stock solution, stored at –20 °C, light sensitive.
4. Growth medium for SE/15 cells: Dulbecco's modified Eagle's medium (DMEM, GIBCO/BRL) containing 10% fetal bovine serum (FBS), 100 μg/mL G418, and 100 μg/mL zeocin.
5. Phosphate-buffered saline (PBS).
6. 0.25% trypsin and 0.02% EDTA in PBS.

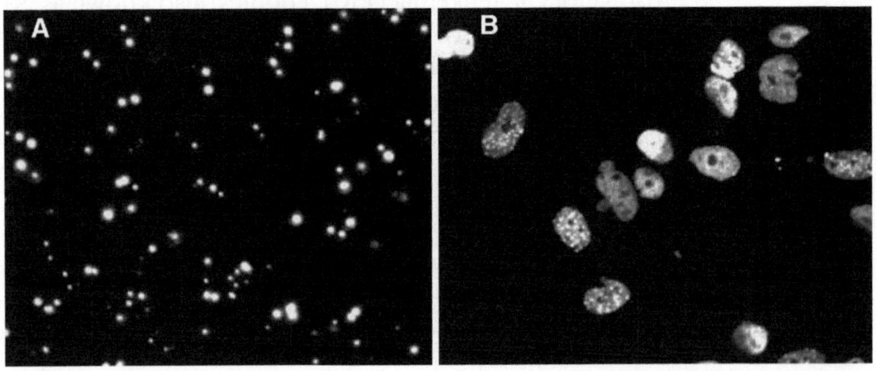

Fig. 1. Fluorescence microscopic images of SE/15 cells showing aberrant localization of GFP-PML upon HCMV infection. SE/15 cells were mock-infected (**A**) or infected with Towne (MOI 3) (**B**). GFP fluorescence micrographs at 18 h p.i. are shown.

2.2. Inoculation of SE/15 Cells

1. Human fibronectin (GIBCO/BRL) solution for coating of dishes. Stock and working concentrations are 1 mg/mL and 10 μg/mL, respectively (*see* **Note 4**).
2. Maintenance medium for HCMV-infected SE/15 cells: F12/DMEM (GIBCO/BRL) containing 5% FBS lacking phenol red. DMEM can be used instead.
3. HCMV solution: HCMV laboratory strains Towne (ATCC, #VR-977) and AD169 (ATCC, #VR-538) are propagated and titrated in HEL cells (or MRC-5 cells, ATCC, #CCL-171) using a standard protocol (*see* **Note 5**).

2.3. Quantification of HCMV-Infected SE/15 Cells by an NPB Fluorescence Assay

1. NPB elution buffer (NPB): 50 mM Tris-HCl, pH 7.4 containing 0.15% NP-40, 150 mM NaCl, 15 mM $MgCl_2$, and 5 mM EDTA.
2. uClear Fluorescence Black P 96-well plates with transparent bottoms (Greiner, #655090).

2.4. Heparin Inhibition Assay

1. Heparin: Stock solution of 1 mg/mL in distilled water. Store at −20 °C.

2.5. Quantification of HCMV-Infected SE/15 Cells by a GFP-PML Imaging Assay

1. Glass Bottom Fluorescence Black Microplates (96-wells, Greiner, #655896).
2. F12/DMEM (GIBCO/BRL) containing 10% FBS.
3. Computer software package for imaging assays: Image J (http://rsb.info.nih.gov/ij).

2.6. Susceptibility Testing of HCMV Clinical Isolates by a GFP-PML Imaging Assay Coupled with Co-Culturing

1. Human diploid fibroblasts (e.g., MRC-5 or HEL fibroblasts).
2. Maintenance medium for HCMV-infected HEL cells: DMEM (GIBCO/BRL) containing 5% FBS, streptomycin, and penicillin.
3. Disposable cell counter: C-chip (DHC-N01, IN CYTO).
4. HCMV clinical isolates propagated in HEL cells.
5. Antiviral stock solutions: 5 mM ganciclovir (GCV) and 100 mM foscarnet (FOS).

3. Methods

Chinese hamster ovary (CHO) cells have been reported to be as sensitive to HCMV as human fibroblasts regarding the initial entry and IE1 expression *(5)*. Since we previously found that HCMV infection frequently causes detachment of CHO cells and that this detachment is efficiently avoided by human p180 overexpression, we generated new cell lines coexpressing GFP-PML and the human host-cell factor p180 *(6)*. Since transfectants containing abnormally large GFP-PML structures appeared to exhibit difficulties in maintaining the structures in a stable manner, we carried out limited dilution to carefully select a suitable clone with respect to the stability and morphogenesis of its GFP-PML bodies, and finally cloned a cell line designated SE/15.

SE/15 cells have functional GFP-PML bodies whose morphology is specifically changed to a diffuse GFP pattern throughout the nucleus upon HCMV infection (**Fig. 1**). The unique appearance of nuclei containing dispersed GFP-PML is never observed in mock-infected cells, begins at postinfection (p.i.) 3–4 h, and continues to increase until postinfection 10–16 h. Thus, visual assessment by fluorescence microscopy is sufficient for sensitive detection of infected cells without the necessity for any further procedures, such as fixation and staining.

3.1. Cultivation of SE/15 Cells

1. For maintenance, culture SE/15 cells in DMEM/10% FBS containing 100 μg/mL G418 and 100 μg/mL zeocin. Cells up to the 25th passage can be used for HCMV quantification assays (*see* **Note 6**).
2. Pass SE/15 cells approaching confluence with trypsin/EDTA to provide new maintenance cultures on 60- or 100-mm dishes using split ratios of 3:1 or 4:1. For assay cultures, prepare fibronectin-coated plates (*see* **Subheading 3.2**).

3.2. Inoculation of SE/15 Cells

1. Add fibronectin solution (10 μg/mL) to the individual wells of tissue culture plates.
2. Incubate the plates overnight at 4 °C. Volumes of 1, 0.5, and 0.1 mL of the solution are sufficient for 12-, 24-, and 96-well culture plates, respectively.
3. Seed SE/15 cells approaching confluence onto fibronectin-coated plates at cell densities of 2×10^5, 1×10^5, and 0.2×10^5 cells/well for 12-, 24-, and 96-well culture plates, respectively (*see* **Note 7**).
4. On the day of the virus inoculation, remove the culture growth medium by aspiration (*see* **Note 8**), and immediately add a small volume (0.05–0.2 mL) of viral stock solution carefully to each well to give an appropriate multiplicity of infection (MOI). Gently rock the plate to achieve even distribution of the virus.
5. To allow adsorption, incubate the cells for 1 h in a CO_2 incubator. The cells may need to be gently tilted every 10 min to avoid cell drying.

6. Remove unabsorbed virus by aspiration and carefully wash three times with prewarmed F12/DMEM/5% FBS (*see* **Note 9**).
7. Add an appropriate volume of F12/DMEM/5% FBS containing no antibiotics to each well and incubate in a CO_2 incubator at 37 °C (*see* **Note 10**).
8. An example of the diffuse GFP-PML in Towne-infected SE/15 cells at 18 h p.i. is shown in **Fig. 1**.

3.3. Quantification of HCMV-Infected SE/15 Cells by an NPB Fluorescence Assay

GFP-PML is present as an insoluble form in normal nuclei *(7)*. In contrast, the diffuse PML in infected nuclei becomes soluble in NP-40-containing buffer (NPB). Therefore, NPB extraction-based fluorescence measurements in SE/15 cells allow quantification of HCMV-infected cells.

1. On the day of the experiment, prepare the following materials: precooled (on ice) NPB buffer, a labeled microcentrifuge tube for each sample, ice container for incubation, PBS.
2. At appropriate times p.i., gently wash HCMV-infected and mock-infected SE/15 cell cultures twice with PBS.
3. Aspirate the residual solution from the cell cultures and place the plates on ice. Add 0.250 mL of ice cold NPB to each well of the 12-well plate (*see* **Note 11**).
4. Incubate the cells for exactly 15 min at 4 °C. Carefully transfer the NPB solutions to 1.5-mL labeled microcentrifuge tubes (*see* **Note 12**), and clarify by centrifugation at 3,000 rpm for 5 min at 4 °C.
5. Transfer the supernatants to a black fluorescence P 96-well plate.
6. Prepare total cell lysates in a parallel experiment. After washing with PBS, scrape the material into appropriately labeled tubes. Add 0.250 mL of prechilled NPB to the 12-well plate samples. Incubate for 15 min on ice and transfer to a black fluorescence P 96-well plate.
7. Measure the NPB fluorescence intensity in a multiwell fluorescence plate reader (e.g., CytoFluor series 4000, Perseptive Biosystems) using an excitation wavelength of 485 nm and an emission wavelength of 510 nm.
8. The data sets of NPB fluorescence intensity required for each sample are NPB extracts of infected cells, mock-infected cells, and total cell lysates.
9. Calculate the net fluorescence intensity of the infected cells as follows: subtract the NPB fluorescence intensity of the mock-infected cells from that of the infected cells.
10. Express the HCMV infectivity as a ratio of the net fluorescence intensity of the NPB extracts to the fluorescence intensity of the total cell lysates.

3.4. Heparin Inhibition Assay Using an NPB Fluorescence Assay

1. Seed SE/15 cells as described in **Subheading 3.2**.
2. On the day of the experiment, prepare the following materials: HCMV (Towne strain) solution which has been thawed and kept on ice, serially diluted solutions of heparin in PBS.
3. Mix HCMV solutions with serial dilutions of heparin to give appropriate concentrations of the virus and heparin.
4. Incubate the mixed virus-heparin solutions at 37 °C for 1 h.
5. Remove the growth medium from the cultures and add the pretreated virus solutions to each well as described in **Subheading 3.2**.
6. Incubate plates in a CO_2 incubator at 37 °C for 18–24 h.
7. Quantify the Towne-infected SE/15 cells by the NPB fluorescence assay as described in **Subheading 3.3**.
8. An example of an inhibition curve for Towne-infected SE/15 cells at 18 h p.i. is shown in **Fig. 2**.

3.5. Quantification of HCMV-Infected SE/15 Cells by a GFP-PML Imaging Assay

Since GFP-PML is dispersed throughout the nuclei of HCMV-infected SE/15 cells and the GFP-PML areas are markedly larger than the GFP-PML dots in uninfected cells (see **Fig. 1**), imaging analysis of fluorescence micrographs enables direct counting of the number of infected nuclei. This protocol allows highly sensitive quantification even in miniscale cultures, and permits assaying low titer clinical isolates (*see* **Note 13**).

As shown in **Fig. 3**, a typical profiling pattern by particle analysis of mock-infected cells shows one peak representing the small GFP-PML foci (**Fig. 3**, left). A second broad peak appears in the profiling pattern of HCMV-infected cells, corresponding to the sizes of nuclei bearing diffuse GFP signals (**Fig. 3**, right). Therefore, HCMV infectivity is expressed as the ratio of the

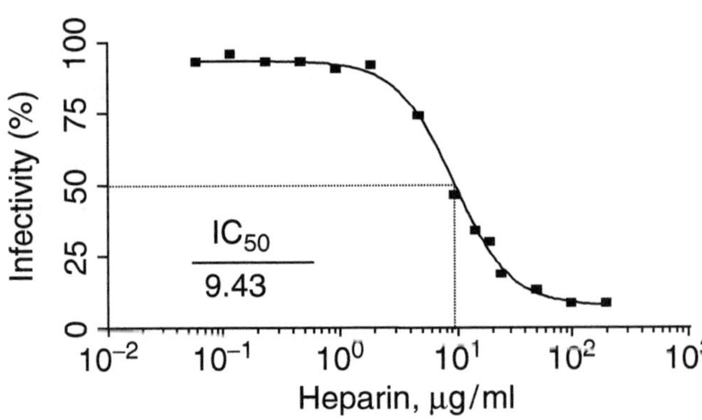

Fig. 2. Heparin inhibition assay using an NPB fluorescence assay. HCMV was preincubated for 1 h at 37 °C with various concentrations of heparin, an entry process inhibitor of the virion into the cells. SE/15 cells were infected (MOI 1) in a 24-well plate for 18 h, followed by the standard NPB fluorescence assay as described in the text. (Reproduced from *(4)* with permission from American Society for Microbiology.).

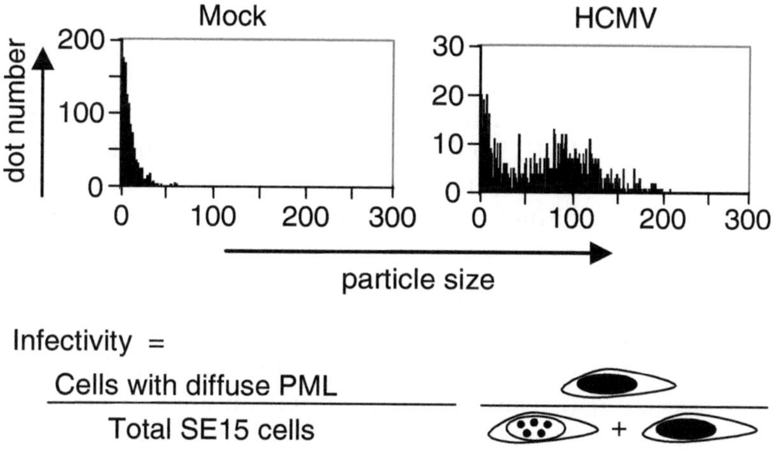

Fig. 3. Profile patterns of the particle sizes of mock-infected and Towne-infected cells. GFP fluorescence images of mock-infected (*left*) and Towne-infected (*right*) cells were captured at 18 h p.i., and analyzed according to the procedure described in the text. (Reproduced from **ref.** *(4)* with permission from American Society for Microbiology).

number of nuclei possessing diffuse GFP signals to the total number of SE/15 cell nuclei.

1. Prepare HCMV-infected and mock-infected SE/15 cells as described above. Growth medium containing no phenol red, G418, or zeocin is highly recommended (*see* **Note 14**).
2. At appropriate times p.i., capture GFP fluorescence images as TIF files using a fluorescence microscope equipped with a cooled charge-coupled device (CCD) camera and a computer (e.g., Olympus model IX71 fluorescence microscope system equipped with an Olympus DP50 CCD camera) (*see* **Note 15**). A 20× objective lens and a fixed data collection time of 0.2 s are used for the conventional assay.
3. When the captured images are not 8-bit gray images, convert them to 8-bit gray images using an image processing software package (e.g., Adobe Photoshop, ImageJ) (*see* **Note 16**).
4. Evaluate the numbers of cells displaying GFP-PML bodies and diffuse GFP-PML using particle counting in ImageJ according to the following procedures.
5. Using the pull-down menu "File" – "Open," open the 8-bit gray images of mock-infected and HCMV-infected cells.
6. Using the pull-down menu "Image" – "Adjust" – "Threshold," set the lower and upper threshold values. When the "apply" button is clicked, the image will be converted into a binary image on the desktop (*see* **Note 17**).
7. Evaluate the total number of mock-infected cells by particle counting as follows.
8. Using the pull-down menu "Analyze" – "Analyze particles," select "outlines" in the show column and 0.00–1.00 in the circularity column.

9. Set the minimum and maximum sizes (pixels) for GFP-PML bodies, where "minimum" refers to the smallest pixels of GFP-PML bodies in mock-infected cells that can segregate electric noise from GFP-PML and "maximum" for GFP-PML bodies is the threshold of diffuse PML, namely the smallest nucleus that can segregate GFP-PML bodies from diffuse PML (*see* **Note 18**).
10. Check "Display results" and "Summarize" and press "OK." The total number of GFP–PML dots is shown in a table of results.
11. Estimate the total number of cells bearing GFP-PML bodies (A) by using the following formula (I): Total number of cells = 0.632 × total number of GFP-PML dots (*see* **Note 19**).
12. Evaluate the cell numbers of infected and noninfected cells using binary images of infected cells as follows.
13. Using the pull-down menu "Analyze" – "Analyze particles," set the "minimum" and "maximum" pixel numbers of diffuse PML using HCMV-infected cells (*see* **Note 20**).
14. "Minimum" is the same value as the maximum pixel number for GFP-PML bodies, while "maximum" is nearly equal to the "upper threshold."
15. Check "Display results" and "Summarize" and press "OK." The number of cells bearing diffuse PML (*B*) is shown.
16. Evaluate the number of cells bearing GFP-PML bodies (C) as described in **Points 7–11**.
17. Evaluate the HCMV infectivity (%) using $B/(B + C) \times 100$ (*see* **Note 21**).

3.6. Susceptibility Testing of HCMV Clinical Isolates by a GFP-PML Imaging Assay Coupled with Co-Culturing

Since SE/15 cells do not support a full replication cycle, co-culturing of SE/15 cells with infected HEL cells is required for susceptibility testing. A schematic representation of the protocol for susceptibility testing is shown in **Fig. 4**.

1. Coat glass-bottom black fluorescence 96-well microplates with fibronectin as described in **Subheading 3.2**
2. Culture HCMV clinical isolates in HEL cells in 24-well plates (1×10^5 cells/well).
3. On the day of the experiment, make serial dilutions of antiviral drugs such as GCV and FOS at tenfold higher concentrations than the final required concentrations.
4. Add antiviral drugs (GCV, FOS, etc.) to the wells, and culture for 48–72 h. Final concentrations of 0.1–100 μM GCV and 1–1,000 μM FOS are recommended.
5. On the day of the co-culturing, harvest the HCMV-infected HEL cells and SE/15 cells with trypsin/EDTA and count the cell numbers.

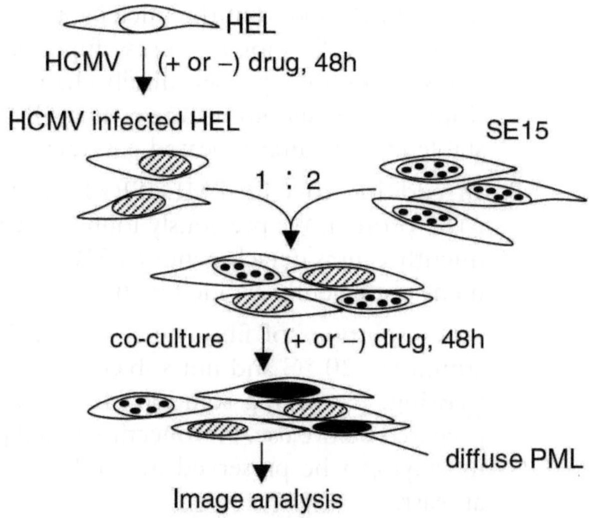

Fig. 4. Schematic representation of the protocol for drug susceptibility testing of clinical isolates using the co-culture system. (Reproduced from **ref.** *(4)* with permission from American Society for Microbiology.

6. Prepare serial dilutions of GCV and FOS.
7. Resuspend the cells in DMEM/5% FBS at a density of 2×10^5 cells/mL.
8. Mix HCMV-infected HEL cells (1×10^4 cells/50 μL) and SE/15 cells (2×10^4 cells/100 μL) at a ratio of 1:2. Seed 150 μL of the mixed cell suspension into glass-bottom black fluorescence microplates precoated with fibronectin.
9. Add antiviral drugs (GCV or FOS) to each sample (*see* **Note 22**).
10. Culture the cells for 24–48 h.
11. Analyze each sample by the GFP-PML imaging assay as described in **Subheading 3.5**
12. Estimate the HCMV infectivity at each drug concentration.
13. Calculate the 50% inhibitory concentration (IC_{50}) values using dose–response curves (*see* **Note 23**).

Notes

1. In this chapter, cultivation of cells is performed at 37 °C in a humidified incubator in an atmosphere of 5% CO_2.
2. The following requirements should be considered when selecting stable cell lines expressing GFP-PML for HCMV assays: (1) GFP-PML morphology should be reminiscent of that of

native PML bodies in the nucleus, that is, not too big or too small; (2) cells should possess functional GFP-PML bodies whose morphology is specifically changed upon HCMV infection; (3) cells should maintain their GFP-PML structures in a stable manner after repeated passages.

3. In addition to GFP-PML, SE/15 cells overexpress human p180 protein. We previously found that HCMV infection frequently causes detachment of CHO cells and that this detachment is efficiently avoided by human p180 overexpression.
4. Stock solutions of fibronectin (0.5–1.0 mg/mL) should be stored at −20 °C and not subjected to repeated freezing and thawing. A working solution (10 μg/mL) should be freshly prepared before use. Fibronectin-coated plates wrapped tightly in vinyl can be preserved in a refrigerator for several weeks at least.
5. HCMV replicates slowly in monolayer cultures of diploid fibroblast cells, and exhibits cytopathic effects in 3–6 days for laboratory strains and 12–30 days for clinical isolates. HCMV laboratory strains, including Towne and AD169, are efficiently shed in culture media, allowing the culture supernatants to be collected as high-titer virus solutions on postinoculation days 4–6. On the other hand, since most of the HCMV clinical isolates are cell associated and only small amounts of virus are shed into the culture media, these virus solutions must be prepared via cell lysis by ultrasonication.
6. Cells at young generations (10–15th passage) should be stocked multiply before propagation of SE/15 cells. Whenever subcultivate, surplus SE/15 cells should be also stocked. Beyond the 30th passage, cells lacking or harboring small amounts of GFP-PML may increase, and these are nonapplicable to quantification assays. Therefore, it is important to regularly check the numbers of cells harboring GFP-PML. In bad or severe culture conditions, nonexpressing cells quickly overgrow, thereby necessary to start with a new cell stock of young generation. Healthy SE/15 cells are a prerequisite for quantification.
7. Even distribution of the cells in each well is required for reproducible assays. After seeding, gently rock the plate to achieve an even distribution and carefully transport the plates to the CO_2 incubator to maintain the distribution.
8. All procedures involving HCMV should be performed in a Class II biological safety cabinet under strict aseptic conditions to minimize the risk of contamination.
9. Since HCMV-infected cells are easily detachable from dishes, wash the cells carefully.

10. Under fluorescent microscopy observation, cells containing diffuse GFP-PML are detectable as early as 4 h p.i., increase until 10–12 h p.i. and then almost reach a plateau.
11. For 96- and 24-well plates, add 30 and 150 μL of NPB, respectively.
12. In the presence of detergent, SE/15 cells are easily detachable from dishes. Add NPB gently via the wall surface of the wells.
13. The NPB fluorescence assay provides a simple and rapid protocol, in which there are only two steps: "extraction" and "measurement by a plate reader." For high titer viruses such as experimental strains Towne and AD169, NPB fluorescence assay is useful. As higher sensitivity was required for assaying clinical isolates, we developed the imaging assay which enables assay of low titer viruses of clinical isolates.
14. Conventional culture plates show a high level of background fluorescence. We recommend using culture plates specified for fluorescence measurements (e.g., glass-bottom black fluorescence microplates, Greiner, #655896 or #662896 are one of the most suitable types of plates for imaging assays).
15. Growth medium containing phenol red makes the *S/N* ratio lower. Cells fixed with paraformaldehyde or formaldehyde can be used in the same protocol.
16. The automatic processing functions in Photoshop (Adobe) may be helpful.
17. Some of the PML signals bearing low intensity may disappear after conversion to a binary file. If the outcome is not appropriate, reset the parameters.
18. In **Fig. 3**, a range of 20–299 is used to represent the sizes (pixels).
19. You can verify your protocol by comparison with data from DAPI-stained images.
20. In **Fig. 3**, a range of 300–1,000 is used to represent the sizes (pixels).
21. Data are presented as mean values from 3 to 5 fields taken from one sample.
22. An even distribution of cells in each well is required for reproducible assays. After seeding, gently rock the plate to achieve an even distribution and carefully transport the plates to the CO_2 incubator to maintain the distribution.
23. The statistical software package Prism 4 (Graphpad) may be useful for fitting of logistic curves and calculating the 50% inhibitory concentrations (IC_{50}) statistically.

References

1. Maul, G. G., D. Negorev, P. Bell, and A. M. Ishov.(2000) Review: properties and assembly mechanisms of ND10, PML bodies, or PODs. *J Struct Biol* 129, 278–287.
2. Ishov, A. M., A. G. Sotnikov, D. Negorev, O. V. Vladimirova, N. Neff, T. Kamitani, E. T. Yeh, J. F. Strauss, G. G. Maul. (1999) PML is critical for ND10 formation and recruits the PML-interacting protein daxx to this nuclear structure when modified by SUMO-1. *J Cell Biol* 147, 221–234.
3. Korioth, F., G. G. Maul, B. Plachter, T. Stamminger, and J. Frey. (1996) The nuclear domain 10 (ND10) is disrupted by the human cytomegalovirus gene product IE1. *Exp Cell Res* 229 155–158.
4. Ueno, T., Y. Eizuru, H. Katano, T. Kurata,T. Sata, S. Irie, and K. Ogawa-Goto. (2006) Novel real-time monitoring system for human cytomegalovirus-infected cells in vitro that uses a green fluorescent protein-PML-expressing cell line. *Antimicrob Agents Chemother* 50, 2806–2813
5. Nowlin, D. M.,N. R. Cooper, andT. Compton. (1991) Expression of a human cytomegalovirus receptor correlates with infectibility of cells. *J Virol* 65, 3114–3121.
6. Ogawa-Goto, K., S. Irie, A. Omori, Y. Miura, H. Katano, H. Hasegawa, T. Kurata, T. Sata, andY. Arao.(2002) An endoplasmic reticulum protein, p180, is highly expressed in human cytomegalovirus-permissive cells and interacts with the tegument protein encoded by UL48. *J Virol* 76, 2350–2362.
7. Zhong, S., S. Muller, S.Ronchetti, P. S. Freemont, A. Dejean, andP. P. Pandolfi. (2000) Role of SUMO-1-modified PML in nuclear body formation. *Blood* 95, 2748–2752.

Chapter 4

Dual-Color Imaging of Tumor Angiogenesis

Robert M. Hoffman

Summary

Angiogenesis is a critical step in the process of tumor metastasis. Many models have been used to study this process, but they have been artificial and do not reflect the actual process that takes place in the human being. Our laboratory has developed realistic models of angiogenesis based on orthotopic transplantation of human tumors in mice. In order to make angiogenesis visible in real time, our laboratory has developed mouse models in which the blood vessels are labeled with green fluorescent protein (GFP) such that they can be visualized by vascularizing tumors expressing red fluorescent protein (RFP). A particularly valuable model is a nude mouse in which the promoter from the stem-cell-marker protein, nestin, drives the expression of GFP. In such transgenic mice, the nascent blood vessels, in contrast to the mature blood vessels, express GFP. This model, in which human tumors expressing RFP are implanted, has been used to test drugs for their antitumor and antiangiogenetic activity. We have observed for the first time the high antiangiogenetic efficacy of cancer drugs such as gemcitabine and doxorubicin. These models should prove very valuable in the discovery of new antiangiogenesis drugs.

Key words: GFP, RFP, Orthotopic transplantation, Nascent blood vessels, Angiogenesis, Antiangiogenesis drugs.

1. Introduction

1.1. Treating Cancer with Antiangiogenesis Drugs

Over a century ago it was observed that the growth of human tumors is often accompanied by increased vascularity *(1)*. The existence of tumor-derived factors responsible for promoting new vessel growth was postulated over 65 years ago *(2)*, and a few years later it was proposed that tumor growth is crucially dependent on the development of a neovascular supply *(3, 4)*.

The concept of treating tumors by inhibiting their ability to recruit new blood vessels (angiogenesis) was pioneered by Folkman *(5)*. It is thought that solid tumors cannot grow beyond

Barry W. Hicks (ed.), *Methods in Molecular Biology, Viral Applications of Green Fluorescent Protein, vol. 515*

DOI: 10.1007/978-1-59745-559-6_4

microscopic sizes without forming new blood vessels, and will remain dormant, or even regress, if angiogenesis is prevented or is not induced *(6)*. Tumors may stimulate angiogenesis *(7)*. Both genetic and environmental events (such as lack of oxygen) appear to contribute to the switching on of new blood vessel growth. Genetic events that could induce angiogenesis include the ras oncogene activation and inactivation of the p53 tumor-suppressor gene *(8, 9)*. The endothelial cells of new blood vessels have receptors that bind to vascular endothelial growth factors (VEGFs), a family of inducers of angiogenesis. These receptors make the endothelial cells of newly formed blood vessels selectively sensitive to agents such as antibodies directed against VEGF or the receptors themselves (see below).

Ferrara and Kerbel *(4)* noted that despite some initial setbacks and negative clinical trial results, major progress has been made over the past few years in targeting angiogenesis for human therapy. In 2004, the US Food and Drug Administration (FDA) approved bevacizumab (Avastin), a humanized anti-VEGF-A monoclonal antibody, for the treatment of metastatic colorectal cancer in combination with 5-fluorouracil (FU). This followed from a phase III study showing a survival benefit *(10)*.

Boehm et al. *(11)* reported that chronic, intermittent therapy of three different mouse tumors with an inhibitor of angiogenesis, endostatin, did not lead to any traces of acquired resistance. For each treatment cycle, the tumor was allowed to grow (or regrow) until it reached ~3% of the body weight of the mouse. Endostatin was then infused daily until the tumors regressed. Tumors regressed even after the sixth successive round of endostatin therapy with no apparent resistance *(6)*. In contrast, standard chemotherapy, using maximum doses of the drug cyclophosphamide, resulted in partial resistance by the third cycle of treatment, and complete resistance by the fourth cycle *(6)*. Resistance to antiangiogenic therapy is emerging *(12)* and thus a better understanding of pathways that may mediate tumor angiogenesis in various circumstances is necessary *(4)*.

Many patients treated with VEGF inhibitors, especially in combination with chemotherapy, may survive longer, but they eventually succumb to their disease. VEGF may be replaced by other angiogenic pathways as the disease progresses. For example, VEGF-unrelated pathways may become activated. Other members of the VEGF family such as the lymphangiogenic factors VEGF-C and VEGF-D, may bind to and activate VEGFR-2 after proteolytic cleavage *(13)*. Other possible mechanisms for acquired resistance to antiangiogenic drugs *(14, 15)* include selection and overgrowth of tumor-cell variants that are hypoxia resistant and thus less dependent on angiogenesis *(16)*. The recruitment of bone-marrow-derived angiogenic cells may also provide a potential mechanism of resistance to antiangiogenic strategies *(4)*.

Furthermore, Ferrara and Kerbel *(4)* noted that recent studies suggest that in some cases endothelial cells associated with tumors are not a genetically stable, nontransformed, compartment (as long assumed) and instead may provide a further mechanism of resistance to antiangiogenic therapies. Streubel et al. *(13)* reported that a significant percentage of the endothelial cells in human B-cell lymphomas harbor lymphoma-specific chromosomal abnormalities, suggesting that endothelial cells in B-cell lymphomas may be in part tumor related. Genetic instability in these endothelial cells could lead to resistance to anti-angiogenesis agents.

1.2. Previous Models Used to Visualize Angiogenesis

The discovery and evaluation of anti-angiogenic substances initially relied on methods such as the chorioallantoic membrane assay *(17, 18)*, the monkey iris neovascularization model *(19)*, the disk angiogenesis assay *(20)*, and various models that use the cornea to assess blood vessel growth *(21–26)*. Although they are important for understanding the mechanisms of blood vessel induction, these models do not represent tumor angiogenesis and are poorly suited to drug discovery.

Subcutaneous tumor xenograft mouse models have been developed to study tumor angiogenesis, but these require cumbersome pathological examination procedures such as histology and immunohistochemistry. Measurements require animal sacrifice and therefore preclude ongoing angiogenesis studies in individual, live, tumor-bearing animals. Moreover, subcutaneous tumor xenografts are not representative models of human disease.

Tumors transplanted in the cornea of the rodents *(27–29)* and rodent skin-fold window chambers have also been used for angiogenesis studies *(30–36)*. The cornea and skin-fold chamber models provide a means for studying tumor angiogenesis in living animals. However, quantification requires specialized procedures, and the sites do not represent natural environments for tumor growth. The cornea and skin-fold window chamber tumor models do not allow metastatic angiogenesis to occur, which may involve mechanisms of angiogenesis *(37)* that are qualitatively different from those occurring in ectopic models.

1.3. Dual-Color Fluorescent Proteins in Tumor–Host Models to Differentially Visualize the Tumor and Blood Vessels

Okabe et al. *(38)* produced transgenic mice with GFP under the control of a chicken β-actin promoter and cytomegalovirus enhancer. All of the tissues from these transgenic mice, with the exception of erythrocytes and hair, fluoresce green. Fluorescent proteins have been shown to be very useful for imaging in tumors including the formation of nascent vessels. We have developed unique mouse models to image tumor angiogenesis with fluorescent proteins, which are described in this chapter.

Tumor cells to be transplanted in the GFP mouse were made visible by transforming them with the red fluorescent protein

(RFP) *(39)*. In order to gain further insight into tumor–host interaction in the living state, including tumor angiogenesis, we have visualized RFP-expressing tumors transplanted in the GFP-expressing transgenic mice under dual-color fluorescence microscopy. The dual-color fluorescence made it possible to visualize the tumor growth in the host by whole-body imaging as well as to visibly distinguish interacting tumor and host cells in fresh tissue. The dual-color approach affords a powerful means of both visualizing and distinguishing the components of the host–tumor interaction *(40)*.

Dual-color images of early events in tumor angiogenesis induced by a B16F10 mouse melanoma in the transgenic GFP-expressing mouse were acquired in fresh tissue preparations. Host-derived GFP-expressing fibroblast cells and endothelial cells form nascent blood vessels were visualized clearly against the red fluorescent background of the RFP-expressing mouse melanoma. Host-derived GFP-expressing mature blood vessels within the RFP-expressing mouse melanoma also became visible. The images were acquired 3 weeks after subcutaneous injection of B16F10-RFP melanoma cells in the GFP mouse *(40)*.

A GFP nude mouse was obtained by crossing nontransgenic nude mice with the transgenic GFP C57/B6 mouse *(41)*. In the nude mice, the organs all brightly expressed GFP, including the heart, lungs, spleen, pancreas, esophagus, stomach, and duodenum. RFP-expressing human cancer cell lines, including PC-3-RFP prostate cancer, HCT-116-RFP colon cancer, MDA-MB-435-RFP breast cancer, and HT1080-RFP fibrosarcoma were transplanted to the transgenic GFP nude mice. All of these human tumors grew extensively in the transgenic GFP nude mouse. Dual-color fluorescence imaging enabled visualization of human tumor–host interaction by whole-body imaging and at the cellular level in fresh and frozen tissues. The GFP mouse model should greatly expand our knowledge of human tumor–host interaction *(41)*.

1.4. Imaging of Nascent Angiogenesis Using Nestin-Driven GFP Transgenic Mice

The intermediate filament protein, nestin, marks progenitor cells of the CNS. Such CNS stem cells are selectively labeled by placing GFP under the control of nestin regulatory sequences in transgenic mice (nestin-driven GFP [ND-GFP]). In the ND-GFP transgenic mice, the hair follicle stem cells express ND-GFP. The hair follicles are linked by a network of ND-GFP-expressing blood vessels which appear to originate from the hair follicle stem cells *(42)*.

We visualized tumor angiogenesis by dual-color fluorescence imaging in ND-GFP transgenic mice after transplantation of the murine melanoma cell line B16F10-expressing RFP. ND-GFP was highly expressed in proliferating endothelial cells and nascent blood vessels in the growing tumor. Immunohistochemical staining showed that the blood vessel-specific antigen CD31 was expressed in ND-GFP-expressing nascent blood vessels. ND-GFP expression was diminished in the vessels with increased blood flow. Progressive angiogenesis during tumor growth was readily visualized by GFP expression. Doxorubicin inhibited the nascent tumor angiogenesis as well as tumor growth in the ND-GFP mice transplanted with B16F10-RFP *(43)*.

The nestin ND-GFP gene was crossed into nude mice on the C57/B6 background to obtain ND-GFP nude mice. ND-GFP was expressed in the brain, spinal cord, pancreas, stomach, esophagus, heart, lung, blood vessels of glomeruli, blood vessels of skeletal muscle, testes, hair follicles, and blood vessel network in the skin of ND-GFP nude mice. Human lung cancer, pancreatic cancer, and colon cancer cell lines as well as a murine melanoma cell line and breast cancer tumor cell line expressing RFP were implanted orthotopically, and a RFP-expressing human fibrosarcoma was implanted s.c. in the ND-GFP nude mice. These tumors grew extensively in the ND-GFP mice. ND-GFP was highly expressed in proliferating endothelial cells and nascent blood vessels in the growing tumors, visualized by dual-color fluorescence imaging. Immunohistochemical staining showed that CD31 was expressed in the ND-GFP-expressing nascent blood vessels *(44)*.

Dual-color fluorescence imaging tumor visualized angiogenesis in the ND-GFP transgenic nude mice after orthotopic transplantation of the MIA PaCa-2 human pancreatic cancer line expressing RFP. ND-GFP was highly expressed in proliferating endothelial cells and nascent blood vessels in the growing tumor. The density of nascent blood vessels in the tumor was readily quantitated. Gemcitabine significantly decreased the mean nascent blood vessel density in the tumor as well as decreased tumor volume. These results demonstrated for the first time that gemcitabine is an inhibitor of angiogenesis as well as tumor growth in pancreatic cancer *(45)*.

In another study, nascent angiogenesis of pancreatic cancer liver metastasis in the ND-GFP transgenic nude mice was imaged. Human pancreatic cancer cells were visualized by RFP. ND-GFP was highly expressed in proliferating endothelial cells and nascent blood vessels in the growing liver metastasis. The density of nascent blood vessels in the tumor was readily quantitated. Gemcitabine significantly decreased the mean nascent blood vessel density in the pancreatic liver metastases *(46)*.

Angiogenesis of sarcoma formed by the HT-1080 human fibrosarcoma cell line expressing RFP was also imaged in the ND-GFP mice. Tumor cells were injected into either the muscle or the bone. Nestin was highly expressed in proliferating endothelial cells and nascent blood vessels in the growing tumors, including the surrounding tissues. Doxorubicin significantly decreased the mean nascent blood vessel density in the tumors as well as decreased tumor volume. Thus, the dual-color model of the ND-GFP nude mouse and RFP sarcoma cells is also useful for the visualization and quantitation of bone and soft tissue tumor angiogenesis and evaluation of angiogenetic inhibitors for such tumors *(47)*.

Thus, fluorescent proteins are very useful for imaging angiogenesis. Their intrinsic brightness, nontoxicity, and multiple colors make them the genetic reporters of choice to image angiogenesis. Particularly powerful are mouse models in which the vessels express one-color protein and the tumor expresses another. It is expected that these models will lead to a deeper understanding of the real-time angiogenesis process and to the discovery of safe and more effective antiangiogenesis agents for metastatic cancer.

2. Materials

2.1. Reagents

1. Restriction enzymes HindIII and NotI
2. RFP cDNA (pDsRed2; Clontech)
3. Plasmid pLNCX2 (Clontech). This plasmid is derived from Moloney murine leukemia virus (MoMuLV) and Moloney murine sarcoma virus (MoMuSV) elements for retroviral gene delivery and expression. Upon transfection into a packaging cell line, pLNCX2 can transiently express, or integrate and stably express, a transcript containing Ψ + (the extended viral packaging signal), and a gene of interest under direction of the human CMV promoter.
4. PT67 packaging cells (Clontech); 3T3 cells for viral titering; cell lines to be transfected with genes encoding fluorescent proteins, such as B16F0 melanoma cells (American Type Culture Collection)
5. Growth medium (normal and selective) appropriate for cell culture, such as DMEM (Invitrogen; Irvine Scientific)
6. Fetal bovine serum (FBS; Gemini Biological Products)
7. Lipofectamine PLUS transfection kit (Invitrogen)
8. G418 neomycin
9. Polysulfonic filter, 4.5 μm
10. Polybrene
11. Trypsin–EDTA and trypsin
12. Mice expressing GFP ("GFP mice"; Jackson Laboratories; Japan SLC, Hamamatsu, Japan)
13. Immunocompetent and immunodeficient mice (Charles River; Taconic; Harlan Teklad)

14. Anesthetic reagents (ketamine HCl, xylazine, acepromazine maleate; "ketamine mixture"; Butler Animal Health Supply)
15. Nair (Carter-Wallace)
16. Doxorubicin
17. NaCl, 0.9%
18. Optimum cutting temperature blocks
19. Antibody to rat immunoglobulin (antirat immunoglobulin) and antimouse immunoglobulin horseradish peroxidase detection kits (BD PharMingen)
20. Monoclonal anti-CD31 (CBL1337; Chemicon)
21. Monoclonal anti-nestin (rat 401; BD PharMingen)
22. Substrate-chromogen 3,3′-diaminobenzidine

2.2. Equipment

1. 60-mm culture dishes; 25-mm culture flasks; 96-well plates
2. Humidified incubator at 37 °C and 5% CO_2
3. Cloning cylinders (Bel-Art Products)
4. 27G2 latex-free syringe, 1 mL (Becton Dickinson)
5. 8-0 surgical sutures
6. Leica fluorescence stereo microscope, model LZ12, with a mercury 50-W power supply, and MZ6 stereo microscope (Leica)
7. D425/60 band-pass filter and 470 DCXR dichroic mirror
8. D470/40 emission filter and GG475 emission filter (Chroma Technology)
9. C5810 three-chip cooled color charge-coupled device (CCD) camera (Olympus) (Hamamatsu Photonics Systems) or DP70 CCD camera (Olympus)
10. Image-Pro Plus 4.0 software (Media Cybernetics)
11. Personal computer (PC; IBM or Fujitsu-Siemens)
12. VCR (Sony, model SLV-R1000)
13. Blue LED flashlight (LDP LLC)
14. Coolpix camera (Nikon)
15. Fluorescent lightbox with fiberoptic lighting at 470nm (FluorVivo Imaging System; INDEC Systems, Inc.)
16. OV100 Small Animal Imaging System (Olympus) with an M20 light source (Olympus Biosystems) and 470-nm excitation light
17. Paint Shop Pro 8 (Corel) and Cell^R (Olympus Biosystems)
18. Olympus BH 2-RFCA fluorescence microscope equipped with a mercury 100-W lamp power supply
19. Leica CM1850 cryostat

2.3. Nude GFP Mice

Use a gene encoding GFP under the control of either the β-actin promoter, resulting in ubiquitous GFP expression, or the promoter of the gene encoding nestin, resulting in selective GFP expression, including expression in nascent blood vessels in tumors. To obtain nude GFP mice, first cross 6-week-old female C57BL/6 GFP mice with 6- to 8-week-old BALB/c homozygous nude (***nu/nu***) or NCR **nu/nu** male mice, then cross male F_1 mice with female F_1 C57BL/6 GFP mice *(48)*. Before working with animals proper ethical documentation must be obtained, and nude mice require special housing facilities and diet (*see* **Note 1**).

2.4. Whole-Body Imaging Equipment

We use an Olympus OV100 Small Animal Imaging System with an MT-20 light source and DP70 CCD camera (Olympus) for whole-body imaging in live mice at variable magnification. The optics of the OV100 fluorescence imaging system have been especially developed for "macroimaging" as well as "microimaging" with high light-gathering capacity. The instrument incorporates a unique combination of high-numerical aperture and long working distance. Four individually optimized objective lenses, parcentered and parfocal, provide a 10^5-fold magnification range for seamless imaging of the entire body down to the subcellular level without disturbing the animal. The OV100 has lenses mounted on an automated turret with a magnification range of ×1.6 to ×16 and a field of view ranging from 6.9 nm to 0.69 mm. The optics and antireflective coatings ensure optimal imaging of multiple fluorescent reporters in small animals. High-resolution images are captured directly on a PC (Fujitsu-Siemens). Images are processed for contrast and brightness and are analyzed with the use of Paint Shop Pro 8 and cell^R.

Many other fluorescence imaging systems can also be used for dual-color tumor–host imaging. For example, a Leica fluorescence stereo microscope (model LZ12) equipped with a mercury 50-W lamp power supply can be used. Selective excitation of GFP is produced via a D425/60 band-pass filter and 470 DCXR dichroic mirror. Emitted fluorescence is collected through a long-pass filter (GG475). Anesthetized animals can be examined with a microscope and images can be acquired with a Hamamatsu C5810 three-chip cooled color CCD camera. Images can also be processed for contrast and brightness and can be analyzed with the use of Image-Pro Plus software. High-resolution images of 1,024 × 724 pixels can be captured directly on an IBM PC or continuously through video output on a high-resolution Sony VCR. Simpler systems such as a light box with appropriate filters and camera or even a blue light LED flashlight with appropriate filters can be used for macroimaging (discussed below) *(48–51)*.

3. Methods

3.1. RFP Retrovirus Production

1. Insert the *Hind*III–NotI fragment from pDsRed2, containing full-length RFP cDNA, into the *Hind*III–*Not*I site of pLNCX2, which contains a neomycin-resistance gene, to establish the pLNCX2-DsRed2 plasmid.
2. Use PT67, an NIH3T3-derived packaging cell line expressing the 10Al viral envelope, to produce retrovirus.
3. Culture PT67 cells in DMEM medium supplemented with 10% heat-inactivated FBS. It takes ~3 days for the cells to reach ~70% confluence after 3×10^5 PT67 cells are seeded in a 25-mm^2 flask with DMEM medium containing 10% FBS.
4. For vector production, when PT67 packaging cells reach 70% confluence in the 25-mm^2 flask, replate cells in a 60-mm culture dish at 60–80% confluence 12 h before transfection.
5. Use 10 μg pLNCX2-DsRed2 DNA and the Lipofectamine PLUS transfection kit. Add 7 μL pLNCX2-DsRed2 DNA to 87 μL serum-free medium in a tube and then add 6 μL Lipofectamine reagent. Mix and incubate for 15 min at 22–26 °C (room temperature).
6. Dilute 4 μL Lipofectamine reagent in 96 μL serum-free medium in a second tube. Mix and incubate for 15 min at room temperature.
7. Combine the DNA prepared in step 3 and diluted Lipofectamine reagent, then mix and incubate for 15 min at room temperature.
8. While the complexes are forming, replace medium on the recently plated cells with 800 μL of serum-free DMEM. Add the DNA–Lipofectamine complex to the dish with cells containing fresh DMEM. Mix the complexes in the medium gently; incubate for 4 h at 37 °C in 5% CO_2.
9. After 4 h incubation, increase volume of medium to 5 mL and incubate for 24 h at 37 °C in 5% CO_2.
10. After 24 h incubation, clone the packaging cells by limit dilution in 96-well plates.
11. For selection of a PT67 packaging cell clone producing large amounts of RFP retroviral vector (PT67-DsRed2), culture the cells first in the presence of 200 μg/mL G418. Every 1–2 days, increase the concentration of G418 to 400, 600, 800, then 1,000 μg/mL (*see* **Note 2**).
12. Clones of PT67-DsRed2 cells with high viral titer production are identified with 3T3 cells used for virus titering. Clones with a titer higher than 1×10^6 plaque-forming units per mL are used for RFP vector production.

3.2. RFP Gene Transduction of Tumor Cell Lines

1. For RFP gene transduction, use cancer cells that are 20% confluent. Plate cancer cells at a density of 1 × 10^5 to 2 × 10^5 cells per 60-mm plate 12–18 h before infection with RFP retrovirus.
2. For retroviral infection, collect conditioned medium from packaging cells (PT67–DsRed2) and filter it through a 0.45-μm polysulfonic filter. Add the virus-containing filtered medium to the target cancer cells.
3. Add polybrene to a final concentration of 8 μg/mL and incubate the cells for 24 h at 37 °C.
4. Replace the medium with DMEM and 10% FBS after 24 h incubation and check for RFP-expressing cells by fluorescence microscopy.
5. Collect tumor cells with trypsin–EDTA and subculture them at a ratio of 1:15 in selective medium, which contains 50 μg/mL G418.
6. To select for the most brightly fluorescent cells, increase G418 to a concentration of 800 μg/mL (in 200 μg/mL increments), by culturing cells for 1–2 days in each concentration of G418 (*see* **Note 3**).
7. Isolate clones expressing RFP with cloning cylinders using trypsin–EDTA and amplify them in DMEM in the absence of the selective agent (*see* **Note 4**). Further select cells for brightness and stability.

3.3. Tumor Models of Fluorescent Protein-Expressing Tumor Cells

Use one of the following options to establish a tumor model of fluorescent protein-expressing tumor cells: i.v. cell injection (a), surgical orthotopic implantation (b), or inoculation of cells by intradermal injection (c).

3.3.1. Cell Injection to Establish an Experimental Metastasis Model

1. Collect fluorescent protein-expressing tumor cells by trypsinization for 3 min at 37 °C with 0.25% trypsin.
2. Wash the cells three times with cold serum-free medium using a tabletop centrifuge at 500 × *g*.
3. Resuspend the cells in ~0.2 mL of serum-free medium.
4. Within 30 min of collecting cells, inject 1 × 10^6 tumor cells in a total volume of 0.2 mL into 6-week-old C57BL/6 GFP mice or nude (*nu/nu*) GFP mice via the lateral tail vein or subcutaneously using a 1-mL 27G2 latex-free syringe (*see* **Note 5**).
5. For liver colonization, inject fluorescent protein-expressing cells directly into the portal vein in anesthetized mice (details on inducing anesthesia are presented below).

3.3.2. Surgical Orthotopic Implantation to Establish a Spontaneous Metastasis Model (See **Note 6***)*

1. Induce anesthesia with a "ketamine mixture" (10 μL ketamine HCl, 7.6 μL xylazine, and 2.4 μL acepromazine maleate) by s.c. injection.
2. Use a microscope (Leica MZ6) with magnification of about ×6 to about ×40 for all procedures of the operation.

3. Isolate fluorescent protein-expressing tumor fragments n mm^3) from subcutaneously growing tumors, formed by injection of RFP-expressing tumor cells by mincing tumor tissue into 1-mm^3 fragments. After proper exposure of the target organ, implant three tumor fragments per transgenic GFP mouse (*see* **Note 7**).
4. With 8-0 surgical sutures, penetrate the tumor fragments and suture the fragments onto the target organ.
5. Keep mice in a barrier facility under high-efficiency particulate air filtration.

3.3.3. RFP-Expressing Cutaneous Melanoma Model

1. Collect the RFP-expressing mouse B16F0 melanoma cells by trypsinization for 3 min at 37 °C with 0.25% trypsin.
2. Wash cells three times with cold serum-containing medium using a tabletop centrifuge at 500 × *g* for 5 min at room temperature, and then keep on ice.
3. Inject 6-week-old male C57BL/6 GFP mice or nude GFP mice with 1 × 10^6 RFP-expressing mouse B16F0 melanoma cells that were collected and washed. This is done by intradermal injection of cells into the dorsal skin of the mouse in a total volume of 50 μL cell culture medium within 40 min of collection.

3.4. Imaging

Use one of the following methods for whole-body imaging of mice: microscopy (a), flashlight imaging (b), light-box imaging (c), or chamber imaging (d).

3.4.1. Microscopy

1. Use a Leica fluorescence stereo microscope (model LZ12) equipped with a mercury 50-W lamp power supply or its equivalent.
2. Produce selective excitation of GFP via a D425/60 band-pass filter and 470 DCXR dichroic mirror.
3. Collect emitted fluorescence through a long-pass filter (GG475) on a Hamamatsu C5810 three-chip cooled color CCD camera or its equivalent.
4. Process images for contrast and brightness with the use of Image-Pro Plus 4.0 software or its equivalent.
5. Capture high-resolution images of 1,024 × 724 pixels directly on an IBM PC or continuously through video output on a high-resolution Sony VCR, model SLV-R1000 or its equivalent.
6. For C57BL/6 mice, remove hair with Nair or by shaving before images are obtained (*see* **Note 8**).

3.4.2. Chamber Imaging

1. Do whole-body imaging with an Olympus OV100 imaging system using 470-nm excitation light originating from an MT-20 light source.

2. Collect emitted fluorescence through appropriate filters configured on a filter wheel with a DP70 CCD camera. Variable magnification imaging can be done with a series of four objective lenses.
3. Capture images on a PC (Fujitsu-Siemens), and process images for contrast and brightness with Paint Shop Pro 8 and cell^R.
4. For C57BL/6 mice, remove hair with Nair or by shaving before images are obtained (*see* **Note 8**).

3.4.3. Tumor Tissue Sampling

1. Obtain tumor tissue biopsies from 3 days to 4 weeks after inoculation of tumor cells. Biopsies of tumor tissue can be obtained from anesthetized mice by removal of a small piece of tumor tissue (1 mm^3 or less) with a scalpel.
2. Staunch bleeding by pressing the wound with sterile gauze. Alternatively, the mouse can be killed and the tissue can be collected and processed for analysis.
3. Cut fresh tissue into pieces of ~1 mm^3 and gently press onto slides for fluorescence microscopy. This procedure is done manually on normal slides.
4. To analyze tumor angiogenesis, digest the tissues with trypsin–EDTA for 5 min at 37 °C before examination.
5. After trypsinization, put tissues on precleaned microscope slides and cover with another microscope slide.

3.5. Fluorescence Microscopy

1. Use an Olympus BH 2-RFCA fluorescence microscope equipped with a mercury 100-W lamp power supply or its equivalent.
2. To visualize both GFP and RFP fluorescence at the same time, produce excitation light via a D425/60 band-pass filter and a 470 DCXR dichroic mirror (*see* **Note 9**).
3. Collect emitted fluorescence light through a GG475 long-pass filter.
4. Capture high-resolution images of 1,024 × 724 pixels with a Hamamatsu C5810 three-chip-cooled color CCD camera or its equivalent and store directly on an IBM PC or its equivalent.
5. Process images for contrast and brightness using Image-Pro Plus 4.0 software or its equivalent.

3.6. Measurement of GFP-Expressing Tumor Blood Vessel Length and Evaluation of Antiangiogenetic Agents

1. Give mice daily i.p. injections of doxorubicin (5 μg/g body weight in a 2-mg/mL solution of 0.9% NaCl) or other drugs or 0.9% NaCl solution (vehicle controls) on days 0, 1, and 2 after implantation of tumor cells.
2. Anesthetize mice with the ketamine mixture and obtain biopsies on days 10, 14, 21, and 28 after implantation using sterile technique (*see* **Note 10**).

3. Biopsies are removed from the tumor with a scalpel.
4. Gently flatten the tumor tissue between the slide and coverslip.
5. Quantify angiogenesis in the tumor tissue by measuring the length of GFP-expressing blood vessels in all fields using fluorescence microscopy.
6. Obtain measurements in all fields at ×40 or ×100 magnification to calculate the total length of GFP-expressing blood vessels.
7. Calculate the vessel density by dividing the total length of GFP-expressing blood vessels (in mm) by the tumor volume (in mm^3).

3.7. Immunohistochemical Staining

1. "Snap-freeze" fresh tissue with liquid nitrogen, then orient and embed the frozen tissue in optimum cutting temperature blocks and store at −80 °C. Cut the frozen sections to a thickness of 5 μm with a Leica CM1850 cryostat.
2. Detect colocalization of GFP fluorescence, CD31, and nestin in the frozen skin sections of mice transgenic for nestin-driven GFP expression using the anti-rat immunoglobulin and anti-mouse immunoglobulin horseradish peroxidase detection kits following the manufacturer's instructions.
3. Use monoclonal anti-CD31 (1:50 dilution) and monoclonal anti-nestin (1:80 dilution) as primary antibodies. To identify the GFP-expressing tumor-infiltrating natural killer cells, macrophages, and dendritic cells, detect localization of GFP together with cell surface markers using immunohistochemical staining with monoclonal antibodies to NK1.1, CD111b, and CD11c, respectively.
4. Use staining with substrate-chromogen 3,3′-diaminobenzidine for antigen detection.

Notes

1. All animal studies are done in accordance with the principles and procedures outlined in the National Institutes of Health National Research Council's *Guide for the Care and Use of Laboratory Animals* (available at http://oacu.od.nih.gov/regs/guide/guidex.htm) under assurance number A3873-1. Nude mice must be kept in a barrier facility under high-efficiency particulate air filtration. Also, these mice must be fed an autoclaved laboratory rodent diet (Tecklad LM-485, Western Research Products).

2. Increasing the amount of G418 in a stepwise way is very important for inducing transgene expression. This procedure ensures high production of RFP+ retrovirus.
3. Increasing G418 in a stepwise way is very important for inducing expression of the transgene. This procedure ensures high expression of RFP.
4. Selecting for cancer cells stably and brightly expressing RFP in the absence of antibiotic, after antibiotic selection, ensures that the cells will stably express RFP in the absence of antibiotic selection. This is necessary since there will be no selection when the cells are transplanted into the mice. It normally takes 7–10 days in culture without selection to obtain a sufficient number of cells for imaging studies.
5. Some cells in suspension may lose viability rapidly over time and therefore should be injected as soon as possible.
6. Orthotopic implantation of tumor fragments results in higher spontaneous metastatic rates than injection of a cell suspension.
7. Bleeding should not occur if the surgery is properly performed. It must be avoided at the surgical site, as hemoglobin will absorb the incident excitation light and interfere with the fluorescent imaging.
8. Hair is highly autofluorescent, so improper removal of hair will result in low-quality images.
9. It is important to minimize autofluorescence of tissue and body fluids by using proper filters. Excitation filters should have a narrow band as close to 490 nm as possible to specifically excite GFP whose peak is distinct from that of the skin, tissues, and fluid of the animal. In addition, proper band-pass emission filters should be used with a cutoff of ~515 nm (Chroma Technology).
10. When using open-biopsy procedures or other surgical procedures that open the animal, it is essential to hydrate the animal by spraying sterile normal saline on the open tissue. Also, when repeating procedures such as an open biopsy or other invasive procedures, it is critical to maintain a properly sterile operation field and sterile instruments to avoid infection.

Acknowledgments

These studies were supported in part by National Cancer Institute grant numbers CA099258 and CA103563.

References

1. Ferrara, N. (2002) VEGF and the quest for tumour angiogenesis factors. *Nat. Rev. Cancer* 2, 795–803.
2. Ide, A. G., Baker, N. H. and Warren, S. L. (1939) Vascularization of the Brown Pearce rabbit epithelioma transplant as seen in the transparent ear chamber. *Am. J. Roentgenol.* 42, 891–899.
3. Algire, G. H., Chalkley, H. W., Legallais, F. Y. and Park, H. D. (1945) Vascular reactions of normal and malignant tissues in vivo. I. Vascular reactions of mice to wounds and to normal and neoplastic transplants. *J. Natl. Cancer Inst.* 6, 73–85.
4. Ferrara, N. and Kerbel, R. S. (2005) Angiogenesis as a therapeutic target. *Nature* 438, 967–974.
5. Folkman, J. (1971) Tumor angiogenesis: therapeutic implications. *N. Engl. J. Med.* 285, 1182–1186.
6. Kerbel, R. S. (1997) A cancer therapy resistant to resistance. *Nature* 390, 335–336.
7. Hanahan, D. and Folkman, J. (1996) Patterns and emerging mechanisms of the angiogenic switch during tumorigenesis. *Cell* 86, 353–364.
8. Rak, J., Filmus, J., Finkenzeller, G., Grugel, S., Marme, D. and Kerbel, R. S. (1995) Oncogenes as inducers of tumor angiogenesis. *Cancer Metastasis Rev.* 14, 263–277.
9. Bouck, N., Stellmach, V. and Hsu, S. C. (1996) How tumors become angiogenic. *Adv. Cancer Res.* 69, 135–174.
10. Hurwitz, H., Fehrenbacher, L., Novotny, W., Cartwright, T., Hainsworth, J., Heim, W., Berlin, J., Baron, A., Griffing, S., Holmgren, E., Ferrara, N., Fyfe, G., Rogers, B., Ross, R. and Kabbinavar, F. (2004) Bevacizumab plus irinotecan, fluorouracil, and leucovorin for metastatic colorectal cancer. *N. Engl. J. Med.* 350, 2335–2342.
11. Boehm, T., Folkman, J., Browder, T. and O’Reilly, M. S. (1997) Antiangiogenic therapy of experimental cancer does not induce acquired drug resistance. *Nature* 390, 404–407.
12. Kerbel, R. and Folkman, J. (2002) Clinical translation of angiogenesis inhibitors. *Nat. Rev. Cancer* 2, 727–739.
13. Streubel, B., Chott, A., Huber, D., Exner, M., Jager, U., Wagner, O. and Schwarzinger, I. (2004) Lymphoma-specific genetic aberrations in microvascular endothelial cells in B-cell lymphomas. *N. Engl. J. Med.* 351, 250–259.
14. Kerbel, R. S., Yu, J., Tran, J., Man, S., Viloria-Petit, A., Klement, G., Coomber, B. L. and Rak, J. (2001) Possible mechanisms of acquired resistance to anti-angiogenic drugs: implications for the use of combination therapy approaches. *Cancer Metastasis Rev.* 20, 79–86.
15. Sweeney, C. J., Miller, K. D. and Sledge, G. W. Jr. (2003) Resistance in the anti-angiogenic era: nay-saying or a word of caution? *Trends Mol. Med.* 9, 24–29.
16. Yu, J. L., Rak, J. W., Coomber, B. L., Hicklin, D. J. and Kerbel, R. S. (2002) Effect of p53 status on tumor response to antiangiogenic therapy. *Science* 295, 1526–1528.
17. Auerbach, R., Kubai, L., Knighton, D. and Folkman, J. (1974) A simple procedure for the long-term cultivation of chicken embryos. *Dev. Biol.* 41, 391–394.
18. Crum, R., Szabo, S. and Folkman, J., (1985) A new class of steroids inhibits angiogenesis in the presence of heparin or a heparin fragment. *Science* 230, 1375–1378.
19. Miller, J. W., Stinson, W. G. and Folkman, J. (1993) Regression of experimental iris neovascularization with systemic alpha-interferon. *Ophthalmology* 100, 9–14.
20. Passaniti, A., Taylor, R. M., Pili, R., Guo, Y., Long, P. V., Haney, J. A., Pauly, R. R., Grant, D. S. and Martin, G. R. (1992) A simple, quantitative method for assessing angiogenesis and antiangiogenic agents using reconstituted basement membrane, heparin, and fibroblast growth factor. *Lab. Investig.* 67, 519–528.
21. Alessandri, G., Raju, F. and Gullino, P. M. (1983) Mobilization of capillary endothelium in vitro induced by effectors of angiogenesis in vivo. *Cancer Res.* 43, 1790–1797.
22. Deutsch, T. A. and Hughes, W. F. (1979) Suppressive effects of indomethacin on thermally induced neovascularization of rabbit corneas. *Am. J. Ophthalmol.* 87, 536–540.
23. Korey, M., Peyman, G. A. and Berkowitz, R. (1977) The effect of hypertonic ointments on corneal alkali burns. *Ann. Ophthalmol.* 9, 1383–1387.
24. Mahoney, J. M. and Waterbury, L. D. (1985) Drug effects on the neovascularization response to silver nitrate cauterization of the rat cornea. *Curr. Eye Res.* 4, 531–535.
25. Li, W. W., Grayson, G., Folkman, J. and D Amore, P. A. (1991) Sustained-release endotoxin. A model for inducing corneal neovascularization. *Invest. Ophthalmol. Vis. Sci.* 32, 2906–2911.
26. Epstein, R. J., Hendricks, R. L. and Stulting, R. D. (1990) Interleukin-2 induces corneal

neovascularization in A/J mice. *Cornea* 9, 318–323.
27. Gimbrone, M. A., Cotran, I. S., Leapman, S. B. and Folkman, J. (1974) Tumor growth and neovascularization: an experimental model using the rabbit cornea. *J. Natl. Cancer Inst.* 52, 413–427.
28. Fournier, G. A., Lutty, G. A., Watt, S., Fenselau, A.and Patz, A. (1981) A corneal micropocket assay for angiogenesis in the rat eye. *Invest. Ophthalmol. Vis. Sci.* 21, 351–354.
29. Muthukkaruppan, V. and Auerbach, R. (1979) Angiogenesis in the mouse cornea. *Science* 205, 1416–1418.
30. Papenfuss, H. D., Gross, J. F., Intaglietta, M. and Treese, F. A. (1979) A transparent access chamber for the rat dorsal skin fold. *Microvasc. Res.* 18, 311–318.
31. Gross, J., Roemer, R., Dewhirst, M. and Meyer, T. (1982) A uniform thermal feild in a hyperthermia chamber for microvascular studies. *Int. J. Heat Mass Transfer* 25, 1313–1320.
32. Dewhirst, M., Gross, J., Sim, D., Arnold, P. and Boyer, D. (1984) The effect of rate of heating or cooling prior to heating on tumor and normal tissue microcirculatory blood flow. *Biorheology* 21, 539–558.
33. Fukumura, D., Xavier, R., Sugiura, T., Chen, Y., Park, E. C., Lu, N., Selig, M., Nielsen, G., Taksir, T., Jain, R. K. and Seed, B. (1998) Tumor induction of VEGF promoter activity in stromal cells. *Cell* 94, 715–725.
34. Li, C. Y., Shan, S., Huang, Q., Braun, R. D., Lanzen, J., Hu, K., Lin, P. and Dewhirst, M. W. (2000) Initial stages of tumor cell-induced angiogenesis: evaluation via skin window chambers in rodent models. *J. Natl. Cancer Inst.* 92, 143–147.
35. Al-Mehdi, A. B., Tozawa, K., Fisher, A. B., Shientag, L., Lee, A. and Muschel, R. J. (2000) Intravascular origin of metastasis from the proliferation of endothelium-attached tumor cells: a new model for metastasis. *Nat. Med.* 6, 100–102.
36. Huang, Q., Shan, S., Braun, R. D., Lanzen, J., Anyrhambatla, G., Kong, G., Borelli, M., Corry, P., Dewhirst, M. W. and Li, C. Y. (1999) Noninvasive visualization of tumors in rodent dorsal skin window chambers. *Nat. Biotechnol.* 17, 1033–1035.
37. Cowen, S. E., Bibby, M. C. and Double, J. A. (1995) Characterization of the vasculature within a murine adenocarcinoma growing in different sites to evaluate the potential of vascular therapies. *Acta Oncol.* 34, 357–360.
38. Okabe, M., Ikawa, M., Kominami, K., Nakanishi, T. and Nishimune, T. (1997) 'Green mice' as a source of ubiquitous green cells. *FEBS Lett.* 407, 313–319.
39. Hoffman, R. M. (2002) Green fluorescent protein imaging of tumour growth, metastasis, and angiogenesis in mouse models. *Lancet Oncol.* 3, 546–556.
40. Yang, M., Li, L., Jiang, P., Moossa, A. R., Penman, S. and Hoffman, R. M. (2003) Dual-color fluorescence imaging distinguishes tumor cells from induced host angiogenic vessels and stromal cells. *Proc. Natl. Acad. Sci. USA* 100, 14259–14262.
41. Yang, M., Reynoso, J., Jiang, P., Li, L., Moossa, A. R. and Hoffman, R. M. (2004) Transgenic nude mouse with ubiquitous green fluorescent protein expression as a host for human tumors. *Cancer Res.* 64, 8651–8656.
42. Amoh, Y., Li, L., Yang, M., Moossa, A. R., Katsuoka, K., Penman, S. and Hoffman, R. M. (2004) Nascent blood vessels in the skin arise from nestin-expressing hair-follicle cells. *Proc. Natl. Acad. Sci. USA* 101, 13291–13295.
43. Amoh, Y., Li, L., Yang, M., Jiang, P., Moossa, A. R., Katsuoka, K. and Hoffman, R. M. (2005) Hair follicle-derived blood vessels vascularize tumors in skin and are inhibited by doxorubicin. *Cancer Res.* 65, 2337–2343.
44. Amoh, Y., Yang, M., Li, L., Reynoso, J., Bouvet, M., Moossa, A. R., Katsuoka, K. and Hoffman, R. M. (2005) Nestin-linked green fluorescent protein transgenic nude mouse for imaging human tumor angiogenesis. *Cancer Res.* 65, 5352–5357.
45. Amoh, Y., Li, L., Tsuji, K., Moossa, A. R., Katsuoka, K., Hoffman, R. M. and Bouvet, M. (2006) Dual-color imaging of nascent blood vessels vascularizing pancreatic cancer in an orthotopic model demonstrates antiangiogenesis efficacy of gemcitabine. *J. Surg. Res.* 132, 164–169.
46. Amoh, Y., Nagakura, C., Maitra, A., Moossa, A. R., Katsuoka, K., Hoffman, R. M. and Bouvet, M. (2006) Dual-color imaging of nascent angiogenesis and its inhibition in liver metastases of pancreatic cancer. *Anticancer Res.* 26, 3237–3242.
47. Hayashi, K., Yamauchi, K., Yamamoto, N., Tsuchiya, H., Tomita, K., Hoffman, R. M. and Bouvet, M. (2007) Dual-color imaging of angiogenesis and its inhibition in bone and soft tissue sarcoma. *J. Surg. Res.* 140, 165–170 [AU3].
48. Hoffman, R. M. and Yang, M. (2006) Color-coded fluorescence imaging of tumor-host interactions. *Nat. Protoc.* 1, 928–935.
49. Hoffman, R. M. and Yang, M. (2006) Whole-body imaging with fluorescent proteins. *Nat. Protoc.* 1, 1429–1438.

50. Hoffman, R. M. (2005) The multiple uses of fluorescent proteins to visualize cancer in vivo. *Nat. Rev. Cancer* 5, 796–806.

51. Yang, M., Baranov, E., Jiang, P., Sun, F.-X., Li, X.-M., Li, L., Hasegawa, S., Bouvet, M., Al-Tuwaijri, M., Chishima, T., Shimada, H., Moossa, A. R., Penman, S. and Hoffman, R. M. (2000) Whole-body optical imaging of green fluorescent protein-expressing tumors and metastases. *Proc. Natl. Acad. Sci. USA* 97, 1206–1211.

Chapter 5

Use of GFP to Analyze Morphology, Connectivity, and Function of Cells in the Central Nervous System

Alan R. Harvey, Erich Ehlert, Joris de Wit, Eleanor S. Drummond, Margaret A. Pollett, Marc J. Ruitenberg, Giles W. Plant, Joost Verhaagen and Christiaan N. Levelt

Summary

We here describe various approaches using GFP that are being used in the morphological and functional analysis of specific cell types in the normal and injured central nervous system. Incorporation of GFP into viral vectors allows phenotypic characterization of transduced cells and can be used to label their axons and terminal projections. Characterization of transduced cell morphology can be enhanced by intracellular injection of living GFP-labeled cells with appropriate fluorescent dyes. Ex vivo labeling of precursor or glial cells using viral vectors that encode GFP permits long-term identification of these cells after transplantation into the brain or spinal cord. In utero electroporation methods result in expression of gene products in developing animals, allowing both functional and morphological studies to be carried out. GFPCre has been developed as a marker gene for viral vector-mediated expression of the bacterial recombinase Cre in the brain of adult mice with "floxed" transgenes. GFPCre-mediated induction of transgene expression can be monitored by GFP expression in defined populations of neurons in the adult brain. Finally, GFP can be used to tag proteins, permitting dynamic visualization of the protein of interest in living cells.

Key words: Viral vectors, Adeno-associated virus, Lentivirus, Electroporation, Olfactory ensheathing glia, Schwann cells, Cre-EGFP, Retina, Semaphorin, Hippocampus, Neocortex, Spinal cord.

Barry W. Hicks (ed.), *Methods in Molecular Biology, Viral Applications of Green Fluorescent Protein, vol. 515*

DOI: 10.1007/978-1-59745-559-6_5

1. Introduction

1.1. Using AAV-GFP to Gauge Transduction Efficiency of Neural Cells In Vivo, to Phenotype Transduced Cells, and to Trace Axonal and Terminal Projections in the Brain

Many groups and facilities are producing purified, replication-deficient adeno-associated viral (AAV) and lentiviral (LV) vectors. These vectors, encoding a range of genes, are now widely used in neuroscience research. Some serotypes of AAV (e.g., AAV2) are known to preferentially transduce neurons and this vector type is currently being tested in a number of clinical trials. In experimental animals studies, incorporation of GFP or some other reporter gene into the vector allows an assessment of in vivo transduction efficiency and, using double-labeling immunohistochemical techniques, it also allows phenotypic characterization of the transduced cell population. AAV-GFP can also be employed to track the axons and terminal projections of transduced neurons, for example, retinal ganglion cells (RGCs) *(1)* (**Fig. 1A**). When using AAV vectors to incorporate specific genes into cells, the reporter genes can also be used to identify transduced neural cells. This is done by using bi-cistronic AAV vectors, in which the gene of interest is linked via an internal ribosome entry site (IRES) to a reporter gene expressing a marker such as GFP. IRES allows expression of the GFP gene after transcription of the upstream neuroprotective transgene thus permitting direct identification of transduced cells. Although IRES-dependent second gene expression is likely to be lower than the expression of the gene placed upstream of the IRES *(2)*, in our bi-cistronic AAV vectors GFP is expressed at sufficiently high levels to allow visualization of GFP immunoreactive axons in the optic nerve, optic tract, and in terminal arbors in central visual target areas *(3)*. Procedures for identifying transduced neurons and their processes are here described for the primary visual pathway in rat.

1.2. GFP Labeling of Neuroglia Ex Vivo and Tracking of Cells After Transplantation

The use of markers to label donor tissue prior to transplantation into the CNS can be of critical importance if the fate of grafted cells and the influence of these cells on the host nervous system are to be accurately assessed. Identification of transplanted material is especially difficult if dissociated cells are used, particularly if the donor cells are migratory or are transplanted into neonatal hosts. As summarized in an earlier review *(4)*, labeling of donor material is essential if the viability of grafted cells is to be quantified in the days, weeks, or months after transplantation, and if additional histological analyses such as immunohistochemistry, retrograde-labeling, or electron microscopy are required in order to assess cellular differentiation and integration into host tissue. Chronic exposure to certain markers may be toxic to cells and the cellular distribution of a particular label may change over time; for example, there is potential transfer of exogenous labels

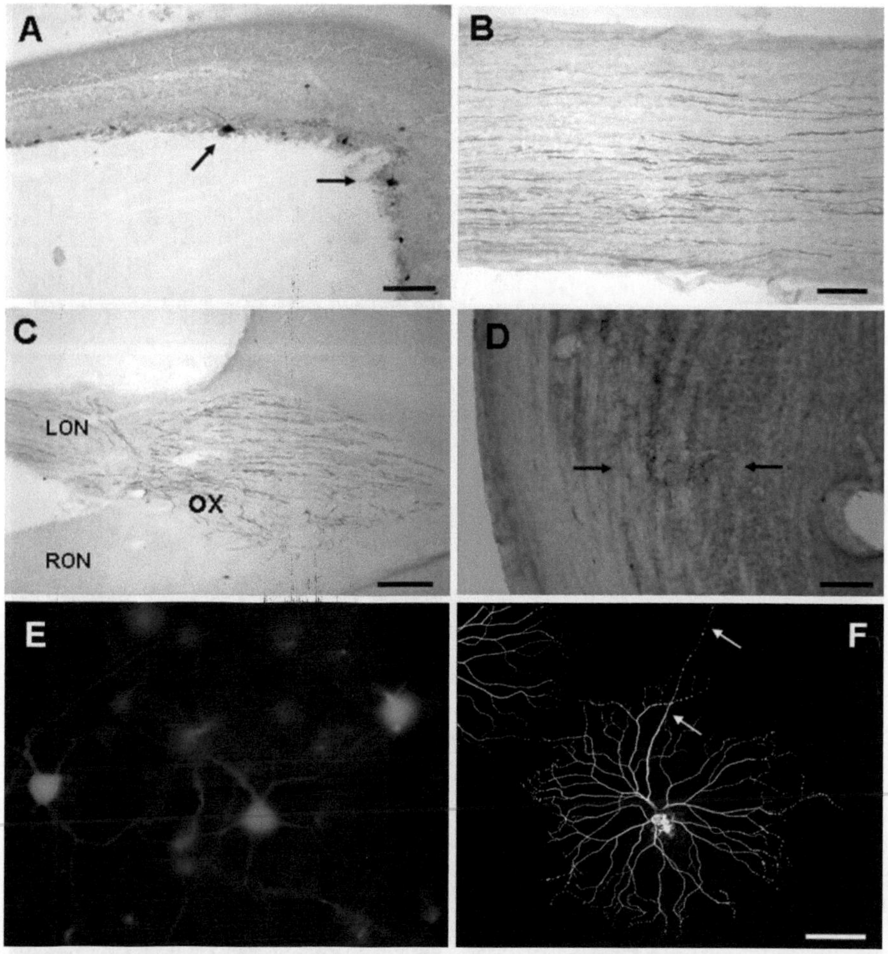

Fig. 1. **A–D** AAV-GFP transduction of adult rat retinal ganglion cells (RGCs) after intravitreal vector injection. All sections were immunostained with an antibody to GFP to amplify the signal. **A** GFP-positive RGCs (*arrowed*) in retina. **B, C** GFP-labeled axons in (**B**) the left optic nerve (LON) and (**C**) at the optic chiasm (OX). Here, the trajectory of individual axons can be followed. **D** The terminal arbor (*arrows*) of a transduced RGC in the thalamus. **E** Living GFP-positive RGCs in a retinal wholemount. The arrowed cell was intracellularly injected with lucifer yellow resulting in a complete fill of the dendritic tree of this transduced neuron. **F** Confocal image of the same cell after fixation and immunohistochemical processing. The axon can also be traced (*arrows* in **F**). Scale bar for all figures = 100 μm.

to host cells, which should always be taken into consideration in graft experiments *(4)*.

We describe here methods for ex vivo labeling of adult Schwann cells (SCs) or olfactory ensheathing glia (OEG) using viral vectors that encode GFP *(5–9)* (**Fig. 2**). The application of viral vector technology to label and track donor cells has many advantages, such as allowing the use of constitutive promoters, selective labeling of characterized cell populations, and – in theory – transduction of cells derived from any animal species. AAV2 does not effectively transduce SCs or OEG but these glial cells are transduced by adenoviral and LV vectors *(6)*. Importantly, when using an appropriate vector such as LV, incorporation of

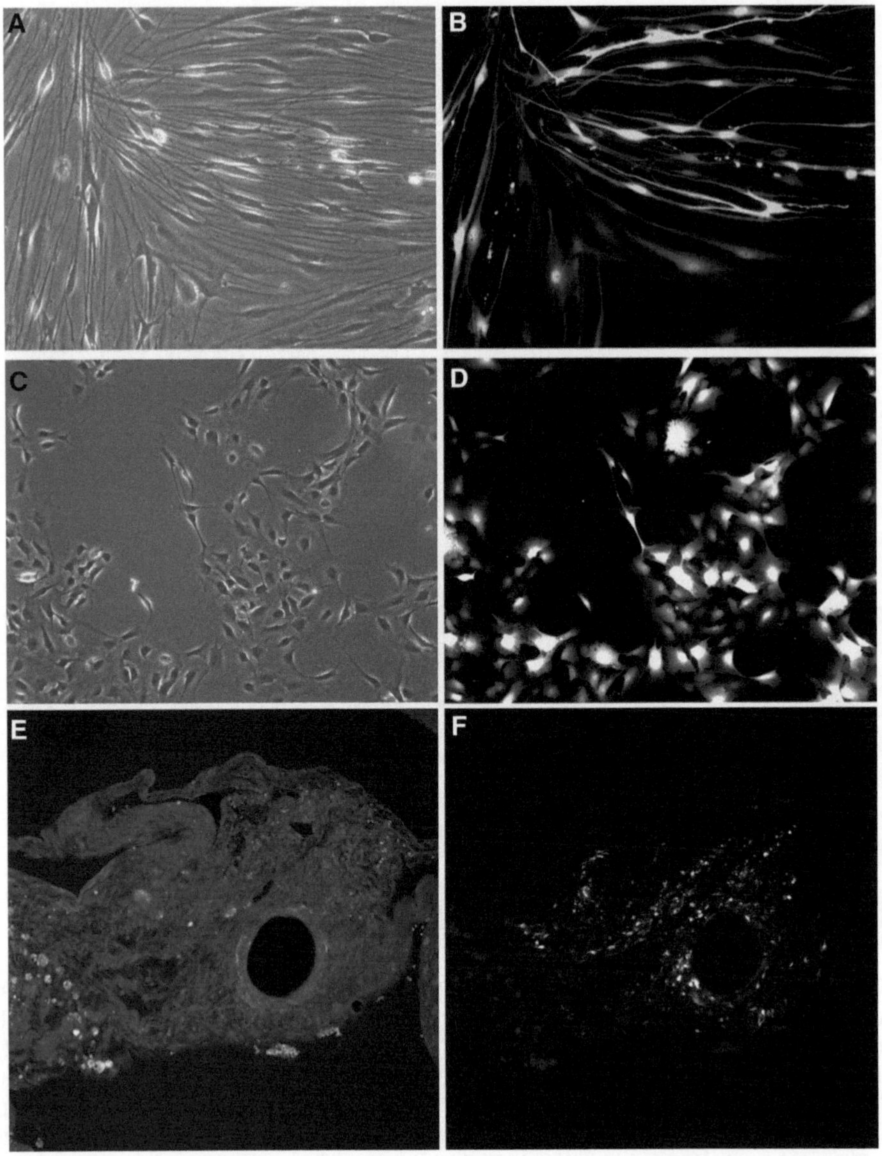

Fig. 2. Ex vivo labeling of purified adult Schwann cells (**A, B**) and adult olfactory ensheathing glia (**C, D**) using LV-GFP. **A, C** Phase contrast images and **B, D** fluorescent images of the same fields showing GFP expression in a large proportion of cells. After transplantation into the injured adult rat spinal cord (**E**), GFP transplanted LV-GFP-labeled ensheathing glia are readily identifiable (**F, A–D**).

the reporter GFP gene into the glial cell genome permits long-term identification of grafted cells (**Fig. 2E, F**) and expression is maintained even if donor cells continue to proliferate after transplantation.

1.3. In Utero Electroporation of Vector-Driven GFP

In utero electroporation is a recently developed technique for transfecting neurons and glia during early stages of development *(10, 11)*. The technique involves the injection of plasmid DNA into the ventricles of mouse embryos through the uterine wall, and electroporating it into the developing neurons and glia lining the ventricle walls using a square wave electroporator. The plasmid DNA will remain episomatic after transfection causing it to dilute out in dividing glia, but to remain present and drive expression effectively in postmitotic neurons throughout life. The main advantages of this technique are that regular plasmid DNA can be used for expression, that in vivo experiments can be performed within days to weeks after preparing the DNA, and that the procedure is rapid and effective in experienced hands. It is possible to transfect specific brain areas, cortical layers, and cell types by changing the placement of the electrodes and the day of electroporation. Sparse or widespread expression of gene products can be mediated allowing both functional and morphological studies to be carried out. In utero electroporation can mediate very high expression levels, permitting visualization of fluorescent proteins with techniques such as in vivo two photon microscopy. Finally, it allows for effective (80%) co-transfection of multiple plasmids. This property makes it possible to express functional proteins of interest and GFP to label the transfected neurons (**Fig. 3**), or to use multiple fluorescent proteins to label both synaptic structures and the outline of the neuron *(12)*.

1.4. AAV-Cre- and LV-Cre-Inducible Transgene Expression

Overexpression of a gene or inactivation of a gene in vivo is an important technique for establishing the function of a specific protein in the nervous system. Conventional overexpression and knock-out strategies, however, suffer from several limitations including embryonic or perinatal death and the compensatory regulation of other genes. Moreover, the temporal and spatial expression or knock-out of a specific gene is difficult to control with conventional transgenic mouse technology. This all hampers the interpretation of gene function studies in adult animals. One approach to producing conditional activation or inactivation of a gene in the nervous system involves the use of the bacterial Cre

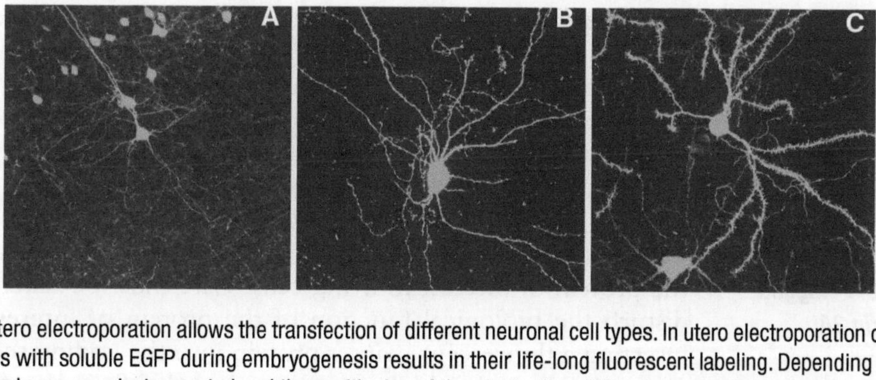

Fig. 3. In utero electroporation allows the transfection of different neuronal cell types. In utero electroporation of developing neurons with soluble EGFP during embryogenesis results in their life-long fluorescent labeling. Depending on the age at which embryos are electroporated and the positioning of the electrodes, different neuronal cell types can be targeted including (**A**) cortical pyramidal neurons, (**B**) cortical interneurons, and (**C**) spiny neurons in the striatum.

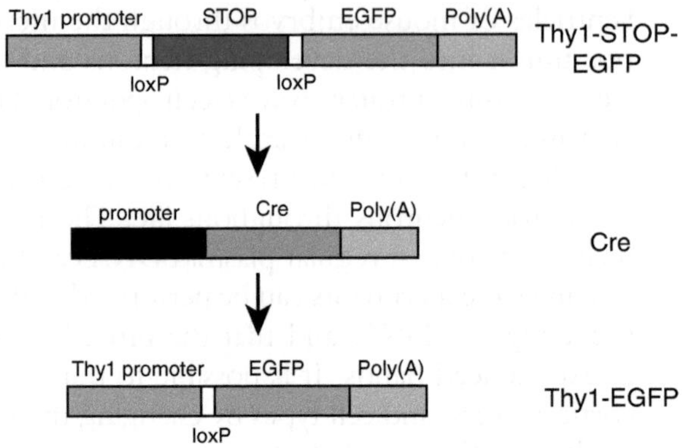

Fig. 4. Expression of membrane-associated EGFP is under the control of the neuronal Thy1-promoter or under the CMV promoter and dependent on Cre-recombinase activity. Mice carry a transgene encoding membrane-associated EGFP driven by either the neuron-specific Thy1-promoter or CMV promoter. A transcriptional "stop-cassette," flanked by loxP sites, is inserted between the promoter and the coding sequence. A similar construct was used driving expression of a dominant negative TrkB.T1-EGFP fusion protein. In the absence of Cre-recombinase, expression of EGFP does not occur. If Cre-recombinase is expressed through lentiviral or adeno-associated viral vector-mediated transduction or by cross-breeding with a Cre-expressing transgenic mouse line, the stop-cassette is excised and EGFP is expressed. Cre expressed by the viral vector is a fusion protein with EGFP (Cre-EGFP). This fusion protein will end up in the nucleus due to the nuclear localization sequence in Cre.

recombinase (Cre), an enzyme that removes DNA-constructs flanked by loxP sites. We have induced transgene expression into discrete regions of the adult mouse brain carrying a transgene that consists of a neuron-specific promoter (Thy1)-a loxP-stop-loxP cassette in front of GFP as a reporter gene (**Fig. 4**) by viral vector-mediated expression of a CreGFP fusion protein. We show that (i) both AAV and LV vectors are capable of delivering CreGFP to discrete regions of the adult brain, (ii) the CreGFP fusion protein is transported to the nucleus which allows the identification of the transduced cells based on well-defined GFP-stained (green) nuclei, and (iii) the CreGFP fusion protein retains its recombinase activity in vivo since we observe the induction of local long-term neuron-specific expression of the reporter gene (in this experiment membrane-associated GFP that labels the whole cell's cytoplasm) in fully mature neurons (**Figs. 5** and **6**). These studies make use of a GFPCre as a marker gene for virally transduced cells and use GFP as a reporter gene to monitor successful recombination resulting in the induction of transgene expression in defined populations of neurons in the adult brain *(13)*.

1.5. Use of GFP to Tag Proteins: GFP-Tagged Semaphorin 3A

Tagging proteins with GFP allows the dynamic visualization of the protein of interest in living cells. Care must be taken not to disturb the biological function of the protein of interest when inserting a fluorescent marker protein into the coding sequence. Control assays should be performed to confirm the function of the tagged protein. Functional GFP-tagged proteins can be expressed in cell lines, primary cells or living animals, and their

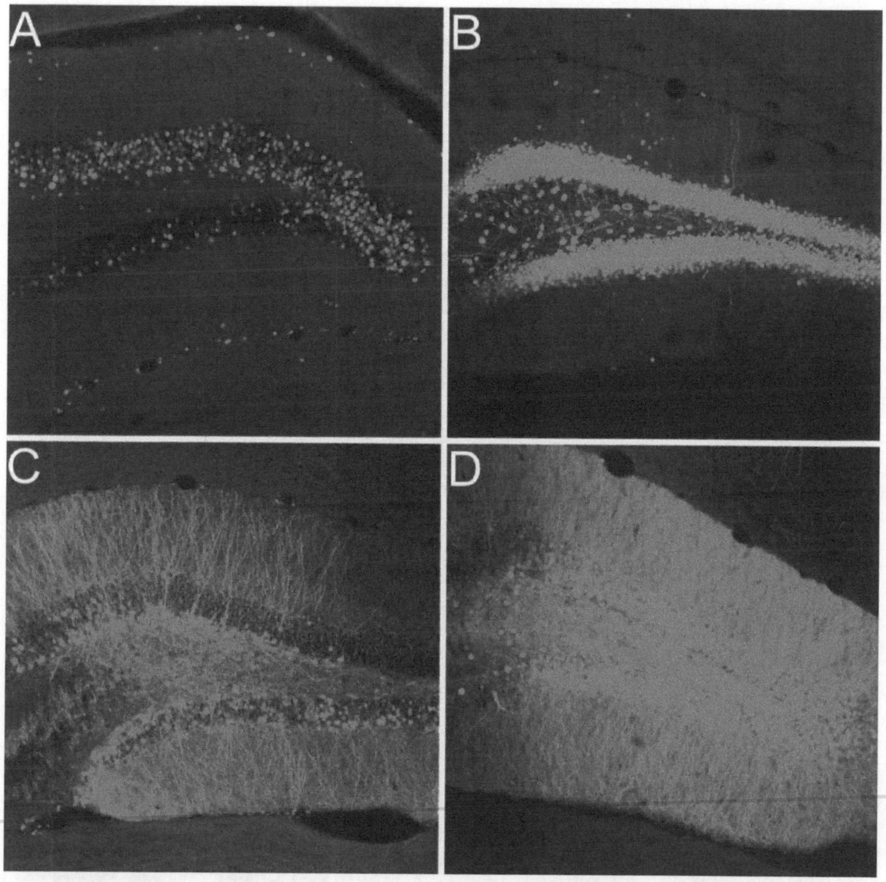

Fig. 5. Conditional expression of eGFP by viral delivery of Cre-GFP in the dentate gyrus. Lentivirus (**A, B**) or adeno-associated virus (**C, D**) expressing Cre-GFP is injected in the dentate gyrus of mice carrying the Thy1-STOP-EGFP expression cassette (**B, D**) or their wild-type litter mates (**A, C**). In wild-type mice, this results in nuclear expression of Cre-GFP, whereas in Thy1-STOP-EGFP mice Cre-GFP also activates expression of EGFP on the plasma membrane.

distribution and function can be monitored in real time. We have used this approach to analyze the trafficking of the secreted chemorepulsive axon guidance cue Semaphorin 3A (Sema3A) in dissociated cortical neurons in which GFP-Sema3A expression is driven by a Thyl promoter *(14)*. Live-cell imaging shows that GFP-Sema3A is transported in highly dynamic vesicles along microtubules in both axons and dendrites of cultured primary cortex neurons (**Fig.** 7 and Movie). While the majority of GFP-sema3A moved along the neuronal processes, a small proportion of the GFP-sema3A remains present in stationary "hot spots" apparently localized at the neuronal cell surface or in diffuse light green areas around the cells. The latter staining most likely represents secreted GFP-sema3A bound to extracellular matrix proteins *(15)*. The application of GFP-tagged proteins and live cell imaging techniques offers a unique way of studying the localization, movement, and function of protein molecules under a variety of experimental conditions.

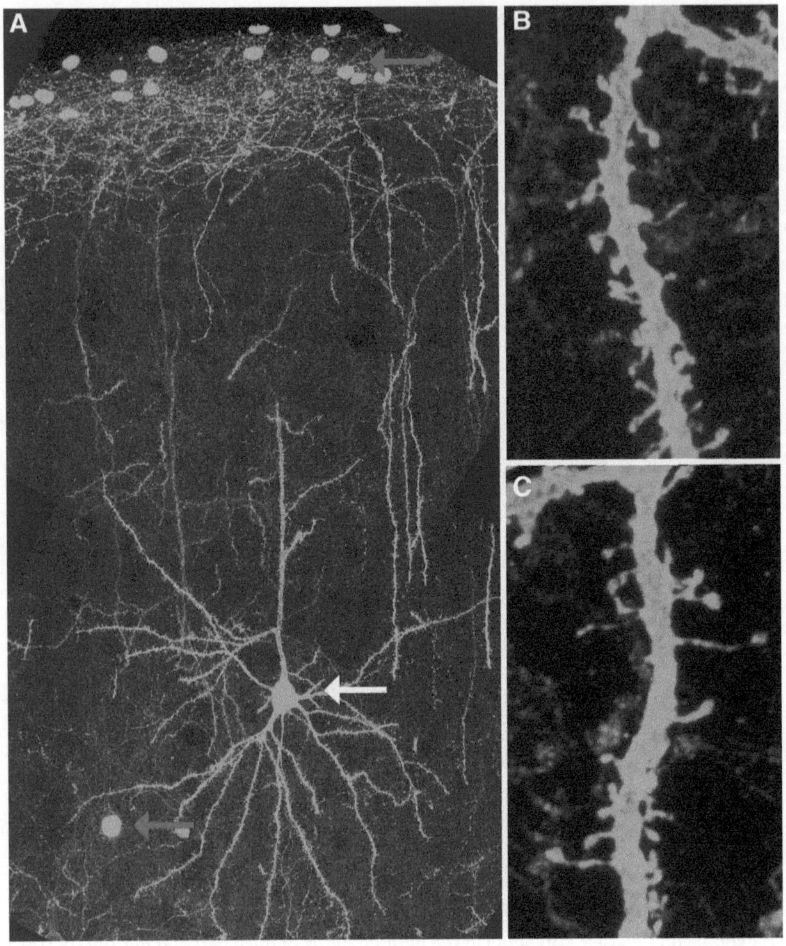

Fig. 6. Expression of membrane-associated EGFP is activated in pyramidal neurons transduced with a lentiviral Cre-EGFP vector. **A** Injection of a lentiviral Cre-EGFP vector results in the transduction of neuronal (*white arrow*) and non-neuronal (*grey arrows*) cells in the neocortex resulting in Cre-EGFP expression. Only in neuronal cells, the Thy1 promoter is active and drives expression of membrane--associated EGFP (*white arrow*). **B** and **C** High magnification images show that expression of EGFP in isolated neurons allows the morphological analysis of dendritic spines.

2. Materials

2.1. Use of AAV-GFP to Gauge Transduction Efficiency of Neural Cells In Vivo, to Phenotype Transduced Cells, and to Trace Axonal and Terminal Projections in the Brain

1. AAV helper plasmid pDG (gift of Dr. J. Kleinschmidt, Heidelberg).
2. HEK 293T cells.
3. Anesthetic: Make a 1:1 mixture of ketamine (100 mg/mL) and xylazine (20 mg/mL) (1–1.5 mL/kg bw, i.p.).
4. Analgesia: Buprenorphine (subcutaneously, 0.02 mg/kg), (Temgesic; Reckitt & Colman, UK).

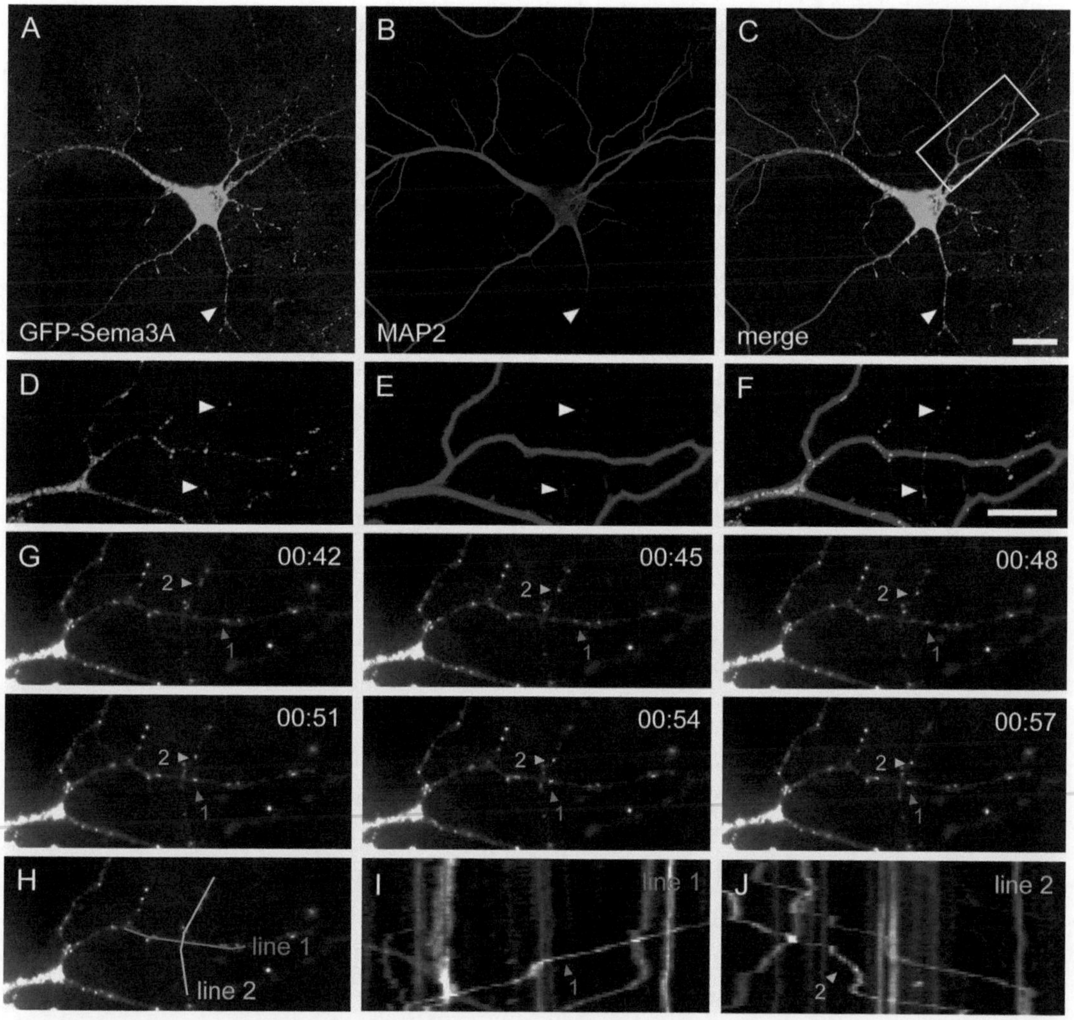

Fig. 7. Trafficking of GFP-Sema3A in a cortical neuron. **A** GFP-Sema3A is transported in secretory vesicles and localizes to discrete puncta in a dissociated cortical neuron (DIV 14). **B** MAP2 staining identifies the somatodendritic compartment of the neuron shown in (**A**). **C** Merged image. GFP-Sema3A containing vesicles localize to dendrites and the MAP2-negative axon (indicated with *arrowheads*). **D–F** Blow-up of boxed area in (**C**). *Arrowheads* indicate two GFP-Sema3A vesicles in a segment of the axon crossing dendrites. **G** Frames from a time-lapse movie lasting 1 min 17 s. Images were taken every second. Numbered *arrowheads* indicate the trajectory of two GFP-Sema3A carrying vesicles. Vesicle 1 moves in the retrograde direction in the dendrite; vesicle 2 moves in the retrograde direction in the axon. **H** Position of lines used to generate the kymographs shown in (**I**) and (**J**). **I** Kymograph plotting the fluorescence along line 1 in (**H**) for each movie frame. The trajectory of vesicle 1 is indicated. Stationary vesicles appear as vertical lines, retrograde vesicles as lines with an upward slope. **J** Kymograph plotting the fluorescence along line 2 in (**H**) for each movie frame. The trajectory of vesicle 2 is indicated. Scale bar in (**C**) 20 μm; in (**F**) 10 μm.

5. Ilium Chloroint (10 mg/g chloramphenicol, 5 mg/g hydrocortisone acetate, to protect corneas; Troy Labs, Smithfield, NSW, Australia).
6. Glass for micropipettes: Blaubrand® intraMARK glass of known internal diameter and 10 μL mark (Brand GMBH, Wertheim, Germany).

7. Cryoprotection agents: Tissue Tek[R] OCT Compound (Miles Inc.) or Jung Tissue Freezing Medium (Leica, Nussloch/Germany).
8. SuperFrost Plus® slides (Menzel-Glaser, Braunschweig, Germany).
9. For double-subbed slides, prepare a 2% solution of gelatin in hot distilled water and a 0.2% solution of chromium potassium sulfate in hot distilled water. When both of them are dissolved, add them together and filter. Dip racks of clean slides into hot solution slowly three times (for a few seconds each time). Leave racks to dry overnight. Repeat the dipping process the next day. The solution may be kept at 4 °C and reheated for reuse. The optimum dipping temperature is 35–40 °C.
10. For poly-L-lysine (PLL) coated slides, make up a solution of 0.1 mg/mL PLL MW > 300,000 (Sigma). Dip slides and dry in 37 °C oven for 30 min. Store at 4 °C. These PLL-coated slides are useable for 1–2 weeks.
11. Cellufine-sulfate chromatography resin.
12. Iodixanol for gradient isolation of viral particles (Nycomed Pharma, Norway).
13. Dulbecco's phosphate buffered saline (PBS).
14. Centricon column (Millipore, MA, USA).
15. Nembutal or Lethabarb (sodium pentobarbitone, Virbac, NSW, Australia) for euthanasia.
16. 4% paraformaldehyde in 0.1 M phosphate buffer (PB) pH 7.4.
17. 30% sucrose.
18. 25, 50, and 100% solutions of Tissue Tek[R] OCT or Jung Tissue Freezing Medium.
19. Immunohistochemistry: Antibody diluent is made by adding 10% normal goat serum (NGS) and 0.2% Triton X-100 to PBS. This solution is also used for the blocking step prior to primary antibody incubation.
20. FITC fluorescent antibody (1:300, Jackson ImmunoResearch Labs or 1:100, Sigma).
21. Citifluor (University of Kent, UK) containing Hoechst 33342 (5 μM, Sigma).
22. Biotin–streptavidin–HRP system antibody amplification (Vectastain Elite ABC kit, Vector Laboratories, CA, USA) and a diamino-benzidine (DAB)-metal complex (Pierce, IL, USA).
23. PBS containing 5–10% normal goat serum (NGS, Hunter Antisera, NSW, Australia) and 0.2% Triton X-100 (Progen Industries, Qld, Australia).
24. PBS containing 1% bovine serum albumin (BSA, Sigma) and 0.2% Triton X-100.
25. 0.6% H_2O_2 in PBS.

26. Ethanol and toluene.
27. DPX (Chem-Supply, SA, Australia).

2.2. Intracellular Dye Injection of GFP-Labeled Neurons

1. Lucifer Yellow (LY) solution: Use a stock 10% solution (1 mg in 10 μL) in 0.1 M Tris (pH 7.6). On injection day, make 2% LY in 0.1 M Tris (4 μL in 16 μL) and store in the dark and in the fridge.
2. Ames solution: Ames media powder (Sigma) and 1.9 g of $NaHCO_3$ is added to 1 L of filtered dH_2O that is bubbled with carbogen (5% CO_2 in oxygen) for 30 min. This is then sterile filtered using vacuum filtration and stored at 4 °C.
3. Injection rig: This rig consists of a fluorescent microscope (Nikon Y-FL) fitted with a 40× immersion objective. It is also attached to a camera capable of taking live, in vivo images (Nikon DS-L1 unit). To allow for constant perfusion of oxygenated Ames media, a plastic slide chamber compartment is used to hold the retina. Attached to the stage is a microelectrode manipulator that is attached to a pulse stimulator and an amplifier.
4. Microelectrodes: Borosilicate glass capillaries (GC120F-15; Harvard) are pulled into microelectrodes using a micropipette puller (P.87; Sutter Instruments) and should have a tip resistance of ~100 MΩ.
5. Antibody diluents: Anti-LY (Molecular Probes) is diluted 1:500 in PBS containing 1% Triton X-100 and 1% bovine serum albumin (BSA).

2.3. GFP Labeling of Neuroglia Ex Vivo and Tracking of Cells After Transplantation

2.3.1. Media and Supplies for Preparation of Sciatic Nerve Explants

1. Culture media: Leibovitz's L-15, DMEM (Gibco), D-10S (DMEM plus 10% heat inactivated fetal bovine serum, FBS), antibiotics (10,000 U penicillin, 10,000 μg streptomycin/mL).
2. Surgical instrument for embryo removal and dissection; fine forceps (sterile).
3. Plasticware: 100-mm, 60-mm, and 35-mm Corning tissue culture dishes.
4. Glass pipettes, graduated pipettes.
5. Halothane/isofluorane.

2.3.2. Dissociation of Explants, Culture, Purification, and Freezing of Schwann Cells

1. Media; D-10S, D-10S/mitogens: D-10S/mitogens comprises DMEM, 10% FBS, 2 μM forskolin (F6886, Sigma), 20 μg/mL bovine pituitary extract (PEX) (cat# 13028-014, Gibco, CA).
2. Dispase/collagenase: To make 50 μL dispase/collagenase, 42.5 μL DMEM (4 °C), 7.5 mL FBS (−20 °C). Make the components below in the DMEM, then sterile filter before bringing up the volume to 42.5 mL: 125 mg dispase (4 °C) (1.25 U/mL), 25 mg collagenase (−20 °C) (0.05%).

3. 35- and 100-mm culture dishes, glass and graduated pipettes, tubes. 100-mm dishes coated with poly-l-lysine (PLL, 200 μg/mL of sterilized water).
4. Trypsin/EDTA (0.05% trypsin/0.02% EDTA).
5. lx Hanks Balanced Salt solution (HBSS) containing no magnesium or calcium (Gibco, Australia).

2.3.3. Medium and Solutions for Schwann Cells

1. 100 mL of D-10S: DMEM (4 °C) 90 mL, FBS (–20 °C) 10 mL, 1 mL penicillin/streptomycin (–20 °C).
2. 100 mL of D-10S/mitogens: 13.3 μL (10 mg in 1.5 μL ethanol or DMSO) forskolin (store at 4 °C), 200 μL PEX (25 mg in 2.5 mL) (store at –20 °C).
3. 50 mL D15S: 42.5 mL DMEM (4 °C), 7.5 mL FBS (–20 °C).
4. 50 mL 10% rat serum (D1ORS): 45 mL DMEM (4 °C), 5 mL rat serum (–80 °C).
5. 50 mL Trypsin/EDTA: 5 mL Trypsin/EDTA(–20° C) (10×), 45 mL lx HBSS.
6. Chemically defined medium; DMEM/F12 50%–50%. 10 μg/mL bovine insulin, 10 μg/mL transferrin, 200 μL putrescine, 30 nM sodium selenite, 1% glutamine, 100 μL/100 mL gentamycin.
7. Thy 1.1 antibody diluted 1:5 in D-10S (supernatant from a Thy 1.1 producing cell line obtained from ATCC, USA).
8. Rabbit complement (diluted 1:8 in 1 mL dH_2O; Cappell CAT# 0012-0620).

2.3.4. Material for Olfactory Ensheathing Glia Isolation

1. Leibovitz's L-15 media (Gibco, Invitrogen, Australia).
2. DF10S: DMEM and Ham's F12 (DF, 1–1 mixture) containing 10% heat inactivated FBS and 50 μg/mL gentamycin.
3. DF10S/mitogens (DF10S supplemented with 20 μg/mL PEX and 2 μM forskolin).
4. Microscissors, fine forceps, 60-mm dishes, 100-mm PLL coated dishes, 15 mL tubes.
5. Nembutal or Lethabarb for euthanasia.
6. 0.25% Trypsin w/v (Worthington Biochemical Corporation, Lakewood, NJ).
7. 50 μg/mL DNase (Sigma) in HBSS.
8. OEG immunopanning dishes: Make 0.05 M Tris buffer (pH 9.5) and sterile filter. Dilute secondary antibody (goat anti-mouse IgG,A,M – Cappel, ION) at 1:100 (100 μg/mL) in Tris buffer. Put about 10 mL into 100-mm nonculture Petri dishes and leave overnight at 4 °C. Wash dishes three times with L-15 media. Incubate with p75NTR antibody (we use supernatant from cultured 192 hybridoma cells that express

the antibody) at a 1:5 dilution in L-15 medium with 5% FBS for 2 h at 4 °C. Wash three times with L-15, then dishes are ready for plating OEG.

2.3.5. Transplantation

1. 1× HBSS, DMEM, 10% rat serum in DMEM, trypsin/EDTA 0.05%.
2. Glass pipettes, graduated pipettes, tubes, hemocytometer.
3. Gel-foam (Pharmacia & Upjohn, Kalamazoo, MI).
4. 5 μL syringe (Hamilton, Reno, NV) mounted in a stereotaxic frame.
5. Benacillin.
6. Analgesic (Temgesic; buprenorphine; subcutaneous; 0.1 mg/kg).
7. 0.9% saline (subcutaneous).

2.4. In Utero Electroporation of Vector-Driven GFP

2.4.1. Supplies and Equipment

1. Square wave electroporator (Electro Square Porator, ECM 830, BTX Harvard Apparatus) with Tweezer electrodes ("tweezertrodes," Model 520, 7 mm, BTX Harvard apparatus).
2. Picospritzer (PLI-100, BTX Harvard Apparatus).
3. Surgery scope (optional, not necessary after E13.5).
4. Heating pad.
5. Electrode puller (p97, Sutter).
6. Surgery tools: 2× small pairs of scissors (for skin and abdominal wall), 1× needle holder, 1× standard size serrated forceps (width 2–3 mm) for suturing, 1× standard size forceps (width 2 mm) with narrow pattern serration to avoid damage to the uterus and 1× small organ holding forceps to manipulate the embryos.

2.4.2. Reagents

1. Prepare 20–50 μL of a fresh 1–4 μg/μL DNA mix containing 10 mM Tris and 0.02% Fast Green. It is advisable to make stock solutions for Tris (10×, 100 mM) and Fast Green (5×, 0.1%, filtered using a 0.22-μm filter). DNA is preferentially dissolved in water and stored frozen.
2. Ethanol 70%.
3. Sterile saline solution.
4. Anesthetics and analgesics: Preferably use isoflurane vaporization or ketamine/xylazine injection and lidocaine analgesic ointment.
5. Electrode gel.
6. Hair remover (Veet or Nair).

2.4.3. Disposables

1. Sutures: For abdominal wall, use silk (Permahand 719H, Ethicon). For skin, use polyester (Ethibond Polyester 6889H, Ethicon).
2. Sterile gauze pads.
3. Cotton tips for applying hair remover and Lidocaine ointment.

4. 20-mL syringe or disposable pipettes for applying saline.
5. Borosilicate glass capillaries (GC120F-15; Harvard) for producing injection needles.
6. Petri dishes for loading the DNA solution into the glass injection needles.

2.5. AAV-Cre- and LV-Cre-Induced Transgene Expression

For materials and methods required to generate transgenic mice, we refer to the manual by Nagy et al. *(16)*. All DNA constructs were generated using standard molecular cloning techniques *(17)*. Schematics of the constructs are shown in **Fig. 4**.

2.5.1. AAV2 and LV Vector Preparation

1. AVB sepharose (Amsterdam Molecular Therapeutics, Amsterdam, The Netherlands).
2. Lysis buffer: 50 mM Tris pH 8.5, 150 mM NaCl, 0.1% Triton X-100, 2 mM $MgCl_2$).
3. HT1080 Human fibrosarcoma cell line (Stratagene).
4. Cellulose acetate filter (0.22 μm).
5. Tubes for ultracentrifuge.
6. 0.1 M Tris pH 8.5.
7. PBS pH 3.5.
8. Dulbecco's PBS containing 5% sucrose.

2.5.2. Viral Vector Injections in Mouse Brain

1. Anesthetic: A 1:0.5:2.5 mixture of Hypnorm:Dormicum:Water (0.1 mL/10 g bw, i.p.).
2. Stereotaxic device (David Kopf instruments).
3. Infusion device: Borosilicate capillaries 1.5/0.86 mm (i.d./o.d.) are pulled into a glass needle using a micropipette puller (P-87; Sutter instruments) and connected to a 10-μL Hamilton syringe using Teflon tubing. The needle tip is adjusted to 80 μm. The infusion speed is controlled by a microinjector pump (PHD2000, Harvard apparatus).
4. AAV particles: AAV2-GFP-cre 1.2×10^9 transforming units (TU)/mL, AAV2-GFP 0.6×10^9 TU/mL.

2.5.3. Tissue Processing and Immunohistochemistry

1. Euthanasia: Nembutal (A.U.V., the Netherlands) 0.1 mL/100 g bw, i.p.).
2. Blocking buffer: PBS, 0.4% Triton X-100, 5% FBS.
3. Rabbit anti-GFP polyclonal antibody (Chemicon).
4. Donkey anti-Rabbit Alexa488 conjugated antibody (Invitrogen).
5. Mowiol mounting media: 0.1 M Tris-HCl, 25% glycerol, 10% Mowiol 4-88, pH 8.5.

2.6. Use of GFP to Tag Proteins: GFP-Tagged Sema3A

2.6.1. Neuronal Cell Culture

1. Hanks' Balanced Salt Solution (HBSS; Gibco BRL, Breda, The Netherlands).
2. Neurobasal medium (Gibco) with supplements: supplemented with B27 (Gibco), 18 mM HEPES, 0.5 mM Glutamax (Gibco), 25 μM β-mercaptoethanol and penicillin/streptomycin (Gibco).
3. 0.25% Trypsin and B27 (Gibco).
4. Glutamax, penicillin, and streptomycin (Gibco).
5. 1 M HEPES.
6. Tyrode's solution (2 mM $CaCl_2$, 2.5 mM KCl, 119 mM NaCl, 2 mM $MgCl_2$, 30 mM glucose, 25 mM HEPES, pH 7.4).
7. Rat glial cells grown on glass cover slips (Menzel Glaser, Braunschweig, Germany).

2.6.2. Live-Cell Imaging

1. Axiovert II microscope (Zeiss, Oberkochen, Germany).
2. Coolsnap HQ camera (Photometrics, Tucson, AZ).
3. Polychrome IV illumination unit (TILL Photonics, Grafelfing, Germany).
4. MetaMorph 6.2 software (Universal Imaging, Downingtown, PA).

3. Methods

3.1. Use of AAV-GFP to Gauge Transduction Efficiency of Neural Cells In Vivo, to Phenotype Transduced Cells, and to Trace Axonal and Terminal Projections in the Brain

In the laboratory at The University of Western Australia, the AAV2 serotype has mostly been used. Generation of the AAV2-GFP vector was performed according to the procedures of Hermens et al. *(18, see* also *19)*. The focus of this chapter is the use of the viral vectors so production is covered only briefly in this section.

1. Co-transfect AAV vector plasmids with the AAV helper plasmid pDG into HEK 293T cells using a calcium phosphate precipitation method.
2. Two days after transfection, harvest the virus-containing cells and release the AAV particles by repeated freeze–thawing.
3. Concentrate the supernatant using cellufine-sulfate chromatography.
4. Purify the viral particles using an iodixanol gradient.
5. Dilute the AAV containing fraction in PBS and reconcentrate using a centricon column (*see* also **Subheading 5.3.5.1**).

6. In our initial studies using AAV, we used the human CMV promoter. GFP was inserted directly downstream of this promoter in the multiple cloning site of pTR-UF. The AAV-GFP stock contained 4×10^9 TU/mL following transfection of HEK 293T cells. More recently we have used both pTR-UF12.1 and pTR-UF12 AAV2 plasmids and the CMV/chicken β actin (CMV-CBA) hybrid promoter.

3.1.1. Intravitreal Eye Injections

1. Pull glass micropipettes on a microelectrode puller. Pulled glass pipettes should have a relatively shallow profile with a shoulder of 3–4 mm. Break them with fine forceps to give a tip of about 200–300 μm in diameter.
2. Attach the micropipette by plastic tubing to a 50-μL Hamilton syringe that is filled with mineral oil.
3. Push the pipette tip through the sclera and insert through peripheral retina, immediately adjacent to the corneal/scleral junction (ora serrata) (*see* **Note 1**).
4. For adult rats, inject about 4.0 μL of AAV-GFP vector into the vitreous of the eye (*see* **Note 2**).

3.1.2. Processing of Retinal and Brain Tissue and Optic Nerves

1. After appropriate survival times (usually 1–6 months), deeply anesthetize rats (Nembutal or Lethabarb, i.p.) and perfuse with a saline wash followed by 4% paraformaldehyde in 0.1 M PB.
2. Dissect out the injected eyes from the orbit, and using fine serrated curved scissors, cut away the cornea just below the ora serrata.
3. Remove the lens, and then using fine forceps carefully pull out any residual vitreous from the eye cup.
4. Post-fix the eye cup containing the retina in the same fixative for 30–35 min.
5. Wash twice with PBS and store in PBS at 4 °C.
6. Carefully remove the brains and optic nerves from the cranium and post-fix for 2 h.
7. Wash as before and store in PBS at 4 °C.
8. Cryoprotect the eyes in 30% sucrose, then in increasing concentrations (25, 50, 100%) of Tissue Tek[R] OCT or Jung Tissue Freezing Medium.
9. Freeze the eyes by placing eye cups face down onto a glass slide with a small amount of cryoprotectant. Put the slide onto a bed of dry ice and cover with more dry ice until block is frozen. Remove the tissue using a sharp razor blade, wrap it in foil, and store dry at −80 °C until ready for cutting.
10. Mount the block onto a chuck and once the block has warmed to about −18 to −19 °C cut horizontal cryostat

sections (20 μm thick) through the eye cup, from dorsal to ventral.

11. Thaw-mount the sections onto double-subbed or PLL coated slides and store at −20 °C until immunoreacted. Every second or third section should be mounted. In most cases, about 9–10 slides can be obtained per adult rat eye, each slide containing 10–12 sections in series across the whole retina.
12. To trace axons and terminal arbors of transduced RGCs, the optic nerves, optic chiasm, and brains of these animals are cryoprotected (30% sucrose in PBS), sectioned, and immunoreacted for GFP.
13. Cut coronal frozen sections of the brain tissue (40 μm thick) through the diencephalon and midbrain. Store the sections in PBS at 4 °C.
14. Cut horizontal 20 μm sections of the optic nerves, optic chiasm, and proximal optic tract on a cryostat, and thaw-mount onto SuperFrost Plus® or double-subbed slides and store at −20 °C.

3.1.3. GFP Immunohistochemistry

GFP label in the eye can be seen under direct fluorescence microscopy; however, visualization of cell bodies and processes is greatly enhanced using immunohistochemical amplification – using either fluorescent secondary antibodies or the biotin–streptavidin–HRP system and a DAB–metal complex.

1. Remove retinal sections from the −20 °C freezer, and immediately fix for 1 min in 1% paraformaldehyde in PB.
2. Wash the sections three times in PBS and block tissue for 20 min in PBS containing 5–10% normal goat serum and 0.2% Triton X-100.
3. Incubate series of retinal sections overnight in a humidified chamber at 4 °C in the above solution containing polyclonal antibodies to GFP (1:200, Chemicon, AB3080). The next day, two different visualization procedures can be used.
4. When double or triple immunofluorescent labeling is to be carried out (*see* **Subheading 3.1.4**), after final washes in PBS visualize GFP by incubating sections in an anti-rabbit secondary FITC fluorescent antibody for 1–2 h at room temperature.
5. Mount sections in Citifluor containing Hoechst 33342 to label cell nuclei. Seal coverslips using nail varnish to prevent evaporation of the aqueous mounting medium.
6. For peroxidase staining, block endogenous peroxidase activity in 0.6% H_2O_2 in PBS.
7. Incubate sections in primary antibody (GFP, 1:200) in diluent consisting of 1% BSA and 0.2% Triton X-100 in PBS at 4 °C overnight.

8. Visualize antibody binding using a biotin–streptavidin–HRP system and a DAB–metal complex according to manufacturer's instructions.
9. Rinse slides in PBS, dehydrate in ethanol, clear in toluene, and mount in DPX.
10. Sections of optic nerve and brain can also be processed in this way to visualize GFP-positive axons and terminals that originate from transduced RGCs (**Fig. 1A–D**) (*see* **Note 3**).

3.1.4. Phenotypic Identification of Retinal Cells

1. Remove retinal sections on double-subbed or PLL-coated slides from the freezer and place in 1% paraformaldehyde in PBS (1 min).
2. After three washes, incubate sections for 1 h in a blocking solution comprising 10% NGS and 0.2% Triton X-100 in PBS.
3. Incubate sections overnight in PBS containing polyclonal anti-GFP (Chemicon) and relevant monoclonal antibodies to identify glia or different retinal neuronal types.
4. On the next day, wash sections three times with PBS and incubate for 1 h at room temperature in anti-rabbit secondary FITC fluorescent antibody and goat anti-mouse Cy3 (Jackson, both 1:300).
5. Wash sections three times with PBS and coverslip in Citifluor (*see* **Note 4**).

3.2. Intracellular Dye Injection of GFP-Labeled Neurons

In this approach, living transduced GFP-positive neurons (the descriptions here relate to RGCs in retina) are injected intracellularly with dyes to fill the cell and allow detailed study of dendritic morphology (**Fig. 1E, F**). When using bi-cistronic vectors, this approach potentially allows comparisons of the long-term effect of different transgenes on neuronal structure, and also may allow concomitant physiological analysis of virally modified neurons.

3.2.1. Retinal Dissection

1. Give animal overdose of lethabarb or other anesthesia and remove whole eye.
2. Dissect out whole retina and put in newly made up (under 2 weeks old) oxygenated Ames solution.
3. Make four cuts from the outside toward the center to make a retina flat.
4. With the RGCs facing downward, slide retina onto a coverslip using a paintbrush, while maintaining it as flat as possible.
5. Using two paintbrushes, uncurl any edges that are folded under the retina (*see* **Note 5**).
6. Place a dry piece of black Millipore paper over the top of the retina.

7. Gently press down on the filter paper to make sure that the retina sticks to the paper and flip the paper over using forceps.
8. Place in oxygenated Ames. If properly attached, the retina should not float off at all.

3.2.2. Injection of Lucifer Yellow (LY)

1. Place the retina in a slide chamber compartment, put on a fine wire retainer of some sort to keep the retina in place, then fill up chamber with Ames solution.
2. Perfuse the chamber with oxygenated Ames at a rate of about 1 drop every 2 s.
3. Pull micropipette electrodes and fill with the LY.
4. Put on alternating voltage at around 4 V and you should be able to see the LY puffing out (resistance of electrode should be between 50 and 300 MΩ).
5. Inject the cells you want by pushing the electrode into the cell gently. If the cell membrane is dimpling then you can lightly tap the micromanipulators or can buzz the capacitance which should push the electrode into the cell (*see* **Note 6**).
6. Leave pipette in for 2–5 min until the fill looks good. Gently remove the electrode (*see* **Note** 7).
7. Fill as many cells as necessary.
8. Remove retina (still on filter paper) from slide chamber and place in oxygenated Ames.

3.2.3. Immunohistochemistry for Anti-LY

1. Fix the retina in 4% paraformaldehyde for 2 h at room temperature in the dark.
2. Wash in PBS, one fast wash, then 3× 10 min.
3. Place in anti-LY antibody solution and incubate overnight at 4 °C.
4. Rinse in PBS for 1 h, then incubate in Cy3 anti-rabbit secondary antibody (1:300 in 1% Triton X-100, 1% BSA in PBS) for 4.5 h at room temperature in the dark.
5. Wash in PBS, one fast wash, and then 3× 10 min.
6. Coverslip in citifluor and seal with nail varnish.
7. Store in the fridge in the dark until use – take photos as soon as possible.

3.2.4. Microscopy and Analysis of Cells

Where possible, a photograph is taken of each cell immediately before it is injected and then straight afterwards (**Fig. 1E**). Confocal images of individual cells can then be taken (**Fig. 1F**) and *Z* projections of each cell generated using ImageJ software (free download from http://rsb.info.nih.gov/ij). Measurements are then made using the ImageJ program. Dendritic fields can also be traced directly from prepared retinal wholemounts using Neurolucida (MBF Bioscience). Parameters measured for each LY filled RGC include:

1. Soma size, measured as the best fitting circle around the cell body
2. Dendritic field area and volume
3. Branching frequency and complexity and the number of terminal tips
4. Whether there are axon-like processes and the trajectory of these processes.

3.3. GFP Labeling of Neuroglia Ex Vivo and Tracking of Cells After Transplantation

3.3.1. Preparation of Sciatic Nerves

1. Prepare a hood by putting 10 mL of L-15 in each of two 100-mm culture dishes and 5 mL of L-15 in one 60-mm dish which has been placed inside a larger 100-mm dish.
2. We usually use adult female Fischer F344 rats. Euthanize rats (halothane/isofluorane is convenient). Shave the dissection site and clean skin with soapy water and then 70% alcohol.
3. Using different sets of forceps and scissors for each tissue layer, dissect out sciatic nerves from each leg. Remove any muscle, fat, or blood vessels and put cleaned nerves in a 100-mm dish containing L-15. Using fine forceps, remove the perineurium from each nerve – start at the end that is most proximal to the spinal cord. Put nerves into the 60-mm dish and cut into ~1-mm segments using fine spring scissors (*see* **Note 8**).
4. Set up three 35-mm dishes in one 100-mm culture dish and add ~2 mL of D-10S to each 35-mm dish. Set up enough 35-mm dishes so that you will only have 10–15 pieces of nerve in each dish. A bent tip pipette is used to transfer 10–15 pieces of nerve into each 35-mm dish.
5. Remove any excess media so that the explants do not float in the dish but are still covered by tissue culture medium.
6. Feed cultures twice a week with D-10S, and transfer them to fresh 35-mm dishes containing fresh media approximately every week. They should be transferred when you see a large amount of outgrowth from the explant. At early stages, these are primarily fibroblasts. The segments will be dissociated when the majority of the cells migrating out of the explants are SCs, usually after about 3 weeks.

3.3.2. Dissociation of Explants, Culture, and Immunopurification of Schwann Cells

1. Set up a 35-mm dish in a 100-mm dish. Transfer all the explants from one preparation into the 35-mm dish and add 1 mL of the dispase/collagenase solution.
2. Put the explants in a CO_2 incubator overnight.
3. The next day, coat dishes with poly-l-lysine for plating out the cells at the end of the procedure.

4. Transfer the solution containing the explants into a 16-mL snap cap tube containing 1 mL of D-10S. Rinse the dish which had the explants with 2 mL of D-10S, until there is nothing left on the dish. Spin the cells at 1,500 rpm for 10 min at 4 °C.
5. Remove the supernatant, resuspend the cells in D-10S (2 mL) with a small bore glass fired pipette, and then add more D-10S to make a total of 5 mL.
6. Spin the cells at 1,500 rpm for 5 min at 4 °C. Repeat steps 5 and 6 until the supernatant looks clear, the cells are then ready to be plated out.
7. Resuspend the cells in D-10S/mitogens. At first, use only 2 mL of D-10S/mitogens, then add sufficient D-10S/ mitogens to put 2 mL/100-mm dish, mix well. Plate cells onto PLL-coated dishes which contain 5 mL of the D-10S/ mitogen feed.
8. Cells are then put into 37 °C 5% CO2 incubator and are fed twice a week with D-10S/mitogen (7 mL/dish).
9. When the cells are ready to be passaged, remove the medium and wash the cells twice with HBSS (containing no magnesium or calcium). Add 3 mL of trypsin/EDTA in HBSS to each dish and wait for 3–5 min until cells start to detach.
10. Remove the cells from the dish and put into a tube containing 1 mL/dish of D-10S, then rinse with 3 mL of D-10S and put this into the tube containing the cells (tube size is determined by the number of dishes being split; for one dish use a 15-mL tube, for two dishes use a 50-mL tube). Spin the cells at 1,500 rpm for 10 min at 4 °C.
11. Remove the supernatant and resuspend cells in 2–3 mL of D-10S with a small bore pipette, then add D-10S to a final volume of 5 or 10 mL (small or large tube, respectively). Spin at 1,500 rpm for 5 min at 4 °C.
12. Remove the supernatant and resuspend in 2–3 mL D-10S/ mitogens, bring to a final volume of 2 mL/dish to be plated onto PLL dishes which contain 5 mL of D-10S/mitogens. Put the dishes into the CO_2 incubator and feed as described in the dissociation protocol.
13. For immunopurification of SCs (*see* **Note 9**), initially proceed as for **steps 9–11** above.
14. Remove the supernatant and resuspend the cells in 1 mL of Thy 1.1 antibody per 100-mm dish. Put the tube with the cells on a shaker within a 37 °C CO_2 incubator for 30 min.
15. Spin the cells at 1,500 rpm for 5 min at 4 °C, remove the supernatant, and resuspend the cells in 1 mL of rabbit complement (diluted 1:8 in 1 mL dH_2O) per 100-mm dish. Again put the tube on the shaker (for 15 min).

16. Add D-10S/mitogens so that you have 2 mL per dish to be plated. Plate 2 mL of cells out into the PLL dishes which contain 5 mL of D-10S/mitogens. Put dishes into the CO_2 incubator.
17. Feed the cells as described in the Dissociation protocol (*see* **Note 10**).

3.3.3. Freezing of Schwann Cells

Initially proceed as for **steps 9–11** above (**Subheading 3.3.2**).

1. Count the cells using a Hemocytometer. Count the number of cells in one of the large blocks. Calculation: # of cells × 10^4 = # of cells/mL. # of cells/mL × volume (mL) = total # of cells. You should freeze about 1.5 million cells/mL/cryovial.
2. Spin the cells at 1,500 rpm for 5 min at 4 °C. Resuspend the cells in the appropriate amount of D-15S, based on hemocytometer counts. When cells are well suspended, add DMSO to get a final concentration of 6%. Quickly aliquot the cell suspension and put in a freezing container at −80 °C or in liquid nitrogen (*see* **Note 11**). Undiluted DMSO can kill cells and should not be allowed to be in contact with cells for any length of time prior to mixing and freezing.

3.3.4. Adult Olfactory Ensheathing Glia Culture and Purification by Immunopanning

1. Primary olfactory bulb cultures are prepared from the 3- to 4-month-old olfactory bulbs of adult female Fischer 344 rats. Deeply anesthetize Fischer rats by injection of Nembutal and then kill by decapitation. Take off dura and carefully dissect out the olfactory bulbs. Place in a 100-mm dish containing 10 mL of Leibovitz's L-15 media. Ensure that at least a portion of the frontal cortex is attached to each bulb.
2. Using microscissors, cut away any pieces of cortex and with fine forceps and carefully remove the pia and blood vessels. Cut away the most ventral portion of the olfactory bulbs to obtain the olfactory nerve fiber layer and some of the glomerular layer.
3. Place ventral parts of the bulbs into one 60-mm dish with 2 mL of HBSS containing 0.25% trypsin (w/v) and 50 μg/mL DNase. Cut the tissue into 0.5–1 mm^3 pieces and incubate in this solution for 60 min at 37 °C in a CO_2-free incubator on a rotating shaker. Stop trypsinization by adding DF10S.
4. Transfer the tissue to a tube after trituration (about 10×) through a fire polished glass pipette. Triturate sufficient times until a cloudy suspension is obtained (not necessarily a single cell suspension but there should be no obvious large clumps). Place the fragments into a 15-mL tube.

5. Add about 10 mL of DF10S medium to make 13–14 mL solution, then spin at 1,500 rpm for 10 min at 4 °C. Discard the supernatant and very carefully resuspend in 3 mL using a fire polished glass pipette and triturate about 15× in DF10S.
6. For five rats (10 bulbs), use three 100-mm PLL coated dishes, which contain 7 mL of DF10S medium. Split suspension into three and distribute an equal volume into each of the dishes. Leave the cultures for 3–4 days in an incubator (37 °C, 5% CO_2) and then change medium, using DF10S/mitogens.
7. Leave for about another 4 days (change medium after 2 days), which in total is about 7 days since isolation from the bulbs. OEG are then sorted using FACS or immunopanning.
8. For immunopanning (to purify OEG from other contaminating cells found in the olfactory bulb such as astrocytes, microglia, oligodendrocytes and precursors, and endothelial cells), detach cells using 0.05% trypsin and 0.02% EDTA, then centrifuge and wash twice with DF10S. Resuspend the cells in L-15 medium and plate on 100-mm dishes that have been prepared for immunopanning.
9. OEG are immunopanned using an antibody against p75 neurotrophin receptor (p75NTR). Make 0.05 M Tris buffer (pH 9.5) and sterile filter.
10. Dilute secondary antibody (goat anti-mouse IgG,A,M – Cappel, ION) at 1:100 (100 μg/mL) in Tris buffer. Put about 10 mL into 100 mm nonculture Petri dishes (Corning) and leave overnight at 4 °C. Use of this type of dish is important to minimize nonspecific cell adhesion.
11. Wash dishes three times with L-15 media. Incubate with p75NTR antibody (we use supernatant from cultured 192 hybridoma cells that express the antibody) at a 1:5 dilution in L-15 medium with 5% FBS for 2 h at 4 °C.
12. Wash three times with L-15, then plate cells onto antibody-treated dishes. Use of the immunopanning dishes is important to minimize nonspecific cell adhesion.
13. For each dish, use 10 mL of L-15 cell suspension. Leave cells for 30 min at 4 °C. Wash dishes five times using L-15 to remove any unbound cells. This is important because fibroblasts can loosely adhere to the dishes.
14. The positive panned cells are maintained in a humidified incubator in DF10S for about 4 days at 37 °C.
15. When cells have reached 70–80% confluence, detach cells using trypsin/EDTA. Papain (~10 U/mL) can also be used.

Incubate for 5 min at 37 °C in humidified incubator. Using a cell scraper, carefully dislodge any cells that have not been removed by the trypsin/EDTA. Add DF10S medium to the dish to inactivate the trypsin.

16. Place cells in a 15-mL tube and make solution up to 10 mL. Centrifuge at 1,500 rpm for 10 min. Resuspend pellet in 1–2 mL of DF10S medium, adjust volume to 6 mL, and plate onto six PLL-coated 100-mm dishes.
17. Feed every 3 days with either DF10S or DF10S/mitogens (*see* **Note 12**).

3.3.5. Ex Vivo Transduction of Primary Cell Cultures (SCs or OEG)

Depending on what vector system is to be used for the introduction of the GFP gene into target cells, different dosages of vector addition to the culture medium may be required for optimal transduction. Optimal multiplicity of infection (m.o.i) needs to be determined for individual cell types via titration. Optimization experiments are usually conducted in 24-well plates. If the required dosage for optimal transduction of your cell culture is already known, use this m.o.i. in **step 4** and disregard titration step.

1. Seed cells at a density of 2×10^4–3×10^4 cells per well, about 12–24 h prior to transduction (incubate cells overnight using standard tissue culture conditions) and cells should reach 70–80% confluency the following day.
2. The next day, visually inspect the cells using normal light microscopy. Ensure that the cells appear healthy and are at the appropriate density.
3. Take the GFP vector from –80 °C storage, thaw at room temperature (*see* **Note 13**).
4. To determine the required dosage for optimal transduction of cells, prepare a titration range starting from an m.o.i. of 1 up to an m.o.i. of 100, which is usually sufficient (higher m.o.i.s can be tested if required). A typical titration range may include m.o.i. 1, 10, 25, 50, 100, and so on. Use a minimum amount of medium. For an m.o.i. of 1, add the same number of transducing units (T.U.) as the original number of cells plated into the well. Incubate cells with viral vector-containing medium under standard tissue culture conditions (*see* **Note 14**).
5. Refresh the medium after 6–14 h (overnight) and grow for an additional 48 h to allow for initiation of transgene expression (*see* **Note 15**).
6. After 48 h, incubate the cells with 10 μg Hoechst 33342 nuclear dye per milliliter of culture medium for 5 min at 37 °C. Then fix cells with 4% paraformaldehyde solution in PBS for 10–15 min at room temperature.

7. Visualize native GFP fluorescence under a conventional fluorescent or confocal laser scanning microscope (**Fig. 2a–d**). The transduction efficiency is determined from the ratio of GFP-expressing cells versus the total cell number (Hoechst nuclear dye). Hoechst nuclear dye can be visualized using a UV light source. Count at least ten fields of view, using a 10–20× objective, at random locations across the well.

3.3.6. Transplantation Using GFP-Labeled Glia

In spinal cord injury studies in rats, there are various injury models including contusion, hemisection, or complete transection. It is beyond the scope of this chapter to describe these methods, but grafting of virally transduced, GFP-expressing cells into the spinal cord is carried out as follows.

Preparation of Cells

1. Rinse the 100-mm dishes containing cells twice with lx HBSS. Add 3 mL of trypsin/EDTA to each dish and wait for 3–5 min until cells start to detach.
2. Remove cells from the dish and put them into a tube (check splitting protocol for size) containing l mL of 10% rat serum/dish. Rinse dish with 3 mL of 10% rat serum and add to cells in the tube. Spin the cells at 1,500 rpm for 10 min at 4 °C.
3. Remove the supernatant and resuspend cells in 2–3 mL of DF with a small bore pipette, then add DF to a final volume of 10 mL. Spin the cells at 1,500 rpm for 5 min at 4 °C.
4. Remove the supernatant and resuspend cells in 2–3 mL of DF, then add DF to a final volume of 10 mL. Now the cells are ready to count.
5. Count the cells using a Hemocytometer. Count the number of cells in one of the large blocks. Calculation: # of cells × 10^4 = # of cells/mL. # of cells/mL × volume (mL) = total # of cells.
6. Spin the cells at 1,500 rpm for 5 min at 4 °C, remove all the supernatant (make sure you remove as much of the DF as possible). The cells are now ready to use.

Surgery and Injection of Cells

1. Adult rats (usually Fischer females) are anesthetized with 5% halothane (in 60:40 O_2:N_2O) then maintained at 2–3%. The skin and muscle layers are cut rostrocaudally to expose the vertebral column, and a T10 laminectomy performed *(20)*.
2. Gel-foam is used to stop bleeding if it occurs.
3. Rats are partially suspended from clamps to the dorsal processes of T8 and T11/12 to circumvent diaphragm-induced dorsoventral movement of the spinal column.
4. SCs or OEG are transplanted after no more than two passages, or 14–21 days in vitro *(20)*. Inject a 5-μL suspension containing 5 × 10^5 SCs or OEG in culture medium, or medium alone

(1:1 DMEM/F-12) into the spinal cord at a depth of 0.8 mm over a period of 3 min using a 5-μL syringe mounted in a stereotaxic frame. Withdraw the needle after 3 min (*see* **Note 16**).

5. Immediately after surgery, give each animal injections of penicillin (Benacillin; intramuscular), analgesic (Temgesic; buprenorphine; subcutaneous; 0.1 mg/kg), and 2–4 mL 0.9% saline (subcutaneous). Place animal boxes partially onto a heating pad for 24 h to assist with maintenance of body temperature. Continue antibiotic on alternate days, analgesic twice daily, and saline once daily for 1 week.

3.4. In Utero Electroporation of Vector-Driven GFP

3.4.1. Preparation for Surgery

1. Pull glass needles and break the end resulting in a sharp, ~50 μm wide tip. Do not pull the needles too long as they will break easily when attempting to inject through the developing fetal skull.
2. Apply 20 μL of the DNA solution onto a Petri dish using a standard pipette and fill two glass needles using the Picospritzer (*see* **Note 17**). Adjust the injection volume to 0.5–2 μL.
3. Prepare a syringe or small beaker with warm (37 °C) saline solution. Keep it warm during surgery.
4. Clean and sterilize surgery tools.
5. Have pregnant mouse at required day of pregnancy ready. For electroporating excitatory neurons in layers 5/6 or inhibitory neurons, use mice at E12.5–E13.5. For electroporating excitatory neurons in layer 2/3, use mice at E14.5–E16.5.

3.4.2. Surgery

1. Anesthetize a pregnant mouse using anesthesia of choice. Keep the mouse on a heating pad with a clean cover. An intramuscular Temgesic injection may be used for extra analgesia.
2. Remove hair from the belly using chemical hair remover. A razor can also be used, but care must be taken not to damage the nipples. Clean and disinfect the skin with 70% ethanol.
3. Make a straight vertical cut in the skin. The cut should be 1.5–2 cm depending on the day of pregnancy. The uterus with embryos should be able to fit through the cut without applying force. Separate the edges slightly from the abdominal wall to make suturing afterwards easier.
4. Make a cut of the same size through the abdominal wall.
5. Place sterile gauze pads around the cut and keep them wet with warm saline.
6. Carefully pull out one uterine horn and place it on one of the pads next to the incision. Make sure not to apply force and never to damage the uterine wall (*see* **Note 18**).

7. Carefully manipulate one embryo at a time to make the head clearly visible. Using the picospritzer with a glass needle inject 0.5–2 μL DNA into one lateral ventricle of each embryo. The ventricles are located half way along one side of the reddish Y-shaped suture of the developing skull. Successful injection will be clearly visible due to the Fast Green dye in the DNA mixture.
8. Electroporate the embryos using the "tweezertrode" electrodes covered with electrode gel. The positive paddle of the tweezertrode has to be placed at the cortical position to be targeted. Try to place the negative paddle at the other side of the head with the ventricle positioned between the two paddles. To hit interneurons and spiny stellate projection neurons in the striatum, place the positive paddle laterally just under the ventricle and the negative paddle again at the other side of the head with the ventricle between the two poles. This will result in targeting developing neurons in the ganglionic eminence *(21)*. Electroporate using 3–5 pulses at 35 V. More pulses result in more neurons being transfected. Avoid hitting the mother's blood vessels or the body of the embryo as this may lead to abortions.
9. Return the uterine horn back into the abdominal cavity (*see* **Note 19**).
10. Repeat the procedure with the other uterine horn. Do not electroporate more than eight embryos as this will often result in abortions.
11. Suture the abdominal wall with silk (Permahand, 719H) and the skin with polyester (Ethibond Polyester, 6889H) (*see* **Note 20**).
12. Apply lidocaine on the wound and let the mother recover on a heating pad.
13. Up to 3–4 days after birth, the success of electroporation of constructs encoding fluorescent proteins can be confirmed by inspecting the pups under a fluorescent microscope with a high numerical aperture, long working distance (>2 cm) objective (*see* **Note 21**).

3.5. AAV-Cre- and LV-Cre-Induced Transgene Expression in the Adult Mouse Brain

3.5.1. AAV-Cre Vector Preparation

AAV2 particles are generated using methods referred to in **Subheading 3.1** using an alternative virus purification method listed below:

1. Two days after transfection, lyse cells in lysis buffer.
2. Incubate lysate with DNAse I (10 μg/mL) for 1 h.
3. Load cleared lysated on AVB sepharose column.
4. Wash column with 10 column volumes (CV) 0.1 M Tris pH 8.5.
5. Elute in 20 CV PBS pH 3.5. Collect eluate in 10% V/V 0.1 M Tris pH 8.5.

6. Change the buffer to Dulbecco's PBS containing 5% sucrose. The elution buffer is exchanged with formulation buffer (PBS/5% sucrose) by means of concentration/diafiltration. This is performed with a hollow fiber processed onto a tangential flow filtration (TFF) unit.
7. Determine viral titer by infecting HT1080 cells (Stratagene) according to manufacturers' guidelines.

3.5.2. Lentiviral Vector Preparation

Lentiviral particle generation involves the transfection of 293T cells with the transfer vector, the envelope vector, and the packing vector using calcium phosphate or lipofectamine. Only high efficiency transfections will result in the generation of lentiviral particle preparations that are active in vivo. The particles are harvested at 2 and 3 days after transfection. After the first harvest, the medium is replaced by fresh medium.

1. Spin the collected medium at low speed (1,000 rpm) for 5 min to remove detached cells.
2. Filter the supernatant through a 0.22-μm cellulose acetate filter.
3. Concentrate the lentiviral particles by ultracentrifugation. Spin at 50,000 × g for 2.2 h at room temperature.
4. Discard the supernatant completely, and resuspend the pellet in 50 mL of PBS by carefully pipetting the pellet up and down. Leave at room temperature for 1–2 h while regularly mixing it.
5. Aliquot and store at –80 °C.
6. Determine the titer by transduction of 293T cells.

3.5.3. Viral Vector Injections

Methods for viral vector injections are basically as described in Ruitenberg et al. *(5)* but with adjustments made for mice as described below and in Ahmed et al. *(13)*.

1. Deeply anesthetize animal and place in a stereotaxical device fixed by ear and tooth bars.
2. Expose skull by making incision in rostrocaudal direction.
3. Remove remaining periosteum by scraping with a scalpel knife.
4. Set stereotaxic coordinates for hippocampus from bregma: AP: –2.0, L: +1.3.
5. Drill small hole and lower capillary needle to z-coordinate (DV: –1.5 from dura).
6. Infuse 1 μL virus using a microinjection pump device (0.2 μL/min).
7. Leave injector in place for 1 min after infusion and slowly raise the injector.

8. Close skin and let animal recover at 37 °C.

3.5.4. Tissue Processing and Immunohistochemistry

1. Euthanize animals by overdose Nembutal.
2. Perfuse transcardially with ice cold 0.9% NaCl solution followed by ice cold 4% paraformaldehyde in 0.1 M phosphate buffer, pH 7.4 (PFA).
3. Fix brains for 16 h in PFA at 4 °C.
4. Prepare 40 μm sections on vibrating blade microtome (VT100S, Leica).
5. Incubate 1 h in blocking buffer:PBS, 0.4% Triton X-100, 5% FBS.
6. Incubate 16 h with rabbit anti-GFP 1:100 in blocking buffer.
7. Wash three times (15 min each) with PBS, 0.4% Triton X-100.
8. Incubate 3 h with donkey anti-rabbit-alexa-488 1:400 in blocking buffer.
9. Wash three times (15 min each) in PBS, 0.4% Triton X-100, and mount in Mowiol mounting media.

3.6. Use of GFP to Tag Proteins: Live Imaging of GFP-Tagged Sema3A

We use cultures of dissociated cortical neurons, prepared according to standard procedures.

3.6.1. Neuronal Cell Culture

1. Dissect cerebral cortices from C57BL/6 mouse E18 embryos free of meninges in Hanks' Balanced Salt Solution containing 7 mM HEPES.
2. Digest the tissue with 0.25% trypsin in HBSS for 15 min at 37 °C.
3. Wash tissue three times in HBSS/HEPES and triturate with a fire-polished glass pipette.
4. Plate dissociated neurons at a density of 50,000 cells/well on top of a layer of rat glial cells grown at a density of 25,000 cells/well on 18-mm glass coverslips in 12-well plates.
5. Maintain neuronal cultures in Neurobasal medium with supplements. Replace half of the medium once a week.
6. Calcium phosphate transfections are done as described in Kohrmann et al. (22). Cells are routinely transfected at 6–8 DIV and imaged between 14 and 17 DIV.

3.6.2. Live-Cell Imaging

1. Place coverslips in a chamber containing Tyrode's solution and image on an Axiovert II microscope equipped with a heated stage.
2. Images are best acquired in MetaMorph 6.2 software using a 40× objective. Stacks from 1 min time-lapse recordings are acquired with an interval of 1 s. Use these to visualize the dynamics of vesicle trafficking.

Notes

1. This type of injection minimizes damage to the retina, but care must be taken to angle the pipette toward the back of the eye. Aim at the posterior retinal surface and at the same time avoid damaging the lens.
2. From a pathway tracing and neural plasticity perspective, the use of vectors to incorporate GFP into neurons potentially allows the isolation and identification of individual axons and terminal arbors in the complex architecture of the mammalian CNS. But the amount of vector delivered and the level of transduction is critical. As with the traditional Golgi silver technique to label neurons and their dendrites, care must be taken not to transduce too many neurons, otherwise tracking of individual axons becomes very difficult, if not impossible.
3. These are relatively straightforward protocols. Note for AAV that there is a delay in transgene expression due to the time required for conversion of recombinant AAV-DNA to a transcriptionally active double-stranded form (23). After ocular administration there is some transgene expression after 3 days and using an antibody to GFP we saw intense expression in some rat retinal RGCs 7 days after intravitreal injection, although the total number of transduced cells was less than that seen at 21 days (1).
4. Post-IRES GFP expression is likely to be weaker and hence enhancing the signal using GFP antibodies is essential. But caution must always be used when attempting to quantify transduction efficiency using this type of bi-cistronic construct.
5. To enable the best cell injections care must be taken to ensure that all of the vitreous is removed and that the retina is completely flat. If the vitreous is not completely removed the microelectrode gets stuck in this layer, making it very difficult to maneuver to the RGCs.
6. To inject specific GFP-labeled cells, the transduction must be good enough to illuminate the cell, preferably the cell body. If only the processes are labeled, this makes injection very difficult as the processes are too thin to have a clean injection. If the transduction is faint, retrograde fluorogold labeling is very useful in revealing the cell body.
7. Care must also be taken when inserting and removing the electrode from cells as if other cells are accidentally punctured the LY will quickly flow into these cells as well, making morphological analysis very difficult.

8. When removing the perineurium from adult sciatic nerves, the nerve end sometimes frays making it difficult to identify the perineurial coat. If this happens, cleanly cut a 1 mm piece off the end of the nerve and try again.
9. Immunopurification of SCs using Thy 1.1 plus complement to lyse contaminating fibroblasts is only usually required if serial explant depletion of fibroblast content is unsuccessful (more than 5% contamination).
10. When feeding SC cultures, make sure that you remove the media from the side of the dish and gently add the media to the side of the dish, to avoid dislodging cells. They can be used for transplant preparation and any other protocols.
11. SCs should be split when the cells have reached confluence and frozen when there are more cells than are needed in the near future. Adult SCs can be passaged about six times, but after that the cells become transformed and no longer behave as primary cultures.
12. For OEG, the yield of cells after immunopanning and approximately 1 week in culture should be 1.5 million cells per dish, thus giving ~9–10 million cells for six dishes.
13. Virus should be carefully resuspended by pipetting up and down (do NOT vortex as this will result in denaturing and loss of stock titer!). Once thawed, most viral vector preparations cannot be refrozen as this will cause a substantial drop in the titer and obscure transduction efficiency. We now only use LV vectors for adult SC and adult OEG transduction *(7–9)*.
14. Accurate determination of viral stock titer is important to obtain reproducible transduction efficiency and consistent results.
15. For nonintegrating vectors and fast proliferating cells, serum and/or mitogen concentration can be reduced.
16. For cell injections into the spinal cord, in contusion injury models the cells are injected toward the center of the injury. However, in hemisection or complete transection models, cells are injected 1 mm distal and 1 mm proximal to the lesion site.
17. For in utero electroporation, a second needle should be prepared and ready in case the first needle breaks during surgery.
18. During surgery, always keep the uterus wet with warm saline.
19. Make sure the uterus is not twisted as this will lead to retention of the embryos and death of the mother.
20. Never suture the skin with silk as this will frequently cause reopening of the wound.

21. Please keep in mind that not all promoters are functional in vivo after electroporation. For example, the neuronal Thy1 and CamKII promoters will not be active. The combined b-actin/CMV promoter/enhancer combination (pCAG) works well in adult mice especially when the woodchuck hepatitis virus posttranscriptional regulatory element is included in the construct.

References

1. Harvey, A. R., Kamphuis, W., Eggers, R., Symons, N. A., Blits, B., Niclou, S. P., Boer, G. J., and Verhaagen, J. (2002) Intravitreal injection of adeno-associated viral vectors results in the transduction of different types of retinal neurons in neonatal and adult rats: a comparison with lentiviral vectors. *Mol. Cell. Neurosci.* 21, 141–157.
2. Mizuguchi, H., Xu, Z., Ishii-Watabe, A., Uchida, E., and Hayakawa, T. (2000) IRES-dependent second gene expression is significantly lower than cap-dependent first gene expression in a bicistronic vector. *Mol. Ther.* 1, 376–382.
3. Leaver, S. G., Cui, Q., Plant, G. W., Arulpragasam, A., Hisheh, S., Verhaagen, J., and Harvey, A. R. (2006) AAV-mediated expression of CNTF promotes long-term survival and regeneration of adult rat retinal ganglion cells. *Gene Ther.* 13, 1328–1341.
4. Harvey, A. R. (2000) Labelling and identifying grafted cells. In: Neuromethods (36), Neural Transplantation Methods. S. B. Dunnett, A. A. Boulton and G. B. Baker (Eds), Humana Press, New Jersey, pp 319–361.
5. Ruitenberg, M. J., Eggers, R., Boer, G. J., and Verhaagen, J. (2002) Adeno-associated viral vectors as agents for gene delivery: application in disorders and trauma of the central nervous system. *Methods* 28, 182–194.
6. Ruitenberg, M. J., Plant, G. W., Christensen, C. L., Blits, B., Niclou, S. P, Harvey, A. R., Boer, G. J., and Verhaagen, J (2002) Viral vector-mediated gene expression in olfactory ensheathing glia implants in the lesioned rat spinal cord. *Gene Ther.* 9, 135–146.
7. Ruitenberg, M. J., Levison, D. B., Lee, S. V., Verhaagen, J., Harvey, A. R., and Plant, G. W. (2005) NT-3 expression from engineered olfactory ensheathing glia promotes spinal sparing and regeneration after injury. *Brain* 128, 839–853.
8. Hu, Y., Leaver, S. G., Plant, G. W., Hendricks, W. T. J., Niclou, S. P., Verhaagen, J., Harvey, A. R., and Cui, Q. (2005) Peripheral nerve constructs genetically engineered to express CNTF enhance the regeneration of adult CNS axons. *Mol. Ther.* 11, 906–915.
9. Hu, Y., Arulpragasam, A., Plant, G. W., Hendriks, W. T. J., Cui, Q., and Harvey, A. R. (2007) The importance of transgene and cell type on the regeneration of adult retinal ganglion cell axons within reconstituted bridging grafts. *Exp. Neurol.* 207, 314–328.
10. Tabata, H., and Nakajima, K. (2001) Efficient in utero gene transfer system to the developing mouse brain using electroporation: visualization of neuronal migration in the developing cortex. *Neuroscience* 103, 865–872.
11. Saito, T., and Nakatsuji, N. (2001) Efficient gene transfer into the embryonic mouse brain using in vivo electroporation. *Dev. Biol.* 240, 237–246.
12. Gray, N. W., Weimer, R. M., Bureau, I., and Svoboda K. (2006) Rapid redistribution of synaptic PSD-95 in the neocortex in vivo. *PLoS Biol.* 4, e370.
13. Ahmed, B. Y., Chakravarthy, S., Eggers, R., Hermens, W. T., Zhang, J. Y., Niclou, S. P., Levelt, C., Sablitzky, F., Anderson, P. N., Lieberman, A. R., and Verhaagen, J. (2004) Efficient delivery of cre-recombinase to neurons in vivo and stable transduction of neurons using adeno-associated and lentiviral vectors. *BMC Neurosci.* 5, 4.
14. De Wit J., Toonen, R. F., Verhaagen, J., and Verhage, M. (2006) Vesicular trafficking of semaphorin 3A is activity-dependent and differs between axons and dendrites. *Traffic* 7, 1060–1077.
15. De Wit, J., De Winter, F., Klooster, J., and Verhaagen, J. (2005) Semaphorin 3A displays a punctate distribution on the surface of neuronal cells and interacts with proteoglycans in the extracellular matrix. *Mol. Cell. Neurosci.* 29, 40–55.

16. Nagy, A., Gertsensstein, M., Vintersten, K., and Behringer, R. (2003) *Manipulating the Mouse Embryo: A Laboratory Manual*, third edition. Cold Spring Harbor Laboratory Press, Cold Spring Harbor, NY.
17. Sambrook, J. F., and Russell, D. W. (Eds). (2000) *Molecular Cloning: A Laboratory Manual*, third edition. Cold Spring Harbor Laboratory Press, Plainview, NY.
18. Hermens, W. T., ter Brake, O., Dijkhuizen, P. A., Sonnemans, M. A., Grimm, D., Kleinschmidt, J. A., and Verhaagen, J. (1999) Purification of recombinant adeno-associated virus by iodixanol gradient ultracentrifugation allows rapid and reproducible preparation of vector stocks for gene transfer in the nervous system. *Hum. Gene Ther.* 10, 1885–1891.
19. Eaton, M. J., Blits, B., Ruitenberg, M. J., Verhaagen, J., and Oudega, M. (2002) Amelioration of chronic neuropathic pain after partial nerve injury by adeno-associated viral (AAV) vector-mediated over-expression of BDNF in the rat spinal cord. *Gene Ther.* 9, 1387–1395.
20. Plant, G. W., Christensen, C. L., Oudega, M., and Bunge, M. B. (2003) Delayed transplantation of olfactory ensheathing glia promotes sparing/regeneration of supraspinal axons in the contused adult rat spinal cord. *J. Neurotrauma* 20, 1–16.
21. Borrell, V., Yoshimura, Y., and Callaway, E. M. (2005) Targeted gene delivery to telencephalic inhibitory neurons by directional in utero electroporation. *J. Neurosci. Methods* 143, 151–158.
22. Kohrmann, M., Haubensak, W., Hemraj, I., Kaether, C., Lessmann, V. J., and Kiebler, M. A. (1999) Fast, convenient, and effective method to transiently transfect primary hippocampal neurons. *J. Neurosci. Res.* 58, 831–835.
23. Hermens, W. T., and Verhaagen, J. (1998) Viral vectors, tools for gene transfer in the nervous system. *Prog. Neurobiol.* 55, 399–432.

Chapter 6

Expression of EGFP by Adenovirus-Mediated Gene Transfer in the Central Nervous System

Ryohei Tomioka

Summary

A new method for investigating cortical microcircuitry uses adenovirus to introduce enhanced green fluorescent protein (EGFP) as a reporter gene to small groups of neurons. Adenovirus in solution containing 600 mM NaCl was injected into the cerebral cortex in anesthetized rats and monkeys, resulting in many EGFP-positive neurons in the interconnected distant brain regions as well as at the injection site. This result suggests that adenovirus with high NaCl concentration will be a kind of retrograde tracer. Thus, I have succeeded in finding a condition of adenovirus injection to retrogradely label cortical neurons to the full extent of their dendritic configurations. This system can be used to study the microcircuitry of central nervous system, and specific mammalian gene function within identified circuits in vivo using RNA interference and/or gene overexpression.

Key words: Adenovirus, EGFP, Golgi-like staining, Immunocytochemistry, Microcircuitry, Retrograde transport, Viral tracer.

1. Introduction

The number of reports using viral vectors in all fields of biology is increasing, including neuroscience. Many commercially available viral vectors can be modified for neuroscience applications (e.g., the neuronal circuit and the gene function in vivo) *(1–4)*. However, at least one inevitable problem arises from the injection of these viruses into the central nervous system in vivo. That is, the localized inflammation induced by the injection of relatively large amounts of viral solution at the site of injection. Since virus normally infects only those neurons near the injection site, this means most of experimental evidence that has been

Barry W. Hicks (ed.), *Methods in Molecular Biology, Viral Applications of Green Fluorescent Protein, vol. 515*

DOI: 10.1007/978-1-59745-559-6_6

published is not representative of normal, healthy brain tissue. Thus, it is difficult to analyze interesting properties of intact neurons because their local neuronal network is perturbed by the harmful inflammatory influence. Recently some recombinant rabies viruses have been used as retrogradely or trans-synaptic infectious tracers *(5, 6)*. Although these viruses do allow analysis of infected neurons in areas distant from the injection site, the infected neurons survived for less than 3 weeks. This suggests that the viruses may be toxic to neurons and that the data from that work is also not indicative of normal, health brain tissue.

I have modified the routine method of adenovirus injection by including a high concentration of NaCl (≥600 mM) that induces retrograde transport in adult rat and in monkey neurons. Thus, you can analyze relatively intact, healthy neurons projecting from interconnected distant brain regions toward the injection site. The low cytotoxicity of adenovirus in the CNS *(7)* allows infected neurons to survive for at least 2 months indicating our data is more representative of normal, healthy brain tissue.

The neocortex consists of glutamatergic neurons as excitatory cells and GABAergic neurons as inhibitory cells in all mammals *(8, 9)*. In our rat and monkey experiments, most of EGFP-labeled neurons are regarded as glutamatergic neurons morphologically because they are principal neurons **Figs. (1 and 2)**. However, adenovirus infection is also confirmed in GABAergic neurons because some infected neurons exhibited GABAergic markers around the injection site, as well as pyramidal cells **Fig. (3)**. Considering that the distribution of retrogradely infected neurons corresponds to previous reports using common retrograde tracers, and adenovirus can infect both glutamatergic and GABAergic neurons, recombinant adenovirus vector injected

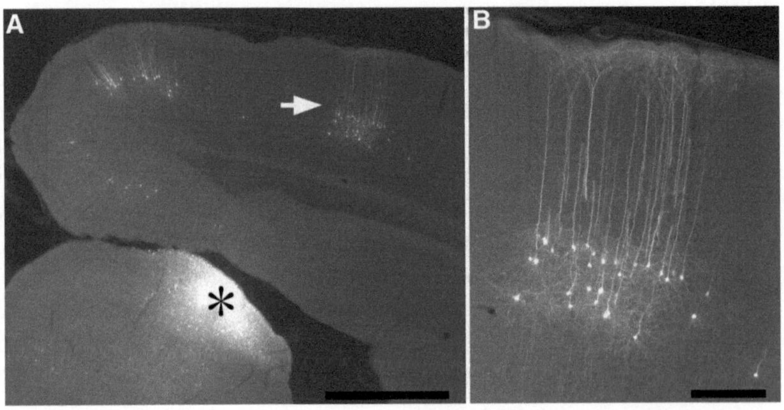

Fig. 1. EGFP expression after adenovirus injection into superior colliculus (survival = 14 days, 600 mM NaCl, processing by immunofluorescence). **a** Injection site (snowflake) in the rat superior colliculus. Many EGFP-expressing neurons were observed in layer 5 of primary visual cortex. **B** Higher magnification of a group of EGFP-expressing neurons in layer 5 from **Fig. 2A** (*arrowhead*). Scale bars: **A** = 1 mm, **B** = 200 μm.

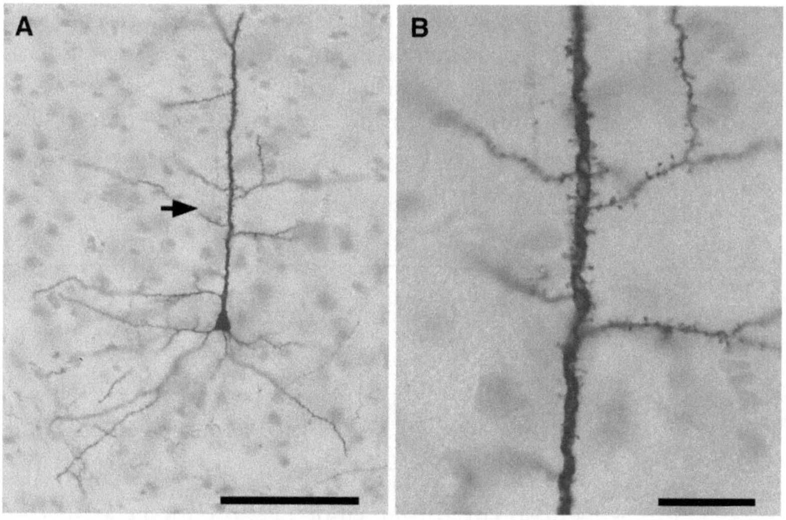

Fig. 2. EGFP expression (immunoperoxidase processing) in the visual cortex. Several EGFP-expressing neurons were observed in layer 5. Region at arrow is rephotographed at higher magnification in (**B**). Dendritic spines are clearly visible. Scale bars = 100 μm (**A**), 20 μm (**B**).

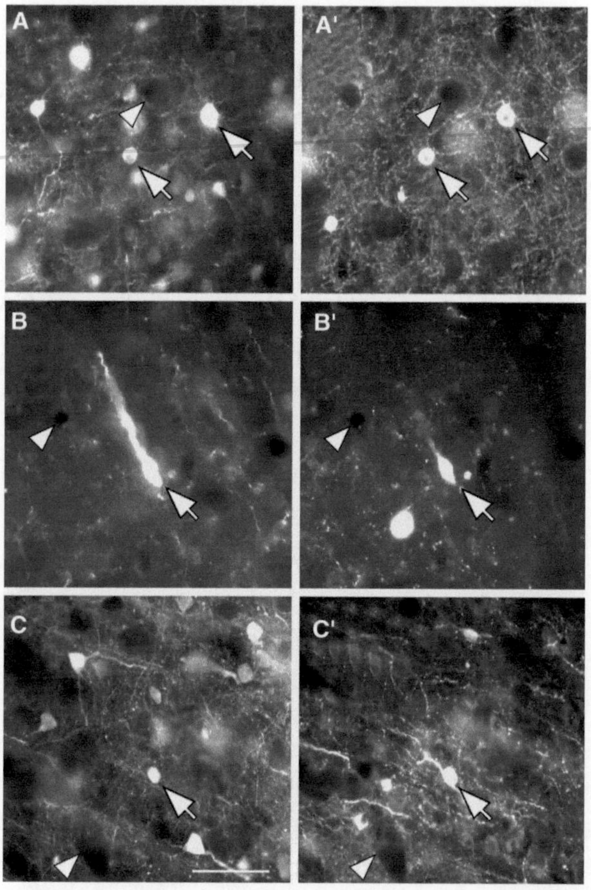

Fig. 3. Fluorescent photomicrographs showing EGFP-labeled cells in the gray matter after injections of adenovirus-expressing EGFP in the monkey neocortex (**A–C**). Each EGFP-positive neuron in the gray matter (*arrow*) is positive for GABAergic markers; parvalbumin (**A**'), somatostatin (**B**'), or calretinin (**C**'), respectively. Scale bar = 50 μm.

with 600 mM NaCl can be regarded as a retrograde tracer without cell-type specificity. This chapter focuses on obtaining high-titer adenovirus solution, injecting it into the CNS to induce trans-synaptic infection, and visualizing infected neurons using antibodies.

2. Materials

2.1. Purification of High-Titer Recombinant Adenoviruses

1. Cell culture media (DMEM; Gibco Carlsbad, CA)
2. HEK293 cell (RIKEN BioResource Center, Tsukuba, Japan)
3. EGFP-encoding adenovirus (TAKARA Bio, Tokyo, Japan)
4. Phosphate buffer saline (PBS; Gibco Carlsbad, CA)
5. Trypsin-EDTA (Gibco Carlsbad, CA)
6. Water bath
7. Tissue culture dish (100 mm)
8. Microplate (48 wells)
9. 100% Ethanol
10. Dry ice
11. Inverted Microscope
12. Incubator
13. P2 room for biosafety level 2 organism (HEPA filter, safety cabinet, and autoclave)
14. Ultracentrifuge (Beckman Fullerton, CA)
15. 2.2 M, 4.0 M, saturated CsCl in 10 mM HEPES (pH 7.4)
16. Centrifuge tube (Ultra-Clear for SW-28 and SW-41)
17. Pump
18. Low speed centrifuge (Tomy Tokyo, Japan)
19. DNase (Sigma-Aldrich, St. Louis, MO)
20. ViraKit AdenoMini (Virapur, San Diego, CA)
21. Vac-Man® Jr. Laboratory Vacuum Manifold, 2-sample capacity (Promega, Madison, WI)
22. Slide-A-Lyzer (Pierce)

2.2. Retrograde EGFP Labeling of Neurons by Adenovirus Injection

1. High-titer recombinant adenoviruses in 50 mM Tris-HCl (pH 8.1) containing 600 mM NaCl
2. Adult Wistar rat (7–12 weeks)
3. 7% Chloral hydrate (Wako, Kyoto, Japan)
4. Stereotaxic apparatus (Narishige, Tokyo, Japan)

5. Microprocessor-controlled syringe pump (KDS 310; Muromachi Kikai Co. Ltd., Tokyo)
6. Hamilton syringe (26 gauge)
7. Glass capillary (Narishige, Tokyo, Japan)
8. Plastic glue (Taiyo ElectricInd. Co., Ltd, Tokyo, Japan). This is necessary to seal the Hamilton syringe with the glass capillary.

2.3. Immunocytochemistry Using Anti-EGFP Antibodies

1. Saline and 4% paraformaldehyde in 0.1 M phosphate buffer for perfusion
2. 30% sucrose for cryoprotection
3. Freezing microtome (Yamato Seiki, Tokyo, Japan)
4. Mount-Quick “AQUEOUS” (option: Daido Sangyo Co., Ltd, Tokyo, Japan)
5. Dry ice
6. Primary anti-EGFP antibody (usually I use rabbit-raised antibody). This antibody was produced in my laboratory
7. Secondary antibody (e.g., Alexa Fluor 488-conjugated anti-rabbit IgG polyclonal goat antibody; Molecular Probes, Eugene, OR)
8. IMMU-MOUNT (Thermo Shandon; Pittsburgh, PA)

3. Methods

3.1. Cell Culture and Production of Recombinant Adenovirus

1. The HEK293 cells should be nearly confluent when infected with adenovirus. To reach 80–90% confluency, plate cells in a 10-cm dish 2–3 days before adenoviral infection.
2. Replace the medium with 2 mL of fresh DMEM containing 1×10^6 to 1×10^7 EGFP-adenoviral particles.
3. Incubate for 1 h at 37 °C in a humidified atmosphere maintained at 5% CO_2.
4. Add 8 mL of fresh DMEM, and incubate for 3–4 days at 37 °C in a humidified atmosphere maintained at 5% CO_2.
5. After most of the HEK293 cells detach, transfer the cell suspension into a 15-mL conical centrifuge tube.
6. Freeze the cell suspension a dry ice/ethanol bath for 5 min, and thaw it by incubation in a water bath at 37 °C for 5 min. Repeat this procedure a second time (*see* **Note 1**).
7. Spin the lysed cells at 3,000 rpm for 10 min in Tomy tabletop centrifuge.

8.Filter the supernatant through 0.45-μm filter unit.

3.2. Purification of Recombinant Adenovirus

Either of the methods below enables one to obtain high-titer recombinant adenovirus. Currently I mainly utilize the "Purification using an ion exchange membrane" because it is easier.

3.2.1. Purification Using Cesium Chloride (CsCl) Density Gradient

1. Generate CsCl gradient in a 40 mL tube (Beckman polyallomer tubes, Catalogue # 337986) by first placing 5.0 mL of 2.2 M CsCl solution and then under laying with 10 mL of 4.0 M CsCl solution.
2. Overlay the 15 mL of CsCl solutions with 15 mL of filtered supernatant containing adenovirus.
3. Spin for 2 h at 26,000 rpm in an SW-28 rotor at 4 °C. After the spin, the viral band will appear between the 2.2 and 4.0 M CsCl layers.
4. Aspirate the media from the top to reach the viral band, and then collect as much of the virus as possible (*see* **Note 2**).
5. Transfer the viral band to a conical tube, and mix each 3.0 mL of virus with 3.0 mL of saturated CsCl, then add the 6.0 mL to a fresh 12-mL centrifuge tube (Beckman, thick wall polycarbonate tube).
6. Gently overlay the 6 mL of viral solution with 2 mL of 4.0 M CsCl followed by 3 mL of 2.2 M CsCl. Centrifuge for 3 h at 4 °C in SW-41 rotor at 35,000 rpm. Adenovirus will float into the interface between the 2.2 and 4.0 M CsCl layers.
7. Collect as small volume of the viral solution as possible (*see* **Note 3**).
8. Transfer the viral solution to a Pierce Slide-A-Lyzer cassette. Dialyze at 4 °C, first for 2 h, then overnight, against 500 and 1,000 mL, respectively, of 50 mM Tris-HCl (pH 8.1) containing 600 mM NaCl.

3.2.2. Purification Using an Ion Exchange Column

1. Add DNase (10 Kunitz Units per 1 mL) to the filtered cell supernatant containing adenovirus (from 3.1). Incubate at 37 °C for 30 min.
2. Add an equal volume of the kit's Dilution Buffer A to DNase-treated adenovirus solution.
3. A Vac-Man® Jr. Laboratory Vacuum Manifold was inserted into a 500-mL side-arm flask and used in conjunction with aspirator-type vacuum pump and the kit's column.
4. Pre wet the column by applying 400 μL of PBS, and apply vacuum to pull the buffer through.
5. Load the viral solution to the column, and apply vacuum to pull the viral solution through. The volume for loading is up to 20 mL of diluted viral solution (*see* **Note 4**).

6. Load 400 μL of Wash Buffer B to the column, and apply vacuum to pull the buffer through the filter. Repeat the wash procedure two times.
7. Place the column on a new, sterile microcentrifuge tube, and load 400 μL of elution buffer. Centrifuge in a microcentrifuge at 4 °C for 10 min at 2,000 rpm.
8. Transfer the eluted solution to a Pierce Slide-A-Lyzer cassette. Dialyze at 4 °C, first for 2 h, then overnight, against 500 and 1,000 mL, respectively, of 50 mM Tris-HCl (pH 8.1) containing 600 mM NaCl.

3.3. Assay for Viral Titer Determination Based on EGFP Expression in HEK293 Cells

1. One day before infection, plate HEK293 cells into two 48-well plates so that plates are ~70% confluent and contain about 0.5 mL of medium per well.
2. Prepare serial dilutions of your virus as follows: Make a 1/100 dilution by adding 10 μL virus solution to 990 μL DMEM. Prepare eight subsequent tenfold dilutions by transferring 100 μL of the previously diluted virus to 900 μL of DMEM. Mix before continuing. Use the serial dilutions of $1/10^3$ to $1/10^{10}$ for titer determination.
3. Add 100 μL of diluted virus to each well of HEK293 cells in columns 1–8.
4. Return the plate to incubator for 3 days at 37 °C (*see* **Note 5**).
5. Using an epifluorescent microscope, count the number of EGFP-positive cells in the most dilute well possible, but where cells are readily quantifiable. Use the following formula to determine the titer (pfu/mL) of your viral stock:

$$\text{pfu/mL} = \frac{\text{Numberof EGFP - positive cells}}{d \times V}$$

where d is the dilution factor (10^{-3} to 10^{-10}) and V is the volume of diluted virus/well (0.1 mL).

3.4. Adenovirus-Mediated Gene Transfer of Central Nervous System

1. Anesthetize a rat using i.p. injection of chloral hydrate (350 mg/kg body weight), and secure it to the stereotaxic frame.
2. Shave the head, and clean the shaved area with 70% ethanol.
3. Make an incision on the skin above the skull and remove connective tissue to expose the top of the skull. Calculate the stereotaxic coordinates for the injection site and drill a hole above it (~2 mm in diameter).
4. Using a surgical microscope to clearly identify the dura mater, make an incision in the dura mater that has been exposed by drilling.

5. Connected a glass capillary (15–20 μm inner diameter) with a 10-μL Hamilton syringe containing adenovirus (1.5 μL of 1.0×10^{12} pfu/mL) and seal the capillary using plastic glue (*see* **Note 6**).
6. Place the Hamilton syringe on the microprocessor-controlled syringe pump that is mounted directly on the manipulator arm of the stereotaxic frame.
7. Calculate your final stereotaxic coordinates for the injection site from the line of Bregma.
8. Pierce the dura at the site of injection, using a 27-gauge needle.
9. Insert the Hamilton syringe with a glass capillary to the predetermined depth from the pia mater.
10. Push the start bottom of the microprocessor-controlled syringe pump to depress the plunger of Hamilton syringe very gently. Usually a flow rate of 0.15 μL/min is acceptable to deliver a total volume of 0.5–1.5 μL of viral solution that provides acceptable delivery at the injection site.
11. Leave the Hamilton syringe in place for 5 min after the viral solution has been injected before withdrawing it.
12. Sew up the incision using surgical stitches.
13. Maintain injected rats for at least 2 weeks.

3.5. Golgi-Like Visualization of Retrogradely Labeled Neurons Using Antibody Against EGFP

1. Reanesthetize rats that have received adenovirus in 600 mM NaCl with chloral hydrate. Perfuse transcardially with 100 mL of saline, followed by 300 mL of 4% paraformaldehyde in 0.1 M phosphate buffer (pH 7.4).
2. Remove the brain by dissection and immerse it overnight at 4 °C in 30% sucrose for cryoprotection.
3. Cut brain tissue into 60-mm thick frontal sections on a freezing microtome.
4. Incubate sections in PBS containing 0.3% Triton X-100 and 1.0% normal goat serum (PBS-TG) for 1 h at room temperature.
5. Incubate sections in PBS-TG containing 0.1–1.0 μg/mL rabbit anti-EGFP antibody overnight at room temperature.
6. Rinse them in PBS three times for 5 min.
7. Incubate them in PBS-TG containing 4.0 μg/mL Alexa Fluor 488-conjugated anti-rabbit IgG polyclonal goat antibody for 2 h at room temperature.
8. Mount them on a glass coverslip, not a slide, because minimal distance between the sample (section) and objective allows use of the highest numerical aperture objectives for improved optical outcome.

Notes

1. Two freeze–thaw cycles are necessary to release adenovirus particles from the cells, but additional freeze–thaw cycles are not recommended as this may lead to degradation of the adenovirus.
2. After first centrifugation step, you may have trouble finding a virus band because it is dilute and there may be several bands. Putting a black background behind the tube facilitates identification of the viral-containing band(s).
3. To obtain high titer virus solution, total amount of them should be less than 0.5 mL. Usually I take only 0.2 mL from the most dense viral band.
4. Do not allow the column membrane to dry while the adenovirus is adhered. Adenovirus should be handled like enzymes and will deteriorate if not properly hydrated.
5. To quantify the viral titer accurately, you should count them within 3 or 4 days after infection. After 7 days post-infection, adenovirus can replicate and be released from the initially infected HEK293 cells leading to an overestimation of the viral titer.
6. When adenovirus is taken into the Hamilton syringe, there should be a large amount of air or a bubble in the syringe above the liquid. This air should be in the syringe to provide a cushion during injection, but it makes it impossible to inject a precise volume.

Acknowledgments

The author thanks Dr. Teiji Chihara and the chemical analysis team of RIKEN for methodological discussion; Dr Mitsuhiro Hashimoto for help with adenovirus purification; and Dr. Kathleen S. Rockland, Dr. Noritaka Ichinohe, and Dr. Toshio Miyashita for helpful discussion. This work was supported by research funds from RIKEN Brain Science Institute.

References

1. Tamamaki, N., et al. (2000) Neurons in Golgi-stain-like images revealed by GFP-adenovirus infection in vivo. *Neurosci. Res.* 38, 231–236
2. Furuta, T., et al. (2001) In vivo transduction of central neurons using recombinant Sindbis virus: Golgi-like labeling of dendrites and axons with membrane-targeted fluorescent proteins. *J. Histochem. Cytochem.* 49, 1497–1508
3. Dittgen, T., et al. (2004) Lentivirus-based genetic manipulations of cortical neurons and their opti-

cal and electrophysiological monitoring in vivo. *Proc. Natl. Acad. Sci. USA.* 101, 18206–18211

4. Stettler, D.D., et al. (2006) Axons and synaptic boutons are highly dynamic in adult visual cortex. *Neuron* 49, 877–887
5. Smith, B.N., et al. (2000) Pseudorabies virus expressing enhanced green fluorescent protein: a tool for in vitro electrophysiological analysis of transsynaptically labeled neurons in identified central nervous system circuits. *Proc. Natl. Acad. Sci. USA.* 97, 9264–9269
6. Wickersham, I.R., et al. (2007) Retrograde neuronal tracing with a deletion-mutant rabies virus. *Nat. Methods* 4, 47–49
7. Tomioka, R. and Rockland, K.S. (2006) Improved Golgi-like visualization in retrogradely projecting neurons after EGFP-adenovirus infection in adult rat and monkey. *J. Histochem. Cytochem.* 54, 539–548
8. DeFelipe, J. and Farinas, I. (1992) The pyramidal neuron of the cerebral cortex: morphological and chemical characteristics of the synaptic inputs. *Prog. Neurobiol.* 39, 563–607
9. Kawaguchi, Y. and Kondo, S. (2002) Parvalbumin, somatostatin and cholecystokinin as chemical markers for specific GABAergic interneuron types in the rat frontal cortex. *J. Neurocytol.* 31, 277–287

Chapter 7

Adenovirus-Mediated Delivery of Short Hairpin RNA (shRNA) Mediates Efficient Gene Silencing in Terminally Differentiated Cardiac Myocytes

Andreas Rinne, Christoph Littwitz, Kirsten Bender, Marie-Cécile Kienitz, and Lutz Pott

Summary

RNA interference (RNAi) represents the most frequently utilized technique to analyze proteins by loss of function assays. Protein synthesis is impaired by sequence-specific degradation of mRNA, which is triggered by short (19–28 nt) silencing RNAs (siRNA). Efficient gene silencing using RNAi has been demonstrated in numerous cell lines and primary cultured cells. Incorporation of siRNA into terminally differentiated mammalian cells, such as adult cardiac myocytes is limited by their resistance to standard transfection protocols. Viral delivery of short-hairpin RNA (shRNA) overcomes these limitations and allows efficient gene silencing in these cells. This chapter describes the generation and characterization of recombinant siRNA-encoding adenoviruses and their application to adult cardiac myocytes, which represent a standard experimental model in research related to cardiac physiology and pathophysiology. Feasibility of this approach is demonstrated by effective ablation (>80%) of both, a transgene encoding for eGFP and the endogenous muscarinic M_2 acetylcholine receptor.

Key words: RNA interference, Silencing RNA (siRNA), Short-hairpin RNA (shRNA), Primary mammalian cells, Cardiac myocytes, Muscarinic M_2 receptor, Silencing adenoviruses, GIRK.

1. Introduction

RNA interference (RNAi) represents an evolutionary conserved biological mechanism that perturbs protein synthesis by efficient degradation of messenger RNA (mRNA) in a sequence-specific manner. RNAi was originally described in plants and *C. elegans* as posttranscriptional gene silencing *(1, 2)*. The discovery of this

Barry W. Hicks (ed.), *Methods in Molecular Biology, Viral Applications of Green Fluorescent Protein, vol. 515*

DOI: 10.1007/978-1-59745-559-6_7

mechanism in mammalian cells *(3)* initiated its recent success in reverse genetic studies. RNAi is induced by short interfering RNAs (siRNAs) of 21–29 nucleotides in length that bind to mRNA in a sequence-specific manner and inhibit protein expression by activation of the mRNA-degrading protein complex, RISC *(4)*. In terminally differentiated cells, the mechanism of RNA interference can be activated by viral expression of short-hairpin RNAs (shRNAs), which are processed to double-stranded siRNAs and activate the RNAi pathway *(5)*. This protocol describes the construction of recombinant adenoviruses encoding for shRNA and their application to adult cardiac myocytes.

2. Materials

2.1. PCR to Generate shRNA Expression Cassettes

1. 5′-H1 primer and 3′-shRNA primer (as described in 7.3.1).
2. Plasmid pH1-siRNA (Biomyx, San Diego).
3. Taq DNA polymerase.
4. 10× Taq DNA polymerase buffer.
5. Dimethyl sulfoxide (DMSO, molecular biology grade).
6. Mineral oil, PCR grade.
7. Equipment for preparing and running agarose gels, Tris-acetate-EDTA (TAE) buffer, and Tris-borate-EDTA (TBE) buffer.

2.2. Cloning of shRNA Cassettes into pCR®2.-TOPO

1. TOPO TA Cloning® Kit, containing pCR®2.-TOPO and *E. coli* OneShot®TOP10 cells (Invitrogen).
2. Restriction endonuclease *Eco*RI (New England Biolabs, Ipswich).

2.3. Bacterial Strains, Media, and Agar Plates

1. Electrocompetent *E. coli* XL1-Blue (American Type Culture Collection).
2. Electrocompetent *E. coli* BJ 5183 (ATCC or Stratagene, La Jolla).
3. Luria Bertani (LB) Medium.
4. LB agar plates containing 50 μg/mL ampicillin.
5. LB agar plates containing 50 μg/mL kanamycin.

2.4. Cloning of shRNA Cassettes into Adenoviral Transfer Plasmids

1. Plasmid pAdTrack (Stratagene).
2. Restriction endonucleases *Bgl*II and *Kpn*I (New England Biolabs, Ipswich).
3. T4 DNA ligase.
4. 10× T4 DNA ligase buffer.

5. Equipment for electroporation of plasmid DNA into bacteria.
6. Kit for purification of DNA from agarose gels (e.g., QIAEX II Gel Extraction Kit, Qiagen).
7. Ethanol.
8. 3 M sodium acetate.

2.5. Homologous Recombination to Obtain Recombinant Adenoviral Genomes

1. Plasmid pAdEasy-1 (Stratagene).
2. Restriction endonucleases *Pac*I and *Pme*I (New England Biolabs).

2.6. Cell Culture and Production of Recombinant Adenoviral Particles

1. HEK 293 cells (ATCC).
2. Dulbecco's Modified Eagle's Medium (DMEM).
3. DMEM medium, supplemented with 10% FCS and 1% penicillin/streptomycin.
4. T25 tissue culture flasks.
5. 1× Trypsin solution.
6. Rubber policemen.
7. Phosphate buffered Saline (PBS) without Na^+ and Mg^{2+}.
8. Dry ice.
9. Methanol.
10. Lipofectamine 2000 (Invitrogen).

2.7. Culture of Adult Rat Cardiac Myocytes

1. Hearts dissected from rats using ethically approved procedures.
2. 35 mm sterile plastic cell culture dishes (BD Falcon, BD Biosciences).
3. Medium 199.
4. Insulin-transferrin-sodium selenite (ITS) liquid medium supplement (100×, Sigma).
5. Antibiotics kanamycin and gentamycin.
6. Tyrode solutions containing Liberase 4 (Roche Diagnostics, Basel, Switzerland).
7. Voltage clamp apparatus, acetylcholine, and adenosine.

3. Methods

The methods described include (1) the selection of specific shRNA sequences, (2) generation of shRNA expression cassettes using PCR, (3) cloning and testing of these cassettes, (4) cloning the cassettes into an adenoviral backbone, (5) generation of gene silencing adenoviruses, and (6) functional tests of the shRNA sequences using primary cardiac myocytes.

3.1. Choice of siRNA Sequences

Effective shRNA sequences are chosen based on criteria derived from the analysis of effective siRNA or shRNA sequences that have been published over the last years. Some common criteria, such as a length of 21–29 base pairs, a G/C content between 30% and 52%, and thermodynamic parameters necessary for RISC-activation have been published (reviewed in *(6)*). These and other criteria led to the development of search algorithms, which are commercially or freely available (e.g., www.ambion.com,www.dharmacon.com, or www.cenix-biosciences.com). Since different shRNA sequences directed against the same gene may vary in their silencing efficiency, it is recommended to test up to four sequences per gene. To avoid silencing of other genes ("cross-target effects"), all shRNA sequences should be compared with known expressed sequence tags (est) of the organism using a blast-search (www.ncbi.nlm.nih.gov/BLAST/, use *blastn* with the database *est*). The design of shRNA oligonucleotides ("3′-shRNA-primer") includes the following steps:

1. Generate up to four shRNAs using your algorithm of choice (*see* **Note 1**).
2. For each shRNA sequence, generate a scrambled shRNA that does not bind to the target mRNA (negative control).
3. Perform a blast search against ests of the organism for all sequences. This step avoids interference of the selected shRNA with other genes.
4. Design 3′-shRNA-primers according to the sequence presented below (see also **Fig. 1a**). The sequence of a 3′-shRNA-primer encoding for a shRNA directed against the muscarinic M_2 receptor (rat) is given as an example below.

Primer sequences:

5′-H1 primer (human H1 promoter). This primer adds a *Bgl*II restriction site (underlined) to the human H1 promoter: 5′-gga<u>agatct</u>gaatcctatgcttcgaacgct-3′

1. **3′-shRNA primer** (targeting the rat M_2 receptor, GenBank Acc. No.: AB017655):5′-cg<u>ggtacc</u>aaaaaa**ggcccaaaaaggtgatgtg**tgtcgacaa**cacatcacctttttgggcc**tgggaaagagtggtctcatacag-3′.

This 3′-primer encodes for the shRNA (bold) and adds a *Kpn*I restriction site to the PCR cassette (underlined). Note the 5′ to 3′ orientation, as used for ordering. In this orientation, the first bold sequence is the complementary antisense sequence of the shRNA (compare **Fig. 1b**).

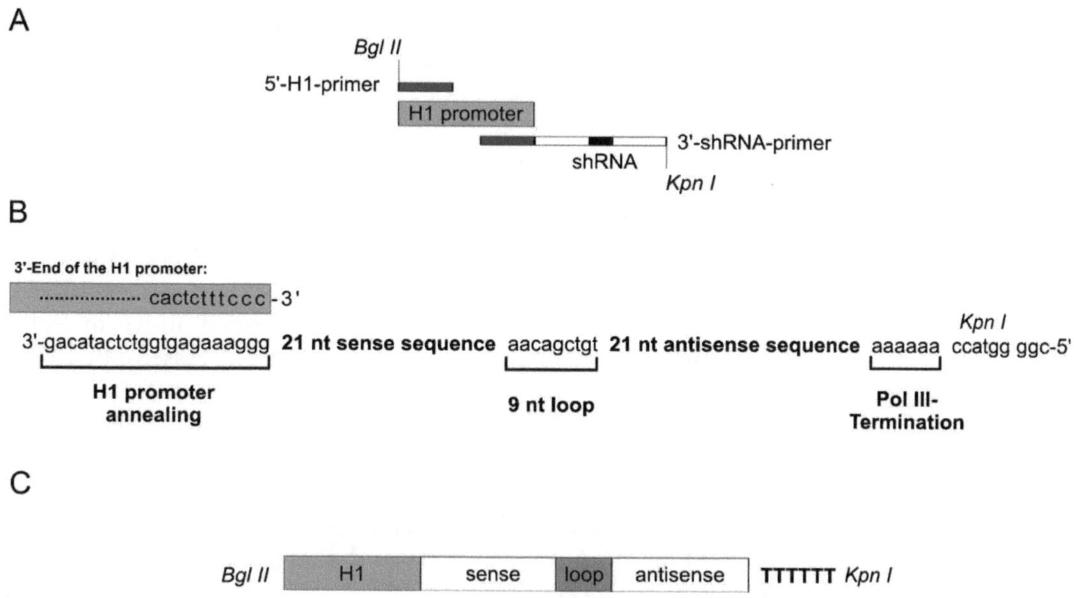

Fig. 1. PCR strategy to generate shRNA cassettes. **a** Amplification of the human H1 promoter by PCR using a short 5′-primer and a long 3′-primer that encodes for the shRNA. **b** Sequence of a 3′ shRNA primer directed against the rat M_2 receptor. Note the 3′ to 5′ orientation. **c** Schematic representation of the resulting shRNA cassette.

2. **3′-shRNA primer (scrambled)**: 5′-cgg*ggtacc*aaaaaa**ctgaatcagtgagcaggag**tgtcgacaa**ctcctgctcactgattcag**tgggaaagagtggtctcatacag-3′.

   This primer encodes for a shRNA with a scrambled sequence that does not bind to M_2-receptor mRNA. To design a 3′-primer targeting your gene of interest, replace the siRNA sequences (bold) in the primer sequence above with your favorite siRNA sequences.

3.2. PCR to Generate shRNA Expression Cassettes

3.2.1. PCR Strategy

1. Consider the promoter to be used. **Figure 1a** shows a schematic outline of the PCR strategy that is used to generate a shRNA expression cassette, which contains a promoter sequence and a siRNA-encoding sequence. Polymerase II promoters (e.g., the CMV promoter) and polymerase III promoters (usually U6 or H1) have been used to express shRNAs. The use of Pol III promoters seems to have some advantages compared to Pol II promoters, including lower toxicity due to moderate shRNA expression levels and activity in nearly all cell types *(7)*.
2. Incorporate endonuclease sites and shRNA into the primer. For generating a shRNA expression cassette by PCR, the human H1 promoter is amplified with a short 5′-primer and a long 3′-shRNA primer, which attaches the information of a short-hairpin RNA (**Fig. 1a**). The latter consists of (1) a sequence for binding to the 3′-end of the H1-promoter, (2) 21–29 nt long shRNA sense and antisense sequences separated

by a loop of 9 nt, and (3) a stretch of six adenines, which encode for six thymine bases to terminate Pol III transcription (**Fig. 1b**). In addition, sites for restriction endonucleases (*Bgl*II, *Kpn*I) are added to the cassette by both primers to allow for subsequent cloning steps (**Fig. 1c**). Since the antisense sequence is in reverse direction to the sense sequence, transcription of this PCR fragment results in mRNA that folds back to form a short RNA hairpin. **Figure 1b** shows an outline of a 3′-shRNA-primer that is used to generate shRNA expression cassettes. The primers described above in **Subheading 3.1** are used to generate a shRNA cassette by PCR, which is schematically represented in **Fig. 1c**. This cassette can be used to test the efficiency of gene silencing and will be incorporated into an adenoviral backbone (*see* **Note 2**).

3.2.2. PCR

Set up a PCR-mix (final volume: 25 µL) containing:

1. 100 ng DNA template (plasmid pH1-siRNA).
2. 100 pM 5′-H1-primer.
3. 100 pM 3′-shRNA-primer.
4. 25 µM dNTPs.
5. 1 U Taq-Polymerase.
6. 2.5 µL 10× reaction buffer.
7. 0.8 µL DMSO (4%, v/v).
8. Sterile H_2O to a final volume of 25 µL.
9. If necessary, overlay the reaction mix with 20 µL PCR-grade mineral oil.
10. Run the PCR using the following conditions: *One initial step* at 94 °C for 5 min, *30 cycles*, consisting of 30 s at 94 °C, 1 min at 60 °C, 1 min at 72 °C, and *one final step*: 72 °C for 10 min to finish the reaction.
11. Verify PCR by agarose gel electrophoresis using a 2% TBE gel. Use 5–10 µL of the reaction to determine the size of the PCR product; the expected fragment size is ~300 bp.

3.3. Cloning and Testing of the shRNA Cassettes

3.3.1. Cloning of shRNA Cassettes into pCR®2.1-TOPO

The shRNA expression cassette from step 3.2 is cloned into the plasmid pCR®2.1-TOPO with a ligase-free T/A-cloning method using a TOPO TA Cloning® Kit, which contains the plasmid pCR®2.1-TOPO and all reagents required for the steps 7.3.3.1 and 7.3.3.2. All plasmids generated this way allow for sequencing to verify the incorporated shRNA sequence and can be used for functional analysis of the incorporated shRNAs.

1. In a sterile tube mix 2 µL PCR reaction, 1 µL salt solution, and 2 µL sterile H_2O (*see* **Note 3**).
2. Add 1 µL of the plasmid pCR®2.1-TOPO (final volume: 6 µL) and mix gently.

3. Incubate at room temperature for 5 min.
4. Place the tube immediately on ice to stop the reaction.
5. Thaw 50 μL of chemically competent *E. coli* OneShot®TOP10 cells on ice.
6. Add 2 μL cloning reaction and incubate on ice for 15 min.
7. Incubate the bacteria at 42 °C for 45 s.
8. Incubate on ice for 10 min.
9. Add 250 μL SOC medium and shake at 37 °C for 45 min.
10. Plate 50–200 μL bacteria on a LB agar plate containing ampicillin (*see* **Note 4**).
11. Incubate the plates over night at 37 °C.
12. Pick 12 colonies and purify the plasmid DNA using your Mini plasmid preparation of choice.
13. Digest the plasmid DNA using *Eco*RI.
14. Run a 2% TBE agarose gel to visualize the restriction fragments.

The restriction enzyme *Eco*RI excises the fragment that is present in the cloning site of pCR®2.1-TOPO. In the case of positive cloning, two fragments should be visible on the agarose gel: the linear plasmid (3.9 kb) and the excised PCR fragment (350 bp).

Prepare a Maxi plasmid preparation using one positive clone to obtain the plasmid pCR2.1[shRNA] (*see* **Note 5**).

3.3.2. Testing the Silencing Potency of the shRNA Cassettes

The plasmid pCR2.1[shRNA] generated in step 7.3.3.1 can be used to test for silencing potency of the encoded shRNA, before the time consuming process of generating recombinant adenoviruses. Two alternative approaches are recommended:

1. Double transfection of a cell line (e.g., HEK 293 or CHO) with pCR2.1[shRNA] and a conventional expression vector, which contains the target cDNA. Subsequent gene silencing is monitored 48–72 h after transfection, either by quantitative RT-PCR (mRNA of the target protein) or by Western blot analysis (protein).
2. Alternatively, a cell line can be co-transfected with pCr2.1[shRNA] and a plasmid that encodes for the target protein fused to GFP. In that case, reduction of protein (and efficacy of the shRNA) can be qualitatively assessed by counting fluorescent cells (*see* **Note 6**).

3.4. Cloning of shRNA Expression Cassettes into pAdTrack

To generate a "gene silencing adenovirus," the shRNA expression cassette (step 7.3.3) is incorporated into an adenoviral backbone. This procedure is divided into two steps: (1) cloning of the shRNA expression cassettes into the transfer plasmid pAdTrack, and (2) incorporation of pAdTrack[shRNA] into the adenoviral backbone pAdEasy-1. A schematic representation of

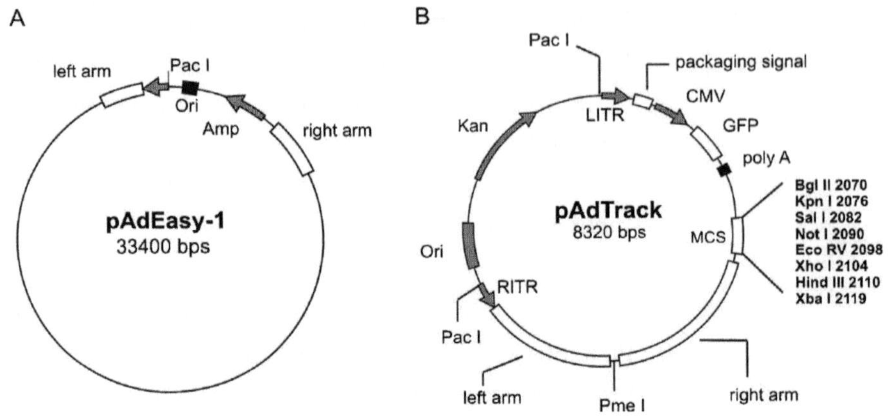

Fig. 2. Plasmids to generate recombinant adenoviruses. **a** The plasmid pAdEasy-1 encodes for the adenoviral backbone (serum type 5). **b** The transfer plasmid pAdTrack provides a multiple cloning site (MCS) to insert shRNA cassettes and encodes for GFP as a reporter protein. pAdTrack is integrated into the adenoviral genome (pAdEasy-1) by homologous recombination of the left and right arms (homology regions). Ori: Origin of replication; CMV: Cytomegalovirus promoter; poly A: polyadenylation signal; LITR: left internal terminal repeat; RITR: right internal terminal repeat. Plasmids adapted from ref. *(7)*.

both plasmids is given in **Fig. 2**. The plasmid pAdTrack encodes for GFP as a transgene product, which is used as reporter to monitor viral production and to identify infected primary cells. A detailed description of the pAdEasy-System for generating recombinant adenoviruses is given in *(8)*.

3.4.1. Cloning of pAdTrack[shRNA]

1. Digest 5 μg pCR2.1[shRNA] (step 7.3.3) and 5 μg pAdTrack using the restriction enzymes *Bgl*II and *Kpn*I (5 U each).
2. Purify the digested plasmid pAdTrack and the PCR fragment by agarose gel electrophoresis and subsequent gel extraction using the QIAEX II Gel Extraction Kit or comparable other method of choice. The expected fragment sizes are: pAdTrack 8.3 kb, shRNA cassette 300 bp (*see* **Note 7**).
3. Ligate both purified fragments using 0.5 μL T4 DNA ligase (corresponds to 200 NEB-Units). Use 500 ng pAdTrack and 20 ng shRNA cassette in a final volume of 20 μL. Incubate at 16 °C for 10 h and inactivate the ligase by incubating at 65 °C for 10 min (*see* **Note 8**).
4. To amplify and analyze the plasmids generated, an aliquot of the ligation reaction is transformed into *E. coli* XL1-Blue cells using electroporation.
5. Add 30 μL TE-buffer to the ligation reaction from **step 3** (final volume: 50 μL) and precipitate the DNA by adding 5 μL sodium acetate (3 M) and 110 μL ethanol (100%). Incubate at –20 °C for 30 min (*see* **Note 9**).
6. Spin down on a tabletop centrifuge for 10 min at 12,000 rpm.
7. Remove the supernatant and wash the DNA pellet twice with ethanol (70%).

8. Dry the pellet on the air and add 20 µL of sterile H_2O to dissolve the DNA pellet.
9. Chill an electroporation cuvette and LB medium on ice. Thaw 40 µL of electrocompetent *E. coli* XL1-Blue on ice.
10. Add 10 µL of purified DNA (**step 4**), mix carefully.
11. Pipette the bacteria/DNA mixture on the bottom of the cuvette and electroporate. Add 500 µL of ice-cold LB medium immediately and shake gently at 37 °C for 1 h.
12. Plate 50–200 µL on a LB-agar plate containing kanamycin.
13. Incubate over night at 37 °C.
14. Pick 12–24 colonies and purify the plasmid DNA using your Mini plasmid preparation of choice.
15. Digest the DNA using the enzymes *Bgl*II and *Kpn*I and analyze the restriction pattern by agarose gel electrophoresis on a 1.5% TAE gel. In case of positive cloning, two fragments (8.3 kb and 300 bp) should be visible on the gel.
16. Prepare a Maxi plasmid preparation from one positive clone to obtain pAdTrack[shRNA].

3.4.2. Incorporation of pAdTrack[shRNA] into pAdEasy-1

Recombinant adenoviral DNA is obtained by homologous recombination of the plasmid pAdTrack[shRNA] (step 7.3.4.1) with the adenoviral backbone vector pAdEasy-1 using the bacterial strain *E. coli* BJ 5183, which provides the enzymes that are required for this type of recombination.

1. Linearize 5 µg pAdTrack [shRNA] using the restriction enzyme *Pme*I (5 U).
2. Purify the linear plasmid by agarose gel electrophoresis (1% TAE gel) and extract the DNA from the gel (size: 8.6 kb). It is important to elute the DNA using sterile H_2O.
3. Thaw 20 µL electrocompetent *E. coli* BJ 5183 on ice.
4. Add 500 ng purified linear pAdTrack[shRNA] and 100 ng plasmid pAdEasy-1 (circular) to the bacteria and transfer to a chilled electroporation cuvette.
5. Electroporate as described above and add 500 µL ice-cold LB medium to the bacteria.
6. Shake at 37 °C for 20 min and plate 250 µL on a LB-agar plate containing kanamycin.
7. Incubate over night at 37 °C.
8. Pick up to 24 colonies and purify the plasmid DNA using your Mini plasmid preparation of choice. It is important to elute or dissolve the plasmid DNA from the Mini preparations in sterile H_2O (*see* **Note 10**).
9. Analyze the plasmids by digestion with the restriction enzyme *Pac*I and prepare a 0.8% TAE agarose gel to visualize the

fragments of the digestion. In case of a positive homologous recombination, two fragments are visible: one large fragment >30 kb and one smaller fragment of either 3 or 4.5 kb in size (*see* **Note 11**).

10. To avoid prolonged exposure of the recombinant adenoviral DNA to the *E. coli* strain BJ 5183, which reverses the homologous recombination, the DNA is retransformed into electrocompetent *E. coli* XL1-Blue cells by electroporation and select on kanamycin plates.
11. Pick a single colony, transfer to LB medium containing kanamycin, and perform a Maxi plasmid preparation to obtain pAdEasy[shRNA].

3.5. Generation and Purification of Recombinant Adenoviral Particles Using HEK 293 Cells

Recombinant adenoviral particles are generated and amplified in the host cell line HEK 293 after transfection of pAdEasy-1[shRNA].

3.5.1. Preparation of the DNA

1. Digest 8 μg of pAdEasy[shRNA] with the enzyme *Pac*I (10 U).
2. Purify the DNA by ethanol precipitation.
3. Dissolve the DNA pellet in 20 μL H_2O.

3.5.2. Transfection of HEK 293 Cells

1. One day before transfection, split a culture of HEK 293 cells by trypsinization and seed the cells into a T25 cell culture flask. The cell density on the day of transfection should be 50–75%.
2. For transfection, dilute 10 μL DNA from step 7.3.5.1 (~4 μg) with 470 μL medium DMEM.
3. Add 20 μL lipofectamine 2000 for a final volume of 500 μL and mix properly. Incubate at room temperature for 30 min.
4. After incubation, add 2.5 mL DMEM to a final volume of 3 mL.
5. Remove the medium from the HEK cells, wash the cells once with PBS, and add the transfection mixture (*see* **Note 12**).
6. Incubate for 3–4 h, then change the medium to normal growth medium (6 mL DMEM, containing 10% FBS and 1% penicillin/streptomycin).
7. To produce recombinant adenoviral particles, cells are cultured until about one-third of the cells start to detach from the bottom of the flask without changing the medium. This step usually requires 7–10 days in culture.

3.5.3. Purification of Adenoviral Particles

1. Prepare a dry ice/methanol cooling mixture and heat a water bath to 37 °C.
2. Remove the HEK cells from the bottom of the flask by scraping with a rubber policeman or cell scraper and transfer the suspension to a 15-mL conical tube.

3. Spin down for 10 min at 1,500 rpm using a tabletop centrifuge at 4 °C.
4. Remove the supernatant and resuspend the pellet in 2 mL of sterile PBS.
5. Freeze the cells rapidly in dry ice/methanol and thaw them immediately in a 37 °C water bath.
6. Vortex the cells rigorously and repeat these freezing/thawing/vortex steps for three additional cycles.
7. After the last step, spin down for 10 min at 1,500 rpm (4 °C).
8. Collect the supernatant and store at −20 or −80 °C.
9. In order to obtain higher viral titers, prepare a new T25-flask with HEK 293 cells (density 50–75%) and infect these cells by using 1 mL supernatant from **step 8**, diluted with 5 mL DMEM medium, containing FBS and antibiotics.
10. Cells are harvested and particles are purified as described above, when one-third of the cells start to detach from the bottom of the flask. Usually, this appears after 2–5 days in culture (*see* **Note 13**).
11. *Optional*: The presence of recombinant viral particles might be verified by PCR. Pipette 10 μL viral supernatant in a PCR tube, add 1 U of Proteinase K (PCR-grade), and incubate at 55 °C for 10 min. Inactivate the enzyme by heating up to 95 °C for 5 min. Spin down and use 1–5 μL as template in a PCR to amplify the shRNA cassette as described in 7.3.1.

3.6. Culture and Infection of Adult Rat Cardiac Myocytes

1. Single myocytes from rat hearts are isolated as described *(9)*. Briefly, hearts of adult Wistar Kyoto rats are excised quickly and mounted on a sterile Langendorff perfusion apparatus.
2. Single myocytes are obtained by retrograde perfusion with enzyme-containing Tyrode solutions (Liberase 4, Roche Diagnostics, Basel, Switzerland).
3. Approximately 10^4 cells are plated on 35 mm plastic culture dishes and cultured using serum-free medium M199 supplemented with 25 μg/mL gentamycin, 25 μg/mL kanamycin, and 1× ITS.
4. Cells are infected 24 h after isolation using ~10^6 viral particles per mL culture medium. As a general rule, viral infections are adjusted to yield about 75% GFP-positive myocytes on day 4 after infection.

3.7. Functional Analysis of shRNA-Induced Gene silencing in Adult Cardiac Myocytes

Feasibility of RNAi in adult cardiac myocytes mediated by adenoviral expression of shRNAs was demonstrated using a U6 promoter-driven shRNA directed against eGFP (constructed with an adenoviral shuttle vector lacking the cDNA encoding for eGFP, for details *see (10)*. As shown in **Fig. 3b**, in myocytes expressing

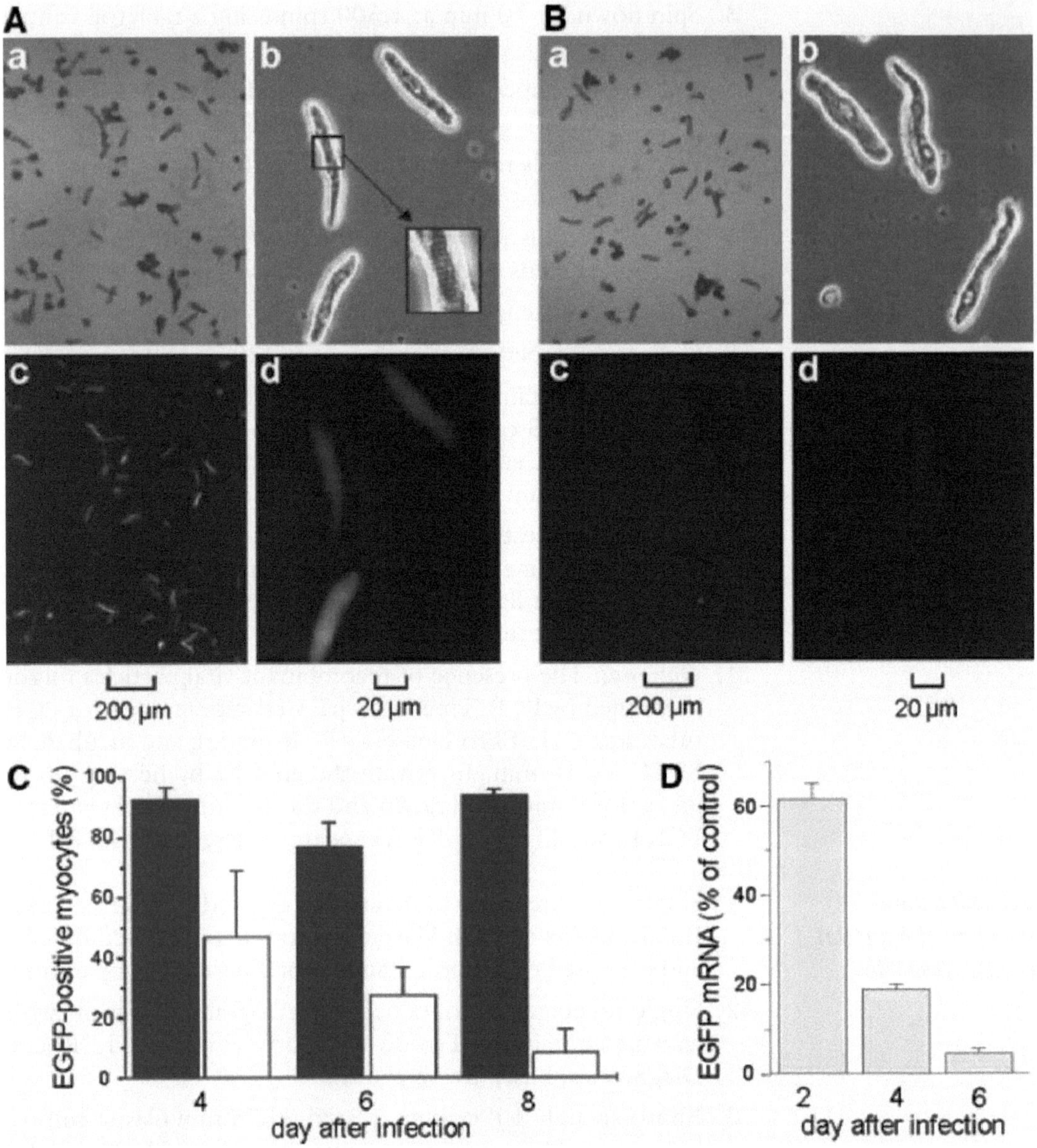

Fig. 3. Silencing of eGFP in adult cardiac myocytes. **a** Transmitted light and fluorescent images of myocytes expressing eGFP and a scrambled shRNA (negative control) in different magnifications. Images were taken on day 8 after infection. **b** Efficient silencing of eGFP in a parallel culture expressing eGFP and a shRNA targeted against eGFP. **c** Time dependence of eGFP silencing assessed by fluorescence (open bars: shRNA expressing cells; black bars: time-matched control cells). **d** Time dependence of siRNA-induced degradation of eGFP mRNA. Taken with permission from **ref.** *(10)*.

eGFP and a shRNA directed against this fluorescent protein, efficient gene silencing was obtained in about 8 days (*see* **Note 14**). Degradation of eGFP-mRNA was identified as underlying mechanism using quantitative RT-PCR, which precedes the ablation of the protein by 2 days (**Fig. 3c, 3d**). Gene silencing of eGFP was never observed in control cells expressing eGFP and a scrambled shRNA (**Fig. 3a, 3c**).

To demonstrate functional silencing of an endogenous protein, a shRNA directed against the muscarinic M_2 receptor for ACh was designed according to this procedure and virally expressed in atrial myocytes. To assess ablation of the muscarinic M_2 receptor in living atrial myocytes, current measurements of G protein activated K^+ channels (GIRK) were analyzed as an experimental readout *(11)*. These channels are activated by direct binding of $G_{i/o}$ derived betagamma-subunits upon stimulation of metabotropic receptors (such as the muscarinic M_2 receptor for acetylcholine (ACh) or the purinergic A_1 receptor for adenosine (Ado) *(12)*. Myocytes were infected with viral particles expressing either a shRNA targeting the M_2 receptor or a scrambled RNA hairpin that does not induce RNAi (negative control). Whole cell GIRK currents were measured 6–8 days after infection using the patch-clamp technique and a holding potential negative to E_K, thus resulting in inward currents. **Figure 4** shows representative measurements of GIRK currents from infected myocytes on day 8 in culture exposed to acetylcholine and adenosine. In time-matched control cells (scrambled shRNA), application of a saturating concentration of ACh (20 μM) resulted in a GIRK current, which is characterized by a large amplitude and fast activation kinetics (**Fig. 4a**). In contrast, myocytes expressing shRNA targeting the M_2 receptor showed acetylcholine-activated GIRK currents with reduced amplitudes

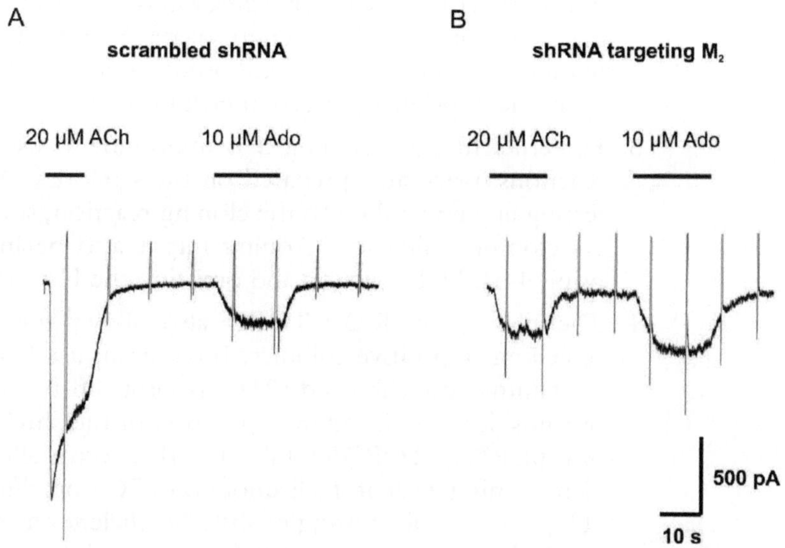

Fig. 4. Functional silencing of muscarinic M_2 receptors in rat atrial cells. **a** Representative electrophysiological measurements of acetylcholine-activated (ACh) GIRK current and adenosine-activated (Ado) GIRK current in a control cell (scrambled shRNA). **b** Silencing of muscarinic M_2 receptors in a myocyte expressing a shRNA targeting the M_2 receptor, as assessed by specific reduction in acetylcholine-activated GIRK current. Note that the current activated by adenosine via A_1 receptors is not affected.

and slower activation kinetics (**Fig. 4b**, 20 μM ACh), indicating functional silencing of the M_2 receptor. This silencing was specific, since both types of cells show typical adenosine-activated GIRK currents when exposed to 10 μM Ado (**Fig. 4a,b**).

Notes

1. The most critical step for a successful RNAi experiment is the choice of siRNA- or shRNA sequences. If no reduction in target protein is observed using several shRNAs, an alternative to sequences chosen with freely available algorithms are customized siRNA sequences, which can be purchased from many companies. These are either calculated on demand or based on databases provided by the companies. Although most companies guarantee a certain efficiency of silencing (e.g., a knockdown >50%) it has to be proven carefully, if these values are based on theoretical calculations or functional tests.
2. Gene silencing by shRNAs represents an established method to induce RNA interference. Advantages of the PCR method among others include (a) long-lasting RNA interference, (b) improved cloning efficacy using adenoviral vectors, and (c) fast generation of several shRNA plasmids in parallel. As described in *(13)*, the generated shRNA cassettes can also be tested by direct transfection of the PCR fragments, without any subsequent cloning steps. Because of the length of the 3′-oligonucleotide (85 nt) and to avoid unspecific PCR products, it is recommended to order the 3′-primers purified in HPLC grade.
3. For efficient T/A cloning it is important to use fresh PCR reactions, preferably prepared on the same day. Avoid transferring any mineral oil to the cloning reaction, since this lowers cloning efficiency. Cloning might also be improved by using 4 μL PCR reaction and omitting the H_2O in **step 1**.
4. The plasmid pCR®2.1-TOPO also allows for blue/white screening of positive colonies. If required, use LB agar plates containing ampicillin and IPTG (isopropyl β-D-1-thiogalactopyranoside). We found incorporation of the shRNA cassettes into pCR®2.1-TOPO useful, since this vector allows for efficient cloning, even if small amounts of PCR products are available, as well as for testing of shRNA efficiency as described in 7.3.3. Alternatively, the PCR fragments might also be cloned directly into any plasmid of choice. For this purpose, the following protocol might be used:
   a. Perform the PCR as described in 7.3.2 and purify 20 μL of the PCR by ethanol precipitation.

b. Dissolve the purified DNA pellet in 16 μL H_2O. Add 2 μL restriction buffer and your enzymes of choice (1 U each), adjust with sterile H_2O to 20 μL, and incubate at 37 °C for a minimum of 1 h.

c. Purify the restriction reaction using agarose gel electrophoresis and clone into your favorite plasmid or pAdTrack following protocol 3.4.1. You might use any combination of restriction enzymes, which are compatible with the multiple cloning site of the plasmid pAdTrack (**Fig. 2**).

5. Alternatively, the restriction sites *Bgl*II and *Kpn*I can be used, generating fragments of 3.9 kb (plasmid) and 300 bp (PCR product) in size.

6. Any standard transfection method can be used. The optimal molecular ratio between the shRNA plasmid and the target gene cDNA has to be tested individually for each shRNA. In our experience, a molecular ratio of plasmid shRNA:plasmid cDNA = 5:1 might be a good starting point. Using HEK cells or CHO cells, a significant reduction in target protein is expected between 48 and 96 h after transfection.

7. A good yield of purified DNA is obtained, if the shRNA cassette is purified using a 2% agarose gel prepared with TBE buffer, while the plasmid pAdTrack is purified using a 1% agarose gel prepared with TAE buffer.

8. In this example, a molecular ratio of pAdTrack:shRNA cassette = 5:1 is used. If the ligation is not successful, vary this ratio in a range from 1:1 to 1:10.

9. Because of the small amount of DNA, a pellet might not be clearly visible during the purification steps. Do not use TE-buffer to dissolve the DNA, because salts interfere with the subsequent electroporation. Voltage and duration of the electroporation step differ with the equipment used and may be optimized individually. We obtained good results using an Eppendorff electroporator and cuvettes (electroporator 2500, cuvettes with 1 mm gap, Eppendorf, Hamburg, Germany) and a voltage of 2,500 V. Alternatively, the ligation reaction can be transformed using chemically competent bacteria cells. In that case, purification by ethanol precipitation is not necessary. However, cloning efficiency might be decreased compared to electroporation.

10. The colonies on the agar plate vary in size. For best results, try to pick the smallest colonies visible.

11. Appearance of any of the following fragments is an indication of incorrect or incomplete homologous recombination: (1) three visible bands in any combination or (2) the presence of a 7-kb band (in combination with any other fragment).

In general, efficiency of the homologous recombination is improved, if fresh prepared plasmids are used, since long-term storage of pAdEasy-1 seems to decrease the efficacy. An example of agarose gels displays the restriction of a positive homologous recombination is given in *(8)*.

12. To ensure high efficient transfection rates, it is necessary to use serum- and antibiotic-free medium DMEM in **steps 1** and **3**. The transfection rate following this protocol should be >50%. This rate as well as the viral production is monitored by GFP fluorescence. In case of poor transfection rates, repeat the transfection step (7.3.5.2) using 8 μg (20 μL) plasmid pAdEasy[shRNA].
13. In some cases, no detachment or lysis of the cells is observed after the first transfection step. In that case, incubation for more than 10 days is considered to be sufficient. After two rounds of re-infection of HEK cells, the viral titers obtained with this protocol (>10^7 viral particles/mL supernatant) are usually high enough for efficient infection of primary cells in culture. Monitor the concentration steps on a daily basis to avoid release of viral particles in the culture medium before the cells are harvested. In the case of advanced lysis of the cells, collect the medium of the flask, which might also be used for subsequent infections.
14. In primary cells, the time-course of protein ablation is also influenced by the turnover of the protein. To distinguish the delay in silencing that is caused by protein turnover, the time courses of mRNA reduction (assessed by quantitative RT-PCR) might be compared with protein ablation (e.g., Western blot analysis). Specificity of the shRNA is typically demonstrated by comparison of the RNAi-phenotype with parallel cultured cells expressing a scrambled shRNA (negative control). This control condition, which should not display reduced amounts of mRNA or protein, is also used to ensure that the cellular phenotype is not altered by the adenoviral infection and/or shRNA expression per se. A recent overview about additional useful negative controls in RNAi experiments is given in *(14)*.

Acknowledgments

The authors thank Dr. B. Vogelstein, Howard Hughes Medical Institute and Molecular Genetics Laboratory, Johns Hopkins Oncology Centre, Baltimore, USA, for kindly providing pAdTrack, pAdEasy-1, and *E. coli* BJ 5183.

References

1. Fire, A. (1999) RNA-triggered gene silencing. *Trends Genet.* 15, 358–363.
2. Tabara, H., Sarkissian, M., Kelly, W. G., Fleenor, J., Grishok, A., Timmons, L., Fire, A., and Mello, C. C. (1999) The rde-1 gene, RNA interference, and transposon silencing in C. elegans. *Cell* 99, 123–132.
3. Elbashir, S. M., Harborth, J., Lendeckel, W., Yalcin, A., Weber, K., and Tuschl T. (2001) Duplexes of 21-nucleotide RNAs mediate RNA interference in cultured mammalian cells. *Nature* 411, 494–498.
4. Dorsett, Y., and Tuschl, T. (2004) siRNAs: applications in functional genomics and potential as therapeutics. *Nat. Rev. Drug Disc.* 3, 318–329.
5. Paddison, P. J., Caudy, A. A., Bernstein, E., Hannon, G. J., and Conklin, D. S. (2002) Short hairpin RNAs (shRNAs) induce sequence-specific silencing in mammalian cells. *Genes Dev.* 16, 948–958.
6. Pei, Y., and Tuschl, T. (2006) On the art of identifying effective and specific siRNAs. *Nat. Methods* 9, 670–676.
7. Snøve, Jr., O., and Rossi, J.J. (2006) Expressing short hairpin RNAs in vivo. *Nat. Methods* 9, 689–695.
8. He, T. C., Zhou, S., da Costa, L. T., Yu, J., Kinzler, K. W., and Vogelstein, B. (1998) A simplified system for generating recombinant adenoviruses. *Proc. Natl. Acad. Sci. USA* 95, 2509–2514.
9. Bechem, M., Pott, L., and Rennebaum, H. (1983) Atrial muscle cells from hearts of adult guinea-pigs in culture: a new preparation for cardiac cellular electrophysiology. *Eur. J. Cell Biol.* 31, 366–369.
10. Rinne, A., Littwitz, C., Kienitz, M. C., Gmereck, A., Bösche, L. I., Pott, L., and Bender, K. (2006) Gene silencing in adult rat cardiac myocytes by adenovirus-mediated RNA interference. *J. Muscle Res. Cell Motil.* 27, 413–421.
11. Bünemann, M., Brandts, B., and Pott, L. (1997) In vivo downregulation of M2 receptors revealed by measurement of muscarinic K^+ current in cultured guinea-pig atrial myocytes. *J. Physiol.* 501, 549–554.
12. Bösche, L. I., Wellner-Kienitz, M. C., Bender, K., and Pott, L. (2003) G protein-independent inhibition of GIRK current by adenosine in rat atrial myocytes overexpressing A1 receptors after adenovirus-mediated gene transfer. *J. Physiol.* 550, 707–717.
13. Castanotto, D., Li, H., and Rossi, J. J. (2002) Functional siRNA expression from transfected PCR products. *RNA* 11, 1454–1460.
14. Cullen, B. R. (2006) Enhancing and confirming the specificity of RNAi experiments. *Nat. Methods* 9, 667–681.

Chapter 8

Microscopic Analysis of Adenoviral Decontamination Using GFP Adenovirus with Comparable Sensitivity to Flow Cytometry

Michael Grusch, Annemarie Losert, Andreas Lackner, Alev Deli, Irene Herbacek, and Klaus Holzmann

Summary

Expression of transgenes from adenovirus vectors has become an extremely important and widely used tool in experimental cancer research and many other areas in the life sciences. It needs to be kept in mind, however, that adenoviruses are human pathogens and avoiding exposure of laboratory personnel to infectious viral particles is therefore an important concern. This issue seems even more important when the transgenes expressed for experimental purposes include oncogenic sequences. Decontamination procedures are thus required, whenever laboratory experiments with adenovirus vectors are performed and the effectiveness of these procedures has to be established. While many reports exist on the decontamination of blood and pharmaceutical products, data on the stability of adenoviruses during experiments performed in most life science laboratories are very scarce. One reason for this is that many of the methods used for assessing viral decontamination are time consuming and laborious and cannot easily be incorporated into the broad range of experimental setups typically performed in the laboratory. In this chapter we describe a reliable, sensitive, and simple method for the assessment of adenovirus decontamination by the use of an adenovirus expressing green fluorescent protein (GFP). The GFP adenovirus is subjected to various test conditions and afterwards susceptible indicator cells are exposed to the recovered virions. GFP expression is detected by a combination of fluorescence microscopy and flow cytometry. The simplicity and flexibility of the method allows one to monitor viral decontamination during the different scenarios occurring in the life science laboratory.

Key words: Adenovirus decontamination, Indicator cells, Green fluorescent protein.

1. Introduction

Development of cancer is generally considered to be a consequence of mutations or altered expression of distinct genes. To study the role of these genes for the growth and survival of cancer cells, gene

Barry W. Hicks (ed.), *Methods in Molecular Biology, Viral Applications of Green Fluorescent Protein, vol. 515*

DOI: 10.1007/978-1-59745-559-6_8

transfer into cancer cell lines and primary cells is an important research tool. Consequently, a broad range of different methods for gene transfer have been developed over the last decades. One of the most efficient and therefore widely used method is gene transfer with recombinant adenoviruses *(1)*. A key advantage of adenoviruses is the broad range of susceptible host cells and the high transfection efficiencies achievable, reaching more than 90% in most tumor cell lines and even in many primary cell types such as human umbilical vein endothelial cells (HUVEC) *(2)*. An important drawback of adenoviral gene transfer is the risk of infection of laboratory personnel. Therefore, appropriate disinfection protocols have to be established, and their effectiveness has to be determined. Many reports have investigated protocols used for the industry-scale decontamination of blood and pharmaceutical products *(3, 4)*. The multitude of different working steps performed in most research laboratories, however, poses a challenge to the establishment of adequate disinfection protocols and a simple yet reliable method to determine their effectiveness is needed. We have shown that replication-deficient adenoviruses expressing GFP can be used to monitor the stability of the viral particles during procedures that are performed with transduced cells for experimental purposes, such as incubation for prolonged periods of time, trypsinization and subculturing of the cells, or isolation of nucleic acids or protein by various methods *(5)*. This protocol can easily be used to test the influence of diverse substances on viral stability, for instance during storage of the viral stocks or the efficacy of procedures specifically performed to decontaminate instruments or the workplace. Here we give a detailed description of all steps required for this method. While we have used a serotype 5 adenovirus-expressing GFP under the control of a CMV immediate early gene promoter that has previously been described by He et al. *(6)*, the principle of the method is applicable to any adenovirus preparation expressing GFP from a strong constitutive promoter. A major advantage of this protocol is that decontamination experiments can be performed with the same type of virus used for the gene transfer experiments. Thus, stability properties are directly transferable from the GFP virus to the virus expressing the respective genes of interest. Hep3B is a robustly growing and widely used hepatoma cell line *(7, 8)* which is very suitable as indicator cell line due to its excellent sensitivity to adenovirus infection. Cell lines other than Hep3B, however, can be used as indicator cells, provided that they are highly sensitive to adenovirus infection. Despite its speed and simplicity, the sensitivity of this GFP-based assay, when performed as described in this chapter, is in the same range as previously reported for other assays assessing viral decontamination *(5, 9)*.

2. Materials

2.1. Cell Culture of Hep3B Indicator Cells

1. Hep3B cells are from the American Type Tissue Collection (ATCC).
2. RPMI-1640 medium (Sigma, St. Louis, MO) supplemented with 10% fetal calf serum (PAA, Linz, Austria).
3. Six-well plates and 100-mm petri dishes.
4. Phosphate buffered saline (PBS): 137 mM NaCl, 2.7 mM KCl, 10 mM Na_2HPO_4, 2 mM KH_2PO_4 in bidistilled H_2O, sterilize by autoclaving, cool to 4 °C before use.
5. Solution of 0.1% trypsin (BD Biosciences, Franklin Lake, NJ) and 0.01% ethylenediamine tetraacetic acid.

2.2. Adenovirus Treatment and Analysis of GFP Expression

1. GFP adenovirus (Ad-GFP) stock with a concentration of at least 10^8 plaque forming units (pfu) per mL. Store at –80 °C (*see* **Note 1**).
2. Control adenovirus (not expressing GFP) stock at a concentration of at least 10^8 pfu per mL (*see* **Note 1**). Store at –80 °C.
3. Adenovirus storage buffer: 5 mM Tris-HCl pH 8.0, 50 mM NaCl, 0.05% bovine serum albumin, 25% glycerol in 0.5× PBS.
4. Dimethyl sulfoxide (DMSO). Store at room temperature and protect from light.
5. 2% Sodium dodecyl sulfate (SDS) in H_2O. Store at room temperature.
6. Laemmli buffer (nonreducing, 60 mM Tris-HCl pH 6.8, 12% glycerol, 2% SDS, 0.005% bromophenol blue). Store at room temperature.
7. RIPA I buffer: 150 mM NaCl, 50 mM Tris-HCl pH 7.4, 0.1% SDS, 1% Nonident P-40, 0.5% sodium deoxycholate, 1 mM Na_3VO_4, 1 mM phenylmethylsulfonyl fluoride (PMSF). Store at –20 °C and add PMSF immediately before use.
8. TRIzol: 900 mM guanidine thiocyanate, 450 mM ammonium thiocyanate, 120 mM sodium acetate pH 4.8, 6% glycerol, 39% water saturated phenol, in RNAse free H_2O. Store at 4 °C.
9. Buffered formalin: 4% formaldehyde in PBS. Store at room temperature and protect from light.
10. Propidium iodide (PI) dissolved in PBS at a concentration of 200 μg/mL. Store at –20 °C and protect from light.

3. Methods

The aim of this protocol is to determine how effectively a given condition, referred to as the test condition, destroys infectious adenovirus particles. The test condition can either be the incubation with a chemical substance or it can be a physical condition like heat treatment, desiccation, or irradiation. **Table 8.1** provides examples of assay results for a range of conditions and reagents to which adenovirus preparations are commonly exposed during experiments in research laboratories.

3.1. Culture and Seeding of Hep3B cells

1. Keep maintenance cultures of Hep3B human hepatoma cells in 100-mm petri dishes in 10 mL RPMI-1640 medium with 10% FCS in a tissue culture incubator at 37 °C, 5% CO_2, and 95% relative humidity.
2. To keep the cells in logarithmic growth, pass them twice weekly at a ratio of approximately 1:10.
3. For seeding Hep3B cells into 6-well test plates, they have to be detached from the petri dish. For that purpose rinse them with cold PBS, incubate them with 1 mL of trypsin/EDTA solution for 5 min at 37 °C and resuspend them in medium with FCS.

Table 1
Examples for the decontaminating effect of frequently used conditions and reagents on adenovirus infectivity determined as percent GFP-expressing cells (% GFP⁺)[a]

	Desiccation		UV irradiation		Temperature		
			Direct	Indirect	37 °C	56 °C	95 °C
Time	45 min	24 h	30 min	30 min	7 days	5 min	5 min
% GFP⁺	3 ± 2	<0.01	0	70 ± 4	38 ± 5	0.2 ± 0.1	0
			Chemicals and reagents				
	Ethanol	Buffered formalin	RIPA I buffer	Laemmli buffer	TRIzol	DMSO	SDS
Concentration	75%	2%	0.5×	0.5×	50%	50%	1%
Time	10 min	10 min	10 min	10 min	10 min	10 min	10 min
% GFP⁺	0	0	85 ± 0.2	0	0	6 ± 0.3	0

[a]Additional examples are provided in the supplementary data of ref. *(5)*

4. For the test experiments, determine the cell concentration with a Neubauer counting chamber and seed 100,000 Hep3B cells in 2 mL RPMI medium with 10% FCS in each well of a 6-well plate.
5. Twenty-four hours after seeding, remove the growth medium and replace it with 2 mL of fresh medium with 10% FCS containing treated or untreated adenovirus (*see* **Subheading 3.2** and **Note 2**).

3.2. Subjecting the GFP Adenovirus to the Test Condition

1. For subjecting the GFP adenovirus to the respective test condition, pipette aliquots of the adenovirus stock solution containing 10^5 infectious virions into 1.5 mL reaction tubes. At least three replicas should be made for every test condition and the assay repeated three times.
2. If the test condition is a physical one, for instance if the impact of desiccation is being tested, then no additional reagents are required. Then add RPMI medium to the mixture to a total volume of 2 mL and transfer it to the indicator cells (**Fig. 1**).
3. In the case of chemical test substances, mix the virus stock in viral storage buffer at the desired ratio with the test substance and incubate for the desired period of time. Then dilute the mixture into 2 mL RPMI medium and transfer it to the indicator cells (*see* **Note 3**).

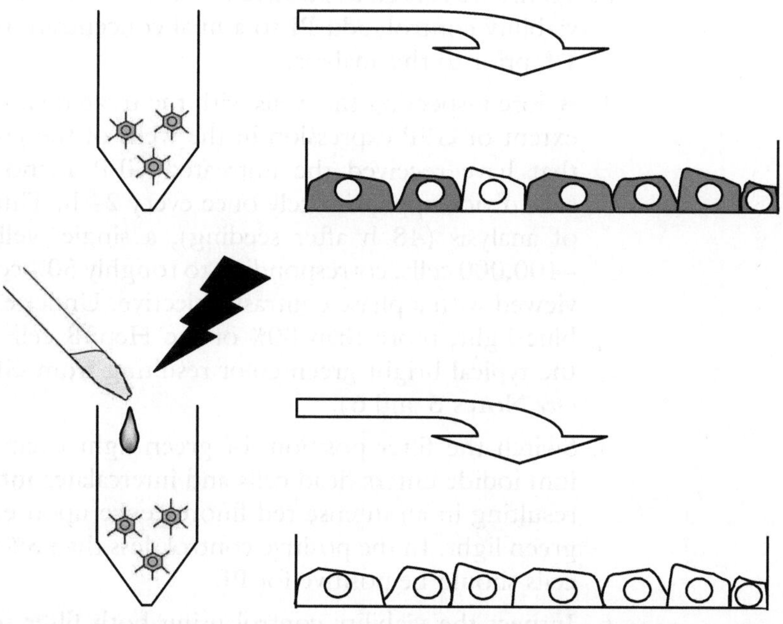

Fig. 1. Schematic diagram of the experimental procedure. Aliquots of virus stock containing 10^5 infectious virions are either left untreated as positive control (*upper part*) or subjected to different physical or chemical treatment conditions (*lower part*) and subsequently diluted in growth medium and transferred to the indicator cells. The presence of GFP-expressing cells (signified by gray cells) is determined 24 h later and is used as readout for the extent of virus elimination by the test condition.

4. It is important to set up an appropriate control for cell viability of the indicator cells. Treat these control cells with the same amount of virus storage buffer plus test substance as the experimental samples, and dilute them into the same amount of RPMI prior to adding to the cells.
5. As a positive control for the quality of the GFP adenovirus stock, transfer 10^5 pfu directly to a separate set of indicator cells.
6. As a negative control, infect one set of indicator cells with 10^5 pfu of the control adenovirus which does not express GFP.

3.3. Analysis by Fluorescence Microscopy

1. For detection of GFP expression and cell viability, we have used a Nikon eclipse TE 300 inverted microscope equipped with an X-cite 120 illumination system, a Spot insight cooled CCD camera, Nikon plan fluor 10× and 40× objectives, and Nikon B-2A (for detection of GFP) and G-2A (for detection of propidium iodide) epifluorescence filter sets. Inverted microscopes of other types and manufacturers can of course be used provided that they are equipped with appropriate epifluorescence illumination, phase contrast objectives, and filter sets for detection of GFP and propidium iodide.
2. Analyze the indicator cells by fluorescence microscopy 24 h after addition of the treated or untreated adenovirus preparations (*see* **Note 4**).
3. To the wells used as positive control and to those used for the viability control, add PI to a final concentration of 2 μg/mL 1 h prior to the analysis.
4. Before inspecting the wells with the treated virions, check the extent of GFP expression in the wells of the positive control that have received the untreated GFP adenovirus. Hep3B cells divide approximately once every 24 h. Thus, at the time of analysis (48 h after seeding), a single well will contain ~400,000 cells, corresponding to roughly 50% confluent when viewed with a phase contrast objective. Under excitation with blue light, more than 80% of the Hep3B cells should show the typical bright green color resulting from GFP expression (*see* **Notes 5** and **6**).
5. Switch the filter position for green light excitation. Propidium iodide enters dead cells and intercalates into their DNA, resulting in an intense red fluorescence upon excitation with green light. In the positive control, less than 5% of the Hep3B cells should be positive for PI.
6. Inspect the viability control using both filter sets. No GFP-positive cells should be detectable, and the fraction of PI-positive cells should not be higher than the positive control. If an increase in PI-positive cells is observed, it is likely to result from a toxic effect of the test substance and you should try to use a smaller volume of the test substance.

7. Take micrographs of the control wells and of the treated virion test wells using the 10× objective. Of each microscope frame, take one picture with GFP settings and one picture in phase contrast.
8. For test conditions which leave a significant fraction of the virions intact, it is not feasible to record all the GFP-positive cells in a well. If every microscope frame contains several GFP-positive cells, take micrographs from three different randomly chosen frames (*see* **Note 7**).
9. From the micrographs (or directly at the microscope), count the number of GFP-positive cells in each frame and divide it by the total number of cells in that frame.
10. Multiply the resulting ratio with 100 and average the results from the three frames per well and from the three wells used for each test condition, in order to get the percentage of GFP-positive cells for that test condition. The resulting percentages should be comparable to those obtained by flow cytometry (*see* **3.4** and **Note 8**).
11. If the number of GFP-positive cells is so small that not every microscopic frame contains a GFP-positive cell, it is important to systematically inspect the whole well and record the total number of GFP-positive cells. The number of total cells in this case can be assumed to be 400,000.

3.4. Analysis of GFP-Positive Cells by Flow Cytometry

1. For analysis of GFP-positive cells by flow cytometry either the same wells as used for fluorescence microscopy can afterwards be used for flow cytometry or alternatively an additional set of identical wells can be analyzed in parallel.
2. Remove the medium and thoroughly wash the cells with chilled PBS at least three times (*see* **Note 9**).
3. Incubate the cells with 0.5 mL trypsin/EDTA solution at 37 °C until all the cells have detached from the plastic surface of the dish. For Hep3B cells, this usually takes 5–10 min.
4. Resuspend the cells in 2 mL fresh medium containing FCS and transfer the suspension to a centrifugation tube.
5. Pellet the cells at 250 × *g* and 4 °C for 5 min, and discard the supernatant.
6. Gently resuspend the cell pellet in 2 mL chilled PBS.
7. Pellet again, discard the supernatant, and finally resuspend the cells in 0.5 mL chilled PBS.
8. For best results, analyze the cells within the next 10–30 min. If the cells are kept in PBS for a longer period of time more of the cells will die, which will increase autofluorescence.
9. These instructions assume the use of a FACS Calibur flow cytometer equipped with a 488-nm excitation laser and Cellquest Pro software (BD Biosciences).

10. Analyze GFP fluorescence using excitation with the 488 nm laser line and fluorescein isothiocyanate (FITC) filter sets. Measuring untreated Hep3B cells, set the GFP gate so that not more than 0.1% of the untreated cells are scored as GFP positive.
11. Measure 30,000 cells from each well and record the percentage of cells scoring as GFP positive. In the wells that have received the control virus, the fraction of GFP-positive cells should not be greater than 0.1%.

3.5. Determination of the Assay Sensitivity

1. For assessment of the effectiveness of viral decontamination, the most relevant question is, whether a substance or a physical treatment completely abolishes viral infectivity. Thus, it is important to determine the sensitivity limit of the assay.
2. For that purpose dilute 1.1×10^5 pfu of the GFP adenovirus stock in 1.1 mL of RPMI and further dilute the virus at a ratio of 1:10 by transferring 100 μL to another tube containing 1 mL medium.
3. Repeat this step until you have tubes containing 1 mL of RPMI and 10^5, 10^4, 10^3, 10^2, and 10^1 pfu of GFP adenovirus, respectively. Prepare these tubes in triplicates.
4. Use the serially diluted virus stock to infect indicator cells and determine the number of GFP-positive cells as described in **Subheadings 3.2–3.4** (*see* **Note 10**). Examples of phase contrast and fluorescence images as well as the corresponding flow cytometry results from serially diluted GFP adenovirus are shown in **Fig. 2.**
5. By flow cytometry, you should be able to clearly distinguish a virus dilution containing 10^2 pfu of the GFP adenovirus from untreated or control adenovirus transduced Hep3B cells, whereas a dilution of 10^1 pfu will not be distinguishable with certainty from the background.
6. By manual inspection of the whole wells under the fluorescence microscope, you should be able to detect a few unambiguously GFP-positive cells even when the cells have received only 10^1 pfu.

Notes

1. These instructions assume that concentrated stocks of replication-deficient GFP-expressing adenovirus and a respective control virus which does not express GFP are available to the user. Detailed instructions for the construction, amplification,

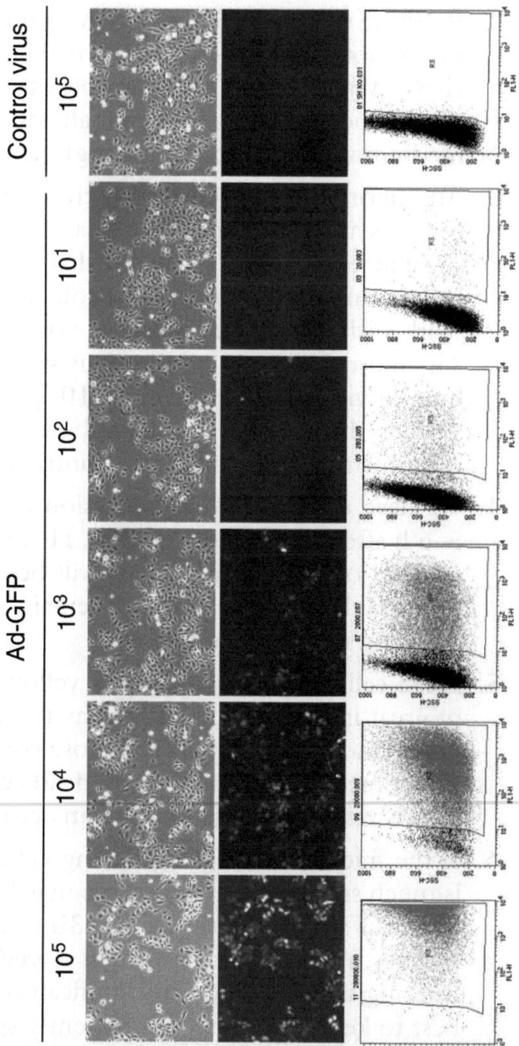

Fig. 2. Representative examples for detection of GFP expression by fluorescence microscopy and flow cytometry of serially diluted virus preparations. Phase contrast (*upper panels*) and fluorescence images (*middle panels*) of the same microscope frames were recorded 24 h after virus addition. Subsequently the cells were washed, removed with trypsin, and analyzed by flow cytometry (*lower panels*). Cells scored as GFP positive by flow cytometry are in gray, and cells scored as GFP negative are in black.

purification, and titer determination of adenovirus stocks can be found in the manuals of commercial suppliers which are available from their websites. The adenoviruses used in our laboratory are constructed according to the instructions of the AdEasy Adenoviral Vector System from Stratagene (La Jolla, CA). Several commercial suppliers also offer ready-to-use adenovirus stocks.

2. Presence of 10% FCS in the growth medium does in our experience not inhibit the infection of Hep3B cells by the adenovirus. While Hep3B cells can tolerate serum-free medium for some time, their viability gradually decreases, and thus we use medium with 10% serum throughout all experiments.
3. An important requirement for the methods is that the indicator cells are alive during the assay. However, many of the test substances are toxic to mammalian cells. Two things need to be kept in mind to avoid this problem. First, it is important to work with the smallest possible volume of the test substance. The more concentrated the virus stock is, the smaller the volume of virus stock containing 10^5 pfu, and consequently the smaller the volume of test substance needed. Second, one must include the appropriate controls for each test substance.
4. We have observed GFP expression in Hep3B cells as early as 5–6 h after addition of Ad-GFP. However, after 24 h, the percentage of GFP-positive cells will be much higher. The fraction of GFP-positive cells will remain almost constant for 2–3 days and then start to decrease.
5. Dead cells may often display a yellow autofluorescence. It is of great importance for this assay to distinguish this from the green fluorescence of GFP. A close comparison with the negative control which will contain dead cells but no GFP-positive cells is very helpful in assuring an accurate discrimination.
6. If the fraction of GFP-expressing cells in the positive control is much smaller than 80%, there may be a problem either with the virus stock or with the Hep3B cells. The virus stock deteriorates within a few days when stored at 4 °C. Also repeated freezing and thawing can significantly reduce infectivity. It is best to keep the stock in sufficiently small aliquots at −80 °C. When Hep3B cells are not passed regularly or seeded at very low density, they will not be in logarithmic growth and may show impaired virus uptake and/or GFP expression.
7. Do not select frames too close to the margin of the culture dish, as phase contrast will not work there making cell counting difficult.
8. Flow cytometry analyzes 30,000 cells per well, whereas three microscope frames will contain only a few hundred cells. Therefore, in such a case, the analysis by flow cytometry will give a more accurate result than manual counting and may be sufficient.
9. In most laboratories, the flow cytometry unit will not be within the biosafety level 2 area. Before leaving the biosafety level 2 laboratory with the cells, it is important to decontaminate them by washing several times, treat them with trypsin, and wash them again.

10. Titer determination of adenovirus preparations is normally done by plaque forming assays in human embryonic kidney (HEK) 293 cells. These cells are used for amplification of recombinant adenoviruses, because they can provide in trans the E1 genes deleted in most replication-deficient adenovirus constructs. We are routinely observing that adenovirus dilutions containing only 10 pfu according to the titer determination in 293 cells are able to transduce a several-fold larger number of Hep3B cells. This is due to the higher susceptibility of Hep3B cells to adenovirus infection compared to 293 cells.

Acknowledgment

The authors thank M. Herlyn for providing the GFP adenovirus.

References

1. Kay, M. A., Glorioso, J. C., and Naldini, L. (2001) Viral vectors for gene therapy: the art of turning infectious agents into vehicles of therapeutics. *Nat Med* 7, 33–40.
2. de Martin, R., Raidl, M., Hofer, E., and Binder, B. R. (1997) Adenovirus-mediated expression of green fluorescent protein. *Gene Ther* 4, 493–5.
3. Celis, P., and Silvester, G. (2004) European Regulatory guidance on virus safety of recombinant proteins, monoclonal antibodies and plasma derived medicinal products. *Dev Biol (Basel)* 118, 3–10.
4. Klein, H. G. (2005) Pathogen inactivation technology: cleansing the blood supply. *J Intern Med* 257, 224–37.
5. Losert, A., Mauritz, I., Erlach, N., Herbacek, I., Schulte-Hermann, R., Holzmann, K., and Grusch, M. (2006) Monitoring viral decontamination procedures with green fluorescent protein-expressing adenovirus. *Anal Biochem* 355, 310–2.
6. He, T. C., Zhou, S., da Costa, L. T., Yu, J., Kinzler, K. W., and Vogelstein, B. (1998) A simplified system for generating recombinant adenoviruses. *Proc Natl Acad Sci USA* 95, 2509–14.
7. Knowles, B. B., Howe, C. C., and Aden, D. P. (1980) Human hepatocellular carcinoma cell lines secrete the major plasma proteins and hepatitis B surface antigen. *Science* 209, 497–9.
8. Vejda, S., Erlach, N., Peter, B., Drucker, C., Rossmanith, W., Pohl, J., Schulte-Hermann, R., and Grusch, M. (2003) Expression of activins C and E induces apoptosis in human and rat hepatoma cells. *Carcinogenesis* 24, 1801–9.
9. Jannat, R., Hsu, D., and Maheshwari, G. (2005) Inactivation of adenovirus type 5 by caustics. *Biotechnol Prog* 21, 446–50.

Chapter 9

Production of Multicistronic HIV-1-Based Lentiviral Vectors

Nachimuthu Chinnasamy, James Shaffer, and Dhanalakshmi Chinnasamy

Summary

In the last decade lentiviral gene transfer vectors have gained significant place both in basic science and gene therapy applications. A number of gene transfer applications would benefit from vectors capable of expressing multiple genes. This chapter focuses on production of bicistronic and tricistronic lentiviral vectors based on the internal ribosomal entry site (IRES) sequence of encephalomyocarditis virus (EMCV) and/or foot-and-mouth disease virus (FMDV) cleavage factor 2A. Multigene vectors produced high titer viral particles and were able to simultaneously express two or three transgenes in transduced cells. The level of expression of individual transgenes varied depending on the transgene itself, its position within the construct, the total number of transgenes expressed, the strategy used for multigene expression, and the number of copies of proviral insertions.

Key words: Multigene lentiviral vectors, Monocistronic lentiviral vector, Bicistronic lentiviral vector, Tricistronic lentiviral vector, IRES, 2A cleavage factor, eGFP, MGMT, HOXB4.

1. Introduction

Lentiviral vectors offer several advantages over other vectors, including stable integration into the host cell genome, lack of transfer of viral genes, and a relatively large capacity for transgene expression. They are efficient in transducing both dividing and nondividing target cells. Many potential gene transfer applications require vectors that express more than one protein. These may include a therapeutic gene plus a selectable marker gene, multiple genes encoding different subunits of a complex protein, or multiple independent genes that cooperate functionally.

Barry W. Hicks (ed.), *Methods in Molecular Biology, Viral Applications of Green Fluorescent Protein, vol. 515*

DOI: 10.1007/978-1-59745-559-6_9

Several strategies are employed in viral vectors to express multiple genes, including mRNA splicing, internal promoters, internal ribosomal entry sites (IRES), fusion proteins, and cleavage factors. Historically, the most popular strategy used in the construction of two gene vectors is the insertion of an IRES element between the two transgenes *(1)*. The IRES of encephalomyocarditis virus (EMCV) has been widely used to link two genes transcribed from a single promoter within recombinant viral vectors. However, there are a number of limitations using IRES elements, including their size and variability in expression of transgenes. In many cases it has been reported that a gene placed downstream is expressed at relatively lower levels when compared to a gene transcribed upstream of an IRES element *(2, 3)*.

Positive strand RNA viruses generally encode polyproteins that are cleaved by viral or host proteinases to produce mature proteins. Among other mechanisms many of these viruses are also known to contain 2A or similar peptide coding sequences to mediate protein cleavage. Foot-and-mouth disease virus (FMDV) is a picarnovirus with an RNA genome that encodes a single polyprotein of ~225 kDa. This polyprotein is cleaved in the host cell to produce different protein products. A self-processing activity in FMDV leads to "cleavage" between the terminal glycine of the 2A product and the initial proline of 2B. The exact mechanism of 2A/2B cleavage is not known. However, it has been hypothesized that the 2A sequence somehow impairs normal peptide bond formation between 2A glycine and 2B proline through a ribosomal skip mechanism without affecting the translation of 2B. The self-processing activity is conferred on heterologous fusion proteins by about 20 amino acids from the 2A region. The cleavage of the polyprotein product occurs at the C-terminal end of the 2A coding region, leaving this peptide fused to the upstream protein and releasing the downstream protein intact (with the addition of an N-terminal proline).

FMDV 2A sequence has been successfully incorporated in adeno-associated *(4)* and retroviral *(5, 6)* vectors to construct multigene vectors. Multigene lentiviral vectors have been developed by other groups using strategies involving inclusion of IRES *(7)*, multiple internal promoters *(8, 9)*, and differential splicing moieties *(10)*. In this chapter we describe the development of HIV-1-based multigene lentiviral vectors using combinations of the FMDV 2A cleavage factor and the EMCV IRES. Bicistronic and tricistronic lentiviral vectors **(Fig. 1)** were able to coexpress two or three different proteins, albeit at levels that depend on the transgene and its location.

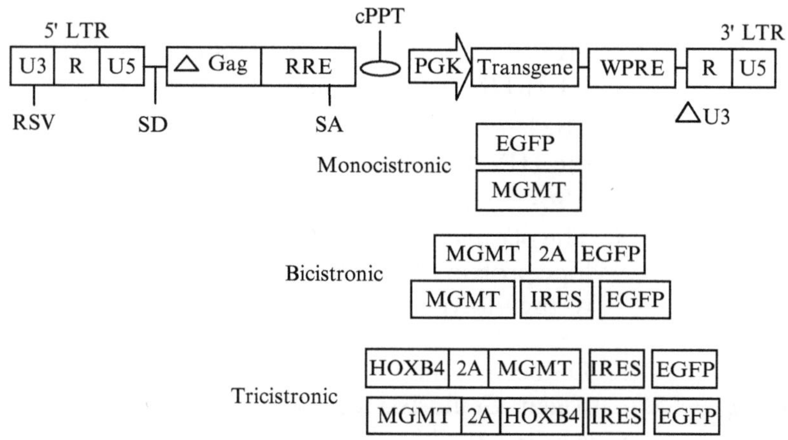

Fig. 1. Schematic illustration of HIV-1-based monocistronic, bicistronic, and tricistronic lentiviral vectors. *LTR*, long-terminal repeat; *Δ Gag*, deleted gag region; *RRE*, Rev-responsive element; *cPPT*, central polypurine tract; *PGK*, human PGK promoter; *WPRE*, posttranscriptional regulatory element of woodchuck hepatitis virus; *RSV*, Rous sarcoma virus; *SD*, splice donor; *SA*, splice acceptor; *2A*, sequence from foot-and-mouth disease virus; *IRES*, internal ribosome entry site sequence from encephalomyocarditis virus (*EMCV*); *eGFP*, enhanced green fluorescent protein; *MGMT*, O^6-methylguanine-DNA-methyltransferase; *HOXB4*, homeobox transcription factor.

2. Materials

2.1. Safety Guidelines

1. Lentiviral vectors should be handled in a laminar flowhood.
2. Use universal precautions when working with human cells and recombinant viral vectors.
3. All the procedures involving viral particles should be carried out according to the requirements of biosafety Level 2 laboratory safety guidelines.
4. Wear protective laboratory clothing including lab coats and gloves.
5. Minimize generating aerosol during handling of viral particles and transduced cells.
6. Collect all aspirated fluids in a 10% bleach solution-containing trap.
7. Decontaminate all wastes generated using autoclave sterilization or soaking in 10% bleach solution.

2.2. Plasmids

1. HIV-1-based lentiviral gene transfer vectors are in pRRL.PPT. PGK.**X**.W.SIN backbone in which **X** denotes the transgene(s) of our interest. Monocistronic vectors expressing eGFP and MGMT; bicistronic vectors expressing MGMT-2A-eGFP

and MGMT-IRES-eGFP; and tricistronic vectors expressing HOXB4-2A-MGMT^{P140K}-IRES-eGFP and MGMT-2A-HOXB4-IRES-eGFP have been described *(11, 12)*.

2. The lentiviral packaging plasmid pCMVΔR8.91 devoid of all the HIV-1 accessory proteins has been described previously *(13)*.
3. The plasmid named pMD-G encodes the heterologous envelope protein VSV-G has been described *(14)* and used widely to pseudotype lentiviral vectors.

2.3. Cell Culture Reagents, Supplies, and Cell Lines

1. Dulbecco's modified Eagle's media (DMEM) (Gibco, Invitrogen, Grand Island, NY).
2. RPMI 1640 media (Gibco).
3. Fetal bovine serum (FBS) (Hyclone, Logan, UT).
4. 0.05% Trypsin–EDTA solution (Gibco).
5. Dulbecco's phosphate buffered saline (DPBS) (Gibco).
6. Glutamine, 200 mM (100×) (Gibco).
7. Penicillin and streptomycin (100×, Gibco).
8. **D10:** DMEM supplemented with 10% FBS (*see* **Note 1**).
9. **R10:** RPMI 1640 supplemented with 10% FBS and 2 mM glutamine.
10. 100-mm tissue culture dishes (BD Falcon, Bedford, MA).
11. 6-, 12-, and 24-well, tissue culture dishes (BD Falcon).
12. T-75 tissue culture flasks with vented caps (BD Falcon).
13. Filter top flasks with 0.22 and 0.45 μm pore size-PES membranes (Nalgene, Rochester, NY).
14. Human embryonic kidney cell line 293T (American Type Culture Collection (ATCC), Manassas, VA) grown in D10 media at 37 °C in a humidified incubator at 5% CO_2.
15. Human cervical cancer cell line HeLa (ATCC) grown in D10 media at 37 °C in a humidified incubator at 5% CO_2.
16. Human erythroleukemia cell line K562 (ATCC), grown in R10 media at 37 °C in a humidified incubator at 5% CO_2.

2.4. Other General Reagents and Supplies and Those for Transfection of 293T Cells and Flow Cytometry

All the reagents used in transfection should be filtrated through a 0.22-μm filter and kept sterile (reagents 1–4).

1. 0.5 M HEPES pH 7.0 (1 M HEPES, Gibco).
2. 150 mM sodium phosphate pH 7.0.
3. 2 M $CaCl_2$.
4. 2 M NaCl.
5. X-VIVO 10 media (BioWhittaker, Walkersville, MD).

6. UltraPure distilled water (Gibco).
7. HIV-1 p24 ELISA kit (Beckman Coulter, Fullerton, CA).
8. Bleach-6.15% sodium hypochlorite (Clorox, Oakland, CA).
9. Poly-D-lysine (Sigma, St. Louis, MO).
10. 1,000× Protamine sulfate (10 mg/mL) (Sigma).
11. Trypan blue (Sigma).
12. Paraformaldehyde (PFA; Sigma).
13. FACS buffer-1% FBS in DPBS.
14. FACS fixative, 1% paraformaldehyde in DPBS from a 4% stock solution.
15. 10% Tween-20 solution (Bio-Rad Laboratories, Hercules, CA).
16. 10× Tris buffered saline (TBS) (Bio-Rad Laboratories).
17. 10× Phosphate buffered saline (PBS) (Bio-Rad Laboratories).
18. Bio-Rad Laemmli sample buffer (Bio-Rad Laboratories).
19. 2-Mercaptoethanol (Sigma).
20. 15-ml Polypropylene conical tube, 17 × 120 mm (BD Labware, Franklin Lakes, NJ).
21. 5-ml Polystyrene round bottom tube, 12 × 75 mm (BD Falcon).

2.5. Instrumentation

1. Beckman centrifuge (Optima XL-100K Ultracentrifuge, Beckman Coulter) with SW28 rotor, or comparable.
2. ELISA plate reader, VERSAmax microplate reader (Molecular Devices, Sunnyvale, CA), or comparable.
3. FACScan or FACSCalibur, Flow cytometer with CellQuest software, or comparable.
4. Inverted fluorescent microscope with appropriate excitation and emission filters to detect fluorescence (Olympus 1X71, Olympus America, Melville, NY), or comparable.
5. GE 750 Ultrasonic Processor with Model CU33 Probe (PGC Scientific, Frederick, MD), or comparable.

3. Methods

3.1. Coating 100-mm Tissue Culture Plates with Poly-D-Lysine

1. Reconstitute poly-D-lysine at 1 mg/mL with DPBS. Poly-D-lysine stock solution can be stored at −20 °C for several months.

2. Dilute 1/10 to 100 μg/mL with DPBS immediately before use.
3. Add ~5–8 mL of the poly-d-lysine to 100-mm tissue culture plate and swirl to coat the bottom of the entire dish. Remove excess and use to coat the next plate.

3.2. Production of HIV-1-Based Lentiviral Vectors

Lentiviral vectors are produced in human embryonic kidney 293T cells by transient transfection of three plasmids carrying different components *(11, 12)* as described below.

3.2.1. Transfection of 293T Cells

1. One the first day, take confluent T-75 flasks of 293T cells, aspirate the media, and wash with 5 mL of DPBS.
2. Add 2 mL of 0.05% Trypsin–EDTA solution to each flask and incubate ~2 min at room temperature. Observe with an inverted microscope making sure cells are detached.
3. Add 5 mL of D10 media and resuspend the cells by pipetting the entire contents of the flask up and down at least five times.
4. Transfer the contents of all flasks into a sterile 250-mL bottle.
5. Pass cells through a 70-μm sieve to eliminate any large clumps of cells.
6. Count the viable cells on a hemocytometer using trypan blue.
7. Plate the cells at 3×10^5 cells/mL (3×10^6 cells/plate) in D10 media in a 100-mm tissue culture dish. The final volume is 10 mL.
8. On the second day, at least 1 hour before the transfection aspirate the media from culture dishes with 293T cells and add fresh D10 media.
9. We usually transfect twenty 100-mm dishes that yield ~200 mL of viral supernatant sufficient to fill six centrifuge tubes for one run in SW28 rotor. Prepare the transfection mixture for every ten 100-mm dishes in 50-mL polystyrene tubes as follows.

   Tube A: 3.825 mL water, 0.5 mL of 0.5 M HEPES pH 7.0, 0.625 mL of 2 M NaCl, and 0.050 mL of 150 mM sodium phosphate pH 7.0.

   Tube B: 4.105 mL water, 0.130 mL of gene transfer plasmid (1 mg/mL), 0.10 mL of pCMVΔR8.91 plasmid (1 mg/mL), 0.040 mL of pMD-G plasmid (1 mg/mL), and 0.625 mL of 2 M $CaCl_2$.

   Plasmid stocks are kept at 1 mg/mL and the final quantity that is transfected to each 100 mm dish will be 13 μg/mL of gene-transfer plasmid, 10 μg/mL of pCMVΔR8.91, and 4 μg/mL of pMD-G plasmid (*see* **Note 2**).
10. Transfer the contents of Tube B to Tube A dropwise, while bubbling with a 2- or 5-mL pipette.

11. Leave the tube at room temperature for 15–20 min. A very fine, white precipitate will begin to form.
12. Vortex the calcium/phosphate/DNA solution briefly to mix the precipitate evenly.
13. Pipette 1 mL of calcium/phosphate/DNA solution dropwise onto each 100-mm dish containing 293T cells. As the solution is added, swirl the dish gently in order to mix the added solution and media evenly.
14. Incubate the culture plates containing transfected 293T cells overnight in a humidified incubator at 37 °C under 5% CO_2.
15. On the third day, gently aspirate the culture media from the transfected 293T cultures.
16. Gently wash each plate two or three times with 5–8 mL of DPBS. Washings should be done gently to avoid dislodging the 293T cells.
17. After washing with DPBS, add 10 mL of warm (37 °C) D10 media, and return the cells to the incubator.

3.2.2. Concentration of Viral Particles from Supernatant Collected from 293T Cells

1. On the sixth day, harvest the virus containing media and filter it through 0.8 μm, then 0.45 μm filters (*see* **Note 3**).
2. Transfer the filtered supernatant to a Beckman polyallomer centrifuge tube.
3. Centrifuge at 50,000 × *g* in a Beckman SW28 rotor for 1.5 h at 4 °C (*see* **Note 4**).
4. Carefully aspirate the supernatant and a small pellet should be visible at the bottom of the tube.
5. Leave ~0.5 mL of the supernatant in bottom of tube to avoid disturbing the pellet.
6. Resuspend the viral pellet into desired volume of the media. We use X-VIVO 10 to resuspend the viral pellet but any medium could be used.
7. The viral pellet must be resuspended homogeneously by pipetting up and down multiple times.
8. Aliquot the viral stock in cryotubes and store at −80 °C until use (*see* **Note 5**).
9. Estimate viral p24 using Beckman coulter™ HIV-1 p24 ELISA kit following the manufacturer's instructions.

3.2.3. Transduction of K562 Cells (See Note 7)

1. One the first day, plate K562 cells at 1×10^4 in 1 mL of R10 media per well in a 24-well plate.
2. Incubate the cells overnight in a humidified incubator at 37 °C under 5% CO_2.

3. On the second day, determine the viable cell count from a representative well using trypan blue.
4. Add known amount of concentrated vector stock to individual wells of K562 cells in duplicate.
5. Add protamine sulfate solution (sterile stock solution of 10 mg/mL in PBS) to control and transduction wells at a final concentration of 10 μg/mL.
6. Mix contents of the wells by gently pipetting the media to distribute the vector particles evenly within the well.
7. Centrifuge the plates at 2,000 rpm for 20–30 min at 32 °C in a tabletop centrifuge.
8. Place the plates at 37 °C in a humidified incubator with 5% CO_2 and allow the transduction to proceed for 48 h.
9. On the fourth day, transfer the cells from individual wells into separate 15-mL polypropylene tubes and spin at 2,000 rpm for 5 min in a tabletop centrifuge.
10. Aspirate the culture media, and wash once with 3 mL of R10 media.
11. Aspirate the wash media, resuspend the cells with 2 mL of R10 media, and place them in a 12-well plate.
12. Incubate the cells in a humidified incubator at 32 °C under 5% CO_2.
13. On the seventh day, remove 1 mL of cells from each well (*see* **Note 8**).
14. Place the cells from each well into separate 5-mL polystyrene round bottom 12 × 75 mm FACS tubes, add 2 mL of FACS buffer, and centrifuge at 1,000 rpm for 5 min in a tabletop centrifuge.
15. Carefully aspirate the supernatant and resuspend the cell pellets in 0.5 mL of FACS buffer or 1% PFA.
16. Keep cells in ice or in a refrigerator until analysis by flow cytometer.

3.2.4. FACS Analysis of Transduced K562 Cells

1. Acquire 10,000 live cell events from each sample using a FACScan or FACSCalibur flow cytometer equipped with CellQuest software.
2. Analyze data obtained from all samples using CellQuest software to determine percentage of eGFP positive cells for transduction efficiency and mean fluorescence intensity (MFI) for the level of transgene expression in each sample **(Fig .2)**.
3. To calculate the titer of the vector stock, multiply the fraction of positive cells in any given sample by the total number

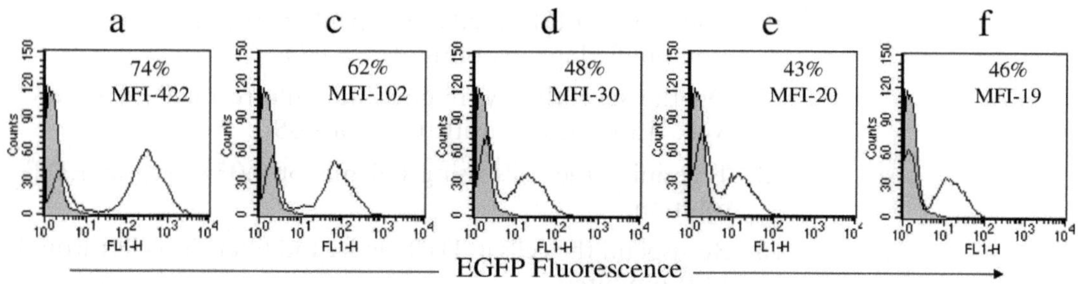

Fig. 2. Comparison of eGFP expression among mono, bi-, and tri-cistronic lentiviral vectors. To compare the level of eGFP expression, K562 cells were transduced with lentiviral vectors expressing the following transgenes, **a** eGFP, **b** MGMT, **c** MGMT-2A-eGFP, **d** MGMT-IRES-eGFP, **e** HOXB4-2A-MGMT-IRES-eGFP, **f** MGMT-2A-HOXB4-IRES-eGFP. Seven days after the transduction, cells were analyzed by flow cytometry for expression of eGFP. Untransduced control cells are shown as gray histograms. Lentivirally transduced cells are shown as open histograms. Percentage of eGFP positive cells is shown as % values on histograms and the mean fluorescence intensity (*MFI*) given as numbers on histograms.

of cells on the infected plate and divide this number by the amount of vector added (*see* **Note 8**).

Example: 20% eGFP^{+} cells by FACS following the addition of 5 ng P24 of viral vector to 1.7×10^5 HeLa cells: Titer = $[(20/100)(1.7 \times 10^5)]/(5 \text{ ng p24})$ = TU/ng of p24.

4. Determine average values for duplicate or triplicate samples using same amount of vector to establish titer.
5. Vector titer should be calculated using more than one ratio of target cell to vector particles within the linear range.

3.2.5. Viral Transduction of HeLa Cells

1. On the first day, plate 1×10^4 HeLa cells in a total volume of 1 mL of D10 media per well in 24-well plates.
2. Incubate the cells overnight in a humidified incubator at 37 °C under 5% CO_2.
3. Perform a viable cell count after harvesting cells from a representative well using trypan blue.
4. Add a known amount of concentrated viral vector stock to duplicate wells of HeLa cells.
5. Add protamine sulfate solution (10 mg/mL stock solution, 1 μL/mL) to a final concentration of 10 μg/mL in the transduction wells.
6. Mix vector with media by gently pipetting the media to distribute the vector particles evenly within the well.
7. Centrifuge the plates at 2,000 rpm for 20–30 min at 32 °C.
8. Place the plates at 37 °C in a humidified incubator with 5% CO_2 and allow the transduction to proceed for 48 h.
9. On day 4, aspirate the culture media from all wells and gently wash the wells twice with 2 mL of warm D10 media.

10. Add 2 mL of D10 media and incubate the plates at 37 °C in a humidified incubator under 5% CO_2.
11. On the seventh day, remove the culture media from each well. Wash the cells with 1 mL of PBS.
12. Trypsinize the cells using 0.5 mL of 0.05% trypsin–EDTA solution.
13. Resuspend the cells in D10 media and filter the cells through a 70-μm filter.
14. Place the cells from each well into a separate 12 × 75 mm polystyrene tubes, add 3 mL of FACS buffer, and centrifuge at 1,000 rpm for 5 min in a tabletop centrifuge.
15. Carefully aspirate the supernatant and resuspend the cell pellets in 0.5 mL of FACS buffer or 1% PFA.
16. Keep cells in ice or in a refrigerator until analysis by flow cytometer.

3.2.6. FACS Analysis of Transduced HeLa Cells

1. Acquire 10,000 live cell events from each sample using a FACScan or FACSCalibur flow cytometer equipped with CellQuest software.
2. Analyze data obtained from all samples using CellQuest software to determine percentage of eGFP positive cells and MFI.
3. To calculate the titer of the vector stock, multiply the fraction of positive cells in any given sample by the total number of cells on the transduced plate and divide this number by the amount of vector added.

Example: 10% $eGFP^+$ cells by FACS following the addition of 10 ng P24 of viral vector 1.7×10^5 HeLa cells: Titer = $[(10/100)(1.7 \times 10^5)]/(10\ \text{ng p24})$ = TU/ng of p24.

4. Determine average values for duplicate or triplicate samples using same amount of vector to establish titer.
5. Vector titer should be calculated using more than one ratio of target cell to vector particles within the linear range.

3.3. SDS-PAGE and Western Blotting (Fig. 9.3)

These instructions are for the use of Bio-Rad mini gel system. This can be easily adapted for other systems. The sample preparation method described below was designed to originally to estimate MGMT activity and Western blotting. Any other sample preparation method could be easily substituted.

1. Add 1.0 mL ice-cold buffer containing 50 mM Tris-HCl pH 8.3, 1 mM EDTA, 10 μg/mL leupeptin, and 3 mM DTT to cells and place on ice.
2. Sonicate samples 8 s at 35% amplitude and return to ice. Repeat the procedure once.

3. Add PMSF to final concentration of 87 μg/mL.
4. Centrifuge sample(s) at 15,000 rpm at 4 °C for 10 min.
5. Transfer supernatant to new tubes.
6. Perform Bio-Rad Protein Assay to estimate protein concentration.
7. For each sample, add desired amount of protein to a microcentrifuge tube followed by 2× Laemmli buffer.
8. Vortex and heat at 95 °C for 10 min.
9. Load 20 μg protein per well in a Bio-Rad 12% Tris-HCl, 10-well Ready Gel.
10. Separate proteins by electrophoresis.
11. Transfer proteins into Bio-Rad Immun-Blot PVDF membranes using Bio-Rad Mini Trans-Blot Cell.
12. Wash membrane 2× 10 min in 1× TBS + 0.05% Tween 20.
13. Block membrane in 5% nonfat milk in 1× TBS + 0.05% Tween 20 (blocking solution) for 1 h at room temperature on a rocking platform (agitation) or at 4 °C overnight.
14. Wash membrane 2× 10 min in 1× TBS + 0.05% Tween 20.

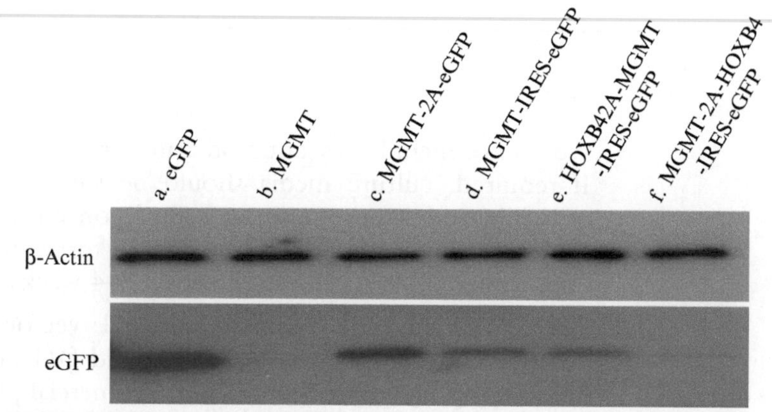

	a	b	c	d	e	f
eGFP %	74		62	48	43	46
eGFP MFI	422		102	30	20	19

Fig. 3. Western blot analysis of β-actin and eGFP expression in lentivirally transduced K562 cells. Cell extracts containing 5 μg of protein were subjected to SDS-PAGE. Blots were probed with mouse monoclonal antibody against GFP (Clontech, Mountain View, CA) or human β-actin (Sigma, St. Louis, MO). Table shows the corresponding % of eGFP positive cells and the mean fluorescence intensity (*MFI*) as analyzed by flow cytometry.

15. Dilute primary antibody in blocking solution and incubate membrane for 1 h at room temperature with agitation or overnight at 4 °C.
16. Wash membrane 4× 10 min in 1× TBS + 0.05% Tween 20.
17. Dilute secondary antibody in blocking solution and incubate membrane for 1 h at room temperature with agitation.
18. Wash membrane 4× 10 min in 1× TBS + 0.05% Tween 20 and 2× 10 min in 1× PBS.
19. Prepare ECL reagent (Pierce Super Signal West Pico Chemiluminescent Substrate), 5 mL peroxide solution, plus 5 mL luminal/enhancer solution.
20. Incubate membrane in ECL reagent for 5 min at room temperature with agitation.
21. Allow excess ECL reagent to drip off membrane and place in plastic sheet protector with TOP facing up.
22. Tape plastic sheet protector in cassette and take to dark room to develop.
23. Expose 30 seconds to Kodak BioMax Light film, develop and determine if more exposure is necessary.

Notes

1. We do not include any antibiotics in our cell culture media. If required, culture media should be supplemented with penicillin and streptomycin at a final concentration of 1×. All the culture media should be filtered through a 0.22-μm filter, stored at 4 °C, and used within 3–4 weeks.
2. Three plasmid HIV-1-based lentiviral vector system is described in this chapter. We use plasmid DNA prepared in house using Qiagen kits and from commercial plasmid production sources.
3. We harvest viral particles once; a second harvest might yield some virus with lower titer.
4. While concentrating the viral particles by centrifugation, tubes should be completely filled. We fill 31–33 mL of viral supernatant per tube. If the volume is lower, the tubes should be completely filled with D10 to avoid deformation of the tube during high-speed centrifugation.
5. Viral stocks should be resuspended homogenously and frozen down at –70 to –80 °C in smaller aliquots. Frozen stocks of VSV-G pseudotyped viral stocks could be thawed

and refrozen at least once without any significant loss of viral titer. However, repeated freezing and thawing of viral stocks results in decrease in titer. Thawing of viral stocks will be best performed at 37 °C and stored in ice.

6. One of our recent publication of Chinnasamy et al. *(12)* compared IRES- and 2A-based multicistronic lentiviral vectors. Following the transduction of K562 cells under similar conditions, the expression of eGFP in a bicistronic vector based on 2A was approximately four times greater than that of an IRES-based vector.
7. We have maintained lentivirally transduced K562 cells over a period of more than 6 months without loss of transgene expression.
8. Viral titer should be measured using a range of vector to target cell ratios to ensure that some of the determinations will fall within the linear range. At higher vector to target cell ratios and as the percentage of transduced cells (e.g., >30% $eGFP^+$ cells) goes up, it becomes unreliable for titer determination. Lower ratios of vector to target cells must be used to estimate a realistic vector titer. Calculate titer using the linear range of transduction, for example, between 2 and 20% $eGFP^+$ cells.

References

1. Ngoi, S.M., Chien, A.C. and Lee, C.G. Exploiting internal ribosome entry sites in gene therapy vector design, *Curr. Gene Ther.*, 4: 15–31, 2004.
2. Mizuguchi, H., Xu, Z., Ishii-Watabe, A., Uchida, E. and Hayakawa, T. IRES-dependent second gene expression is significantly lower than cap-dependent first gene expression in a bicistronic vector, *Mol. Ther.*, 1: 376–382, 2000.
3. Zhou, Y., Aran, J., Gottesman, M.M. and Pastan, I. Co-expression of human adenosine deaminase and multidrug resistance using a bicistronic retroviral vector, Hum. *Gene Ther.*, 9: 287–293, 1998.
4. Furler, S., Paterna, J.C., Weibel, M. and Bueler, H. Recombinant AAV vectors containing the foot and mouth disease virus 2A sequence confer efficient bicistronic gene expression in cultured cells and rat substantia nigra neurons, *Gene Ther.*, 8: 864–873, 2001.
5. de Felipe, P., Martin, V., Cortes, M.L., Ryan, M. and Izquierdo, M. Use of the 2A sequence from foot-and-mouth disease virus in the generation of retroviral vectors for gene therapy, *Gene Ther.*, 6: 198–208, 1999.
6. Klump, H., Schiedlmeier, B., Vogt, B., Ryan, M., Ostertag, W. and Baum, C. Retroviral vector-mediated expression of HoxB4 in hematopoietic cells using a novel coexpression strategy, *Gene Ther.*, 8: 811–817, 2001.
7. Stripecke, R., Cardoso, A.A., Pepper, K.A., Skelton, D.C., Yu, X.J., Mascarenhas, L., Weinberg, K.I., Nadler, L.M. and Kohn, D.B. Lentiviral vectors for efficient delivery of CD80 and granulocyte-macrophage-colony-stimulating factor in human acute lymphoblastic leukemia and acute myeloid leukemia cells to induce antileukemic immune responses, *Blood*, 96: 1317–1326, 2000.
8. Reiser, J., Lai, Z., Zhang, X.Y. and Brady, R.O. Development of multigene and regulated lentivirus vectors, *J. Virol.*, 74: 10589–10599, 2000.
9. Yu, X., Zhan, X., D'Costa, J., Tanavde, V.M., Ye, Z., Peng, T., Malehorn, M.T., Yang, X.,

Civin, C.I. and Cheng, L. Lentiviral vectors with two independent internal promoters transfer high-level expression of multiple transgenes to human hematopoietic stem-progenitor cells, *Mol. Ther.*, 7: 827–838, 2003.

10. Zhu, Y. and Planelles, V. A multigene lentiviral vector system based on differential splicing, *Methods Mol. Med.*, 76: 433–448, 2003.

11. Chinnasamy, D., Fairbairn, L.J., Neuenfeldt, J., Treisman, J.S., Hanson, J.P., Jr., Margison, G.P. and Chinnasamy, N. Lentivirus-mediated expression of mutant MGMTP140K protects human CD34+ cells against the combined toxicity of O6-benzylguanine and 1,3-bis(2-chloroethyl)-nitrosourea or temozolomide, *Hum. Gene Ther.*, 15: 758–769, 2004.

12. Chinnasamy, D., Milsom, M.D., Shaffer, J., Neuenfeldt, J., Shaaban, A.F., Margison, G.P., Fairbairn, L.J. and Chinnasamy, N. Multicistronic lentiviral vectors containing the FMDV 2A cleavage factor demonstrate robust expression of encoded genes at limiting MOI, *Virol. J.*, 3: 14, 2006.

13. Zufferey, R., Nagy, D., Mandel, R.J., Naldini, L. and Trono, D. Multiply attenuated lentiviral vector achieves efficient gene delivery in vivo, *Nat. Biotechnol.*, 15: 871–875, 1997.

14. Naldini, L., Blomer, U., Gallay, P., Ory, D., Mulligan, R., Gage, F.H., Verma, I.M. and Trono, D. In vivo gene delivery and stable transduction of nondividing cells by a lentiviral vector, *Science*, 272: 263–267, 1996.

Chapter 10

Using an EGFPmeter to Evaluate the Lentiviral Vector Production: Tricks and Traps

Claude Bagnis, Pascal Bailly, and Sylvie Chapel-Fernandes

Summary

An efficient viral vector containing supernatant is the first key issue to address for gene therapy or basic research projects relying on viral gene transfer. When present in a lentiviral vector backbone as a reporter gene, the expression of the enhanced green fluorescent protein (EGFP) shows strong interests to assess the production of lentiviral vector supernatants. The immediate nondestructive visual analysis of EGFP expression helps to predict the quality of the viral vector supernatant while it is produced. It also largely facilitates the quantitative evaluation of the number of viral particles present in the collected supernatants. Although some traps should be considered when analyzing the expression of a lentivirally transduced EGFP expression unit, the EGFP is the most powerful tool to consider when designing new gene transfer and expression strategies.

Key words: EGFP, Lentiviral vector, Viral vector production, Titration assay.

1. Introduction

Gene therapy and numerous basic research projects rely on the design of efficient gene transfer strategies. In this respect, the testing of gene transfer vectors requires gene reporters which allow the vectorologist to visualize both the event of transduction and the strength of expression. Numerous genes coding for proteins showing traceable features have been used as reporter genes coding for proteins inducing antibiotic resistance, proteins able to metabolize chromogenic chemical agents, or cell surface proteins detected by immunological techniques *(1)*. Among them, the GFP gene has rapidly dominated the field with several interesting features: (i) immediate detection without cell destruction

Barry W. Hicks (ed.), *Methods in Molecular Biology, Viral Applications of Green Fluorescent Protein, vol. 515*

DOI: 10.1007/978-1-59745-559-6_10

that allows further analyses of genetically modified cells; (ii) GFP expression can be detected without manipulation of the cells that prevents from biases due to uncontrolled parameters associated with the detection process; (iii) GFP expression can be quantified using fluorocytometry; and (iv) of importance, the small size of the cDNA encoding the green fluorescent protein allows its insertion in all gene transfer vectors envisaged in cancer gene therapy. Taking into consideration these features, strong efforts have been made by biotech companies to provide scientists with fluorescent reporter genes coding for proteins exhibiting enhanced autofluorescence or emitting in alternate wavelengths, with chimeric proteins able to locate in subcellular compartments, or with fluorescent proteins with reduced half-life. Most of the experiments dedicated to the design of gene transfer vectors have been performed with EGFP, an "enhanced" version of the GFP gene which encodes for a protein with enhanced fluorescence capacity *(2, 3)*.

The emergence of lentiviral vectors at the end of the nineties largely took benefit of the increasing use of EGFP as a reporter gene *(4–6)*. This chapter highlights the interest of using EGFP during the production of a monocistronic HIV type 1 lentiviral vector and the titering of the subsequent lentiviral vector containing supernatant.

Lentiviral vectors are usually produced by transient transfection of three plasmid vectors into 293T cells *(4, 7, 8)*. A first plasmid expresses the gag, pol, tat, and rev HIV-1 proteins as trans-complementing proteins. A second plasmid expresses the g protein of the Vesicular Stomatitis Virus (VSV) to pseudotype viral particles with an envelope protein able to confer a wide spectrum of infectivity to the lentiviral vector particles. A third plasmid containing the transgene of interest produces genomic lentiviral vector mRNAs which interact with HIV-1 gag, pol, tat, rev, and VSV-g, and cellular components to generate new infectious viral particles. Viral particles are then collected in the supernatant of producing cells.

In the transfected producing cells, the presence of vector mRNAs leads to expression of the transgene of interest. Thus, to a certain extent, when an EGFP expression unit is present in the lentiviral vector genome or when a control EGFP lentiviral vector is produced in parallel, both the level of fluorescence and the percentage of EGFP-positive cells reflect the transfection efficiency. Even if EGFP expression reflects the expression of only one over the three plasmids transfected, a simple analysis of the producing cells provides useful information about the production efficiency.

2. Materials

2.1. Cell culture

1. RPMI 1640 medium with l-glutamine (BioWhittaker, Emerainville, France) supplemented with 10% fetal bovine serum (FBS), EU approved origin (Gibco, Invitrogen Corporation).
2. Cells: 293T, a human embryo kidney cell line, is cultured in RPMI 1640-based medium supplemented with 10% FBS and 1% antibiotic stock solution.
3. Trypsin–EDTA 1× in HBSS without calcium and magnesium and with EDTA (Gibco, Invitrogen Corporation, Carlsbad, CA).
4. Antibiotic stock solution: Penicillin 10,000 IU/mL and streptomycin 10,000 µg/mL (BioWhittaker).

2.2. Transfection

1. Sterile 100 mM NaCl. Dilute 1 mL of a 5-M NaCl stock in 49 mL water and sterilize by filtration using a 0.22-µm filter (Millex filter, Millipore Corporation, Bedford, USA). Store at +4 °C.
2. Polyethylenimine (PEI) working solution (10 mM). Prepare a 100-mM stock solution with 45 mg PEI (MW 25,000, Sigma-Aldrich) in 10 mL water and sterilize by filtration. Adjust to pH 7 with HCl. The stock solution is stored frozen in 1 mL aliquots at −20 °C until use. Working solution (10 mM) is obtained by dilution of a 1-mL aliquot with 9 mL sterilized water and stored at 4 °C.
3. Phosphate buffered saline (PBS; 6.7 mM phosphate) without calcium and magnesium (BioWhittaker).
4. Viral DNA vectors: p8.91 which expresses the HIV-1-derived gag, pol, Tat, and Rev proteins and pMDG which expresses the VSV-g env glycoprotein (kindly provided by D. Trono, Ecole Polytechnique Fédérale de Lausanne, Switzerland).
5. EGFP lentiviral vector plasmid (kindly provided by D. Trono, Ecole Polytechnique Fédérale de Lausanne, Switzerland).

2.3. Concentration

1. Polyethylene glycol (PEG) solution: 100 g of PEG (MW 8000, Sigma-Aldrich) is dissolved in 100 mL PBS and sterilized by autoclave. The final volume of the PEG solution is ~0.56 g PEG/mL of solution.
2. Hepes buffer: 1 M in 0.85% NaCl (BioWhittaker).
3. 0.22 µm filters (Millex filter, Millipore Corporation).

2.4. Detection of EGFP-Positive Cells

1. Fluorescent cells are detected using a IX-70 inverted microscope (Olympus Corporation, Tokyo, Japan) equipped with a U-RFL-T module for UV emission and a filtration cube

dedicated to the detection of EGFP fluorescence (excitation filter, 460–480 nm; emission filter, 495–540; dichromatic mirror 485 nm).

2. FACS analyses are performed with a FACScalibur apparatus (Becton Dickinson, Franklin Lakes, NJ, USA) equipped with a 15-mW 488 nm, air-cooled argon-ion laser, the EGFP-positive cells being detected in the FL-1 channel (band pass filter: 525 ± 15 nm).

2.5. Titration of Lentiviral Vector Production

1. Supplemented RPMI 1640.
2. Trypsin–EDTA 1× in HBSS without calcium and magnesium and with EDTA.
3. Polybrene solution: polybrene (hexadimethine bromide, Sigma-Aldrich) is dissolved in PBS and filter-sterilized to obtain a 1 mg/mL solution to be stored at 4 °C.
4. HuH7 is a human hepatocarcinoma cell line (ATCC).

3. Methods

3.1. Transfection

1. Seed the 293T cells into 15 cm tissue dishes to obtain an 85–95% confluent culture at the day of transfection. In the conditions described herein, about 25 × 10^6 cells can be seeded into 15 cm dishes the day prior to the transfection to reach 85–95% confluency. Alternatively, a 1/4 split of a confluent culture provides cell cultures ready for use after ~72–96 h.
2. Use a three-plasmid transient co-transfection strategy to generate HIV-derived retroviral vector particles. Cells are transfected using a polyethylenimine—based procedure *(9)*. A total of 87.5 μg of plasmid DNA is used for the transfection of a 15-cm diameter plate of 293T cell in the following proportions: 25 μg of p8.91, 12.5 μg of pMDG, and 50 μg of the EGFP lentiviral vector plasmid (*see* **Note 1**).
3. Mix the three DNA samples in tube A with 1.75 mL of 150 mM NaCl.
4. In tube B, mix 1.5 mL of 150 mM NaCl with 250 μL of PEI working solution.
5. Immediately add the contents of tube B to tube A and vortex briefly.
6. After 1 h at room temperature, add 10 mL of supplemented RPMI and replace the culture medium of 293T cells with this transfection medium.

7. Leave the transfection medium on the cells for 17 h ± 2 h (Day 1). Then replace the transfection medium with 10 mL of fresh culture medium supplemented with 25 mM Hepes (*see* **Notes 2** and **3**).
8. Viral supernatants can be collected after 24 h (Day 2). Filter the viral medium through a low protein-binding 0.22 μm filter.
9. The supernatants harvested can be directly used to transduce target cells or stored at –80 °C for further use. If a high transduction efficiency is required, in either case it would be prudent to concentrate the viral particles before use.

3.2. Concentration and Storage of Lentiviral Vector

1. Using a syringe, add PEG solution to the viral supernatant up to obtain a final concentration of 10%.
2. Vigorously homogenize with a vortex mixer.
3. Use 50-mL polypropylene tubes to process large volumes of viral supernatant. Cool the tubes for 20–30 min at 4 °C, then dispense the viral supernatants into the necessary number of tubes.
4. Centrifuge at 9,500–10,000 × *g* for 1 h at 4 °C in a fixed-angle rotor.
5. Remove the tubes and place them into an ice bath. All of the following steps should be performed on ice.
6. Aspirate and completely discard the supernatant. Carefully resuspend the pelleted viral particles in 1% of the original supernatant volume using cold PBS. Avoid frothing the solution when resuspending the pellet.
7. Store aliquots of this viral stock solution at –80 °C until use. Alternatively short storage at –20 °C, up to 1 month, does not affect the viral titer.

3.3. EGFP Monitoring to Predict the Quality of the Lentiviral Vector Production

1. While culturing cells, obtain fluorescent micrographs of the live cells using an inverted fluorescence microscope. For a more quantitative analysis, perform flow cytometry. **Figure 1** shows the kinetics of EGFP gene expression in transfected 293T cells.
2. The expression of EGFP correlates with the viral production efficiency (*see* **Notes 4–7**). The EGFP expression can be detected as early as 24 h after transfection, but this varies with the promoter selected to drive the EGFP expression. Normally, the fluorescence intensifies and reaches a plateau after 3–4 days (*see* **Notes 5**, **6**, and **14**). In a typical experiment such as that shown (**Fig. 1**), a viral titer ranging from 5×10^5 to 1×10^6 viral particles/mL is observed.

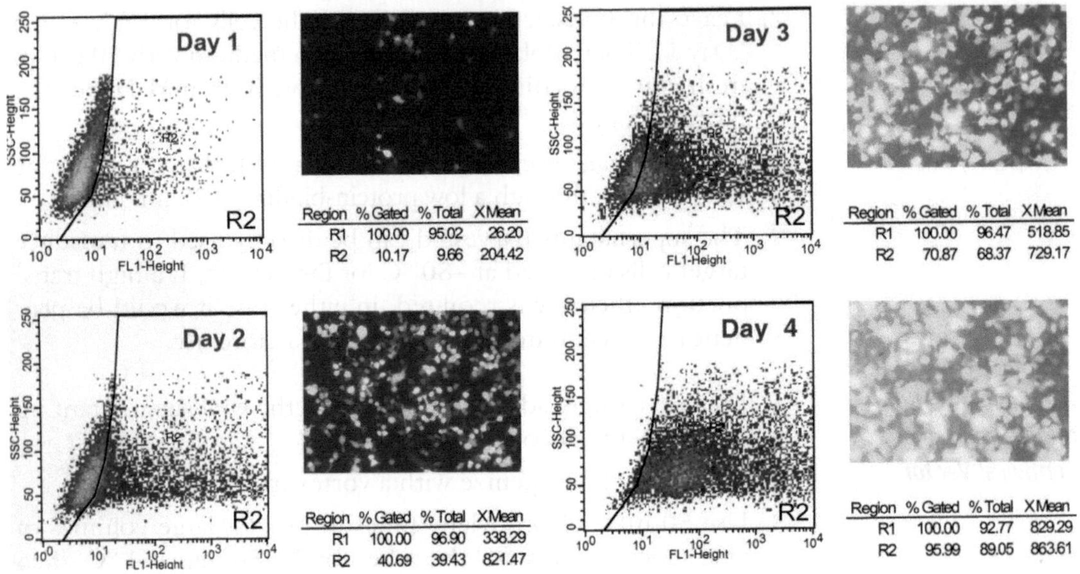

Fig. 1. Kinetics of EGFP expression during lentiviral vector production. 293T cells were transfected as indicated in the text. At Days 1, 2, 3, and 4 after the beginning of the transfection, cells were analyzed under fluorescence microscopy, then trypsinized, and EGFP expression was evaluated by flow cytometry. Micrographs were taken using the same magnification and sensitivity parameters using a fluorescence microscope equipped with a DP-50 camera and the DP-Soft software. A marked increase of EGFP expression can be observed during the lentiviral vector production as assessed by visual analysis and flow cytometry taking into consideration the percentage of EGFP-positive cells (% Total) detected in the R2 region and the global mean fluorescence intensity of the cell population (× mean of the R1 region). As indicated in the **Subheading 10.3** and in the **Note 2**, viral supernatants are collected at Days 2, 3, and 4.

3.4. Titration of Supernatant Containing EGFP Lentiviral Vectors

The titration assay estimates the number of functional viral vector particles in a cell culture supernatant. Estimation of the viral titer relies on the quantification of viral proteins, genomic RNA, the detection of integrated proviral vectors in the genome of the target cell, or the detection of transduced cells expressing the transgene *(10, 11)*. Although it restricts the viral titer to both functional viral particles able to transfer the EGFP and cells able to express the transgene, the viral titration assay based upon detection of EGFP fluorescent transduced cells is a common and classical procedure now. Titration assay and viral titer are classical matters of stormy debates between vectorologists since numerous parameters including the quality of the serum, the quality of the polybrene, the target cell type, the subclone of the target cell used, the volume of transduction, the number of target cells, and so on can all profoundly affect the results of the titration assay. It is important to remember that even in the case of similar technical processes, a one log variation in the number of viral particles detected can be observed between two expert laboratories with the same viral supernatant batch. A constructive strategy is to standardize the technical procedure in the laboratory and to approach the viral titer as a parameter for internal use. Both nonfunctional viral particles which may account for more than

50% of the viral vector preparation and which may compete with functional viral particles for cell transduction, and transduction events with no or low level of expression are not detected using this assay.

HuH7 is a human hepatocarcinoma cell line which has been shown very sensitive to transduction with HIV type 1-derived lentiviral vectors pseudo-typed with the VSV-g protein. Alternate pseudo-typing of the viral particle may require the titration assay to be performed with relevant cell lines. Numerous cell lines can be used to titrate a viral supernatant and the titration assay can be performed with adherent and nonadherent cells. An "absolute" titration assay requires the most sensitive cell line to transduction in order to detect the maximum of functional viral particles. A "relative" titration assay can be performed with the cells of the study, that is, CD34+ hematopoietic stem cells for experiments dealing with gene transfer into hematopoietic progenitors, or with model cells close to the final target cells.

1. Trypsinized HuH7 cells are seeded into 24 well-plates in supplemented RPMI 1 day prior to analysis with about 1×10^6 cells in 1.5 mL of culture medium per well. Of note, HuH7 cells show a strong contact inhibition.
2. The viral supernatant to be tested has to be thawed in ice. Aliquots can be freeze–thawed up to three times without altering the viral titer of VSV-g protein pseudotyped lentiviral vectors.
3. Experiments should be performed in triplicate with at least four serial dilutions of the pure supernatants and should include a unique negative control with mock HuH7 cells. Use sufficient viral stock so that at least 300 μL per point is available to add to cells Perform serial dilutions of the viral stocks using HuH7 culture medium supplemented with 10 μg/mL of polybrene. Positive control, that is, EGFP positive HuH7 cells previously transduced can be included in a titration assay but a routine procedure usually takes into consideration the presence of EGFP positive cells in the sample transduced with nondiluted supernatant (*see* **Notes 8–10**).
4. One day after infecting cells, replace the culture medium containing the viral particles with fresh culture medium
5. On the third day after infecting HuH7 cells, remove the medium, and wash the cells with PBS.
6. Detach the adherent cell by adding 100 μL of Trypsin-EDTA solution. After 5 min at 37 °C, the whole cell population is seeded into 35-mm dishes in fresh culture medium.
7. On the fifth day, wash the HuH7 cells with PBS and add 500 μL of trypsin–EDTA solution. After 5 min at 37 °C, detached cells are collected in a 5-mL FACS tube containing 1 mL of culture medium.

8. Analysis of the genetically modified cell population can be performed by FACS immediately or up to 24 h later taking into consideration the detection of positive events in the FL-1 channel (band pass filter: 525 ± 15 nm). Visual scoring of EGFP positive cells using an inverted microscope equipped with the fluorescence light technology introduces very strong biases and is not recommended (*see* **Note 11**).
9. Assuming (i) the number of HuH7 cells does not vary significantly between the day of plating and day 0, and (ii) the proliferation of transduced cells is not different from the proliferation of nontransduced cells, the viral titer can be calculated using the following formula:

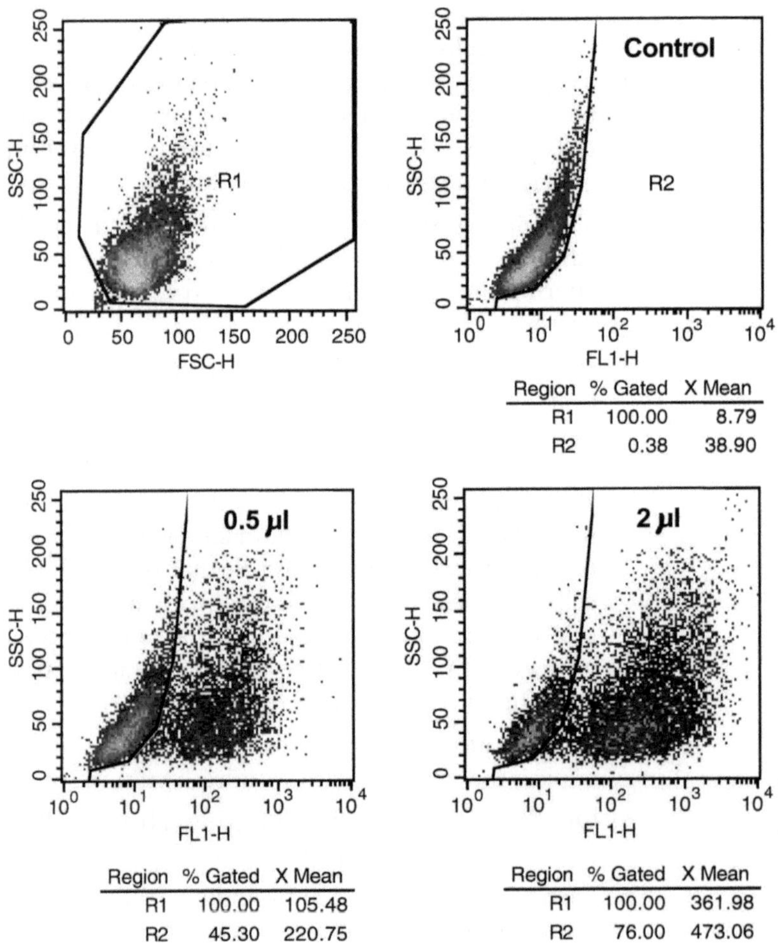

Fig. 2. EGFP expression in HuH7 cells transduced with an EGFP containing lentiviral vector and estimation of the viral titer. 1×10^6 HuH7 cells were loaded with 0.5 and 2 µL of concentrated viral supernatant and processed as indicated in the text. According to the percentage of EGFP-positive cells, a viral titer of 9.06×10^8 vp/mL ($45.3\% \times 10^6 \times (1{,}000/0.5) \times 1$) and 3.80×10^8 vp/mL ($76\% \times 10^6 \times (1{,}000/2) \times 1$) were obtained for experiments performed with 0.5 and 2 µL, respectively. Future experiments have to be designed considering the viral titer calculated from the assay performed with 0.5 µL.

Mean % of EGFP-positive cells × 1 × 10^6 × (1,000/starting volume) × dilution factor = number of viral particles/mL of pure supernatant (*see* **Notes 12–14** and **Fig. 2**).

Notes

1. It is highly recommended to pay particular attention to the quality of the preparation of the plasmids used considering pure DNA yields better transfection efficiency.
2. The transfected cells can be incubated for two further rounds of 24 h that allows a second (Day 3) and a third (Day 4) harvesting of the viral-containing supernatant. Transfected 293T cells tend to fuse with neighboring cells due to the fusogenic properties of the VSV-g envelope protein that impairs the collection of efficient viral supernatant beyond Day 4. Difficulties can be encountered when filtrating the third collect due to the presence of an increased number of cellular debris.
3. The control of the differentiation status and the proliferation kinetics of the target cells is a key parameter to consider. In this respect, serum-free or more adequate culture medium can be used to produce the viral vector supernatant that will be applied to the target cells. It is recommended to compare the production performed with the selected culture medium with the production performed with the standard RPMI-based culture medium to evaluate their respective impact on both the producing cells and the target cells.
4. EGFP can be expressed in 239T and HuH7 cells at very high level. Most of the studies performed with the EGFP does not report toxic effect associated with EGFP gene expression. It has to be reported that it may exert toxic effects in hematopoietic cells transduced with EGFP containing vectors *(12)* but these data has never been confirmed nor reproduced. EGFP accumulates in the cytoplasm of the 293T and HuH7 cells. Surprisingly, in some target cells *(13)*, EGFP-positive transduced cells can be detected as nucleus-bright cells suggesting that a cryptic nuclear localisation signal in the protein can be specifically used in some cell types.
5. Efficient co-expression of two genes using the same retroviral vector can be obtained but remains a major challenge. Whatever the co-expression strategy *(14, 15)*, when the EGFP is used as a reporter gene the cell population transduced with a bicistronic retroviral vector may contain various cell types: (i) cells expressing the two genes at equivalent level, cells expressing only one of the two genes, (ii) cells expressing undetectable

level of EGFP but adequate level of the accompanying gene, and (iii) cells expressing EGFP but not at adequate level or without expression of the accompanying gene. The relative proportion of the transduced cell populations described above and/or the sorting of EGFP-positive cells to study the accompanying gene may strongly bias the results of the study. In addition, it is of note that according to the type of co-expression strategy, the EGFP gene may exhibit a weaker expression than in cells transduced with a control monocistronic vector, due to interference at the translational level (for vectors containing two promoters, ...) or at the transcriptional level (for vectors containing internal ribosomal entry site, ...), that constricts the detectable EGFP-positive cell population to cells highly expressing EGFP. At last, even moderate toxic effects due to the protein encoded by the accompanying gene can profoundly unbalance the relative proportion of the transduced cell populations described above. Thus, use of EGFP as a reporter in a bicistronic retroviral vector can be a powerful tool but this strategy should be approached very carefully even during the production of the viral vector.

6. EGFP expression during the first 24 h of production only reflects the expression of EGFP by cells that have taken up DNA. After this period, EGFP expression observed in positive cells corresponds not only to those cells transfected by the vector plasmid, but also to cells transduced by newly produced viral vectors. Both the level of expression, as assessed by the fluorescence intensity and the percentage of EGFP-positive 293T cells, and the increase in EGFP expression provide useful information to predict the quality of the ongoing lentiviral vector production.
7. Since this strategy relies on the expression of a gene encoded by one of the three plasmids used, in some instances a strong expression of the EGFP gene can hide a low efficiency of lentiviral vector production. Several reasons including either the quality of the transcomplementing plasmids that could affect the transfection efficiency and/or unexpected side effects of the accompanying gene when a bicistronic lentiviral vector is produced, can explain such a discrepancy.
8. When the selected lentiviral vector does not contain the EGFP gene or an easily detectable marker gene, it is possible to run in parallel a production with an EGFP control vector. Of evidencc, the analysis of these concomitant productions should always take into consideration the possibility that unexpected features such as the biological impact of the expression of the transgene of interest in producing cells can profoundly affect the production of lentiviral particles.

9. The transduction volume should be minimal because only the Brownian motion promotes the contacts between viral particles and target cells in this procedure *(16)*.
10. According to the literature, the concentration of polybrene may vary from 2 to 10 μg/mL. This range of concentration may be due to the quality of the batch of polybrene and/or the quality of the batch of serum used. The optimal concentration of polybrene to be used should be tested prior to standardize the titration assay. Using some cell types, such as cord blood CD34+ hematopoietic cells, the absence of polybrene does not modify the transduction efficiency. Of importance, a three- to tenfold excess of polybrene concentration may be toxic for the target cells.
11. When possible, the direct detection of the protein encoded by the transgene of the study is a good tool to estimate the viral titer and to predict the transduction efficiency. For a bicistronic vector which contains EGFP, transduced cells strongly expressing EGFP emit a bright fluorescence which makes impossible the use of antibodies coupled with fluorochromes emitting in the Fluorescence-1, FL-1 channel (band pass filter: 530 nm). It also makes very difficult the use of antibodies coupled with fluorochromes emitting in the FL-2 channel (band pass filter: 585 nm). The co-detection of EGFP and a protein of interest can be performed using the FL-4 channel (band pass filter: 661 nm; this detection requires the FACS apparatus to be equipped with a second red diode laser) with antibodies coupled to a fluorochrome emitting in an alternate wavelength such as the AlloPhycoCyanin.
12. Using the production protocol presented above and according to the resuspension volume used during the concentration process, the viral titer may range from 10^5 viral particles per mL (vp/mL) for nonconcentrated viral supernatant to 10^9 vp/mL for concentrated stocks. These results can be easily obtained with monocistronic lentiviral vector in which the EGFP gene is driven by so-called constitutive promoter such as the promoter derived from the elongation factor 1 alpha gene, the phosphoglycerate kinase gene, or from the cytomegalovirus. For a concentrated viral supernatant batch (10^8–10^9 pv/mL), up to a 1,000-fold dilution of the stock can be tested. For a nonconcentrated viral supernatant batch (10^5–10^7 vp/mL), up to a 100-fold dilution sample starting from 100 μL of the pure supernatant should be tested. As an alternative and easier method which allows a gross evaluation of the viral titer, it is possible to load the culture volume with 0.5, 1, and 2 μL of the pure supernatant in the presence of polybrene (**Fig. 2**).

13. A graph of the percentage of EGFP-positive cells versus dilution plateaus for viral supernatants with a high concentration of viral particles, thus viral titer should be estimated using the results of transduction experiments showing 5–50% of EGFP-positive cells. Viral titer estimated using HuH7 cell population showing more than 50% of EGFP-positive cells may be underestimated. Conversely, viral titer estimated using HuH7 cell population showing less than 5% of EGFP positive cells may be overestimated. **Figure 2** shows a typical experiment.
14. Cell coating with free EGFP proteins present in the supernatant or EGFP proteins coated to or tightly associated with nonfunctional viral particles may account for a temporary fluorescence called pseudo-transduction *(17, 18)*. Protein degradation and target cell proliferation reduce the amount of unexpected EGFP protein per cell and make this fluorescence disappears. Although not frequent, this phenomenon is a common threat. Pseudo-transduction has never been detected in manipulated HuH7 cell population in the conditions depicted here. Of note, in the case of high viral load, the culture medium of the target cells may show a weak fluorescence under fluorescent microscopy due to a high content of free EGFP proteins present in the viral supernatant tested.

References

1. Bagnis, C., Uch, R. and Mannoni, P. (1999) Viral vectors and gene markers: tools and challenges for scientists working blind on genetic modification of hematopoietic cells. In:. *Recent Research Developments in Virology.* (Pandalai, S. G., ed.). Transworld Research Network: , Trivandruml, pp. 217--228.
2. Bierhuizen, M. F., Westerman, Y., Visser, T. P., et al. (1997) Green fluorescent protein variants as markers of retroviral-mediated gene transfer in primary hematopoietic cells and cell lines. *Biochem. Biophys. Res. Commun.* 234, 371–375.
3. Zhang, G., Gurtu, V. and Kain, S. R. (1996) An enhanced green fluorescent protein allows sensitive detection of gene transfer in mammalian cells. *Biochem. Biophys. Res. Commun.* 227, 707–711.
4. Naldini, L., Blomer, U., Gallay, P., et al. (1996) In vivo gene delivery and stable transduction of nondividing cells by a lentiviral vector. *Science* 272, 263–267.
5. Trono, D. (2000) Lentiviral vectors: turning a deadly foe into a therapeutic agent. *Gene Ther.* 7, 20–23.
6. Vigna, E. and Naldini, L. (2000) Lentiviral vectors: excellent tools for experimental gene transfer and promising candidates for gene therapy. *J. Gene Med.* 2, 308–316.
7. Zufferey, R. (2002) Production of lentiviral vectors. *Curr. Top. Microbiol. Immunol.* 261, 107–121.
8. Tiscornia, G., Singer, O. and Verma, I. M. (2006) Production and purification of lentiviral vectors. *Nat. Protoc.* 1, 241–245.
9. Demeneix, B. and Behr, J. P. (2005) Polyethylenimine (PEI). *Adv. Genet.* 53, 217–230.
10. Sastry, L., Johnson, T., Hobson, M. J., et al. (2002) Titering lentiviral vectors: comparison of DNA, RNA and marker expression methods. *Gene Ther.* 9, 1155–1162.
11. Geraerts, M., Willems, S., Baekelandt, V., et al. (2006) Comparison of lentiviral vector titration methods. *BMC Biotechnol.* 6, 34.
12. Hanazono, Y., Yu, J. M., Dunbar, C. E., et al. (1997) Green fluorescent protein retroviral vectors: low titer and high recombination frequency suggest a selective disadvantage. *Hum. Gene Ther.* 8, 1313–1319.

13. Roche, C., Zamora, A. J., Taieb, D., et al. (2004) Lentiviral vectors efficiently transduce human gonadotroph and somatotroph adenomas in vitro. Targeted expression of transgene by pituitary hormone promoters. *J. Endocrinol.* 183, 217–233.

14. Ngoi, S. M., Chien, A. C.Lee, C. G. (2004) Exploiting internal ribosome entry sites in gene therapy vector design. *Curr. Gene. Ther.* 4, 15–31.

15. Hildinger, M., Schilz, A., Eckert, H. G., et al. (1999) Bicistronic retroviral vectors for combining myeloprotection with cell-surface marking. *Gene Ther.* 6, 1222–1230.

16. Palsson, B. and Andreadis, S. (1997) The physico-chemical factors that govern retrovirus-mediated gene transfer. *Exp. Hematol.* 25, 94–102.

17. Liu, M. L., Winther, B. L. and Kay, M. A. (1996) Pseudotransduction of hepatocytes by using concentrated pseudotyped vesicular stomatitis virus G glycoprotein (VSV-G)-Moloney murine leukemia virus-derived retrovirus vectors: comparison of VSV-G and amphotropic vectors for hepatic gene transfer. *J. Virol.* 70, 2497–2502.

18. Haas, D. L., Case, S. S., Crooks, G. M., et al. (2000) Critical factors influencing stable transduction of human CD34+ cells with HIV-1 derived lentiviral vectors. *Mol. Ther.* 2, 71–80.

Chapter 11

Use of a Macrophage-Tropic GFP-Tagged Human Immunodeficiency Virus Type 1 (HIV-1) to Study Viral Reservoirs

Amanda M. Brown

Summary

The human immunodeficiency virus type 1 (HIV-1) predominantly infects two main cell types: T-lymphocytes and monocyte-derived cells such as macrophages, dendritic cells, and Langerhans cells. Studies in patients aimed at uncovering and understanding the molecular mechanisms for viral persistence and latency in macrophages in particular are challenging because these cells reside in tissues and can only be extracted using invasive methods. The in vitro human monocyte-derived macrophage (MDM) model provides an amenable system to study HIV–host cell interactions at the molecular level. Infection of macrophages with a recombinant M-tropic virus that contains the green fluorescent protein (GFP) gene within its genome allows the detection of cells with integrated viral DNA that are producing viral particles. The infected cells can be studied at the single-cell level using a variety of fluorescence-based technologies such as flow cytometry, conventional and confocal microscopy, and laser capture microdissection.

Key words: Monocyte-derived macrophages, Green fluorescent protein, Human immunodeficiency virus type 1, HIV-1, Laser capture microdissection, Flow cytometry.

1. Introduction

Since its detection over 25 years ago, human immunodeficiency virus type 1 (HIV-1) infection has reached epidemic proportions particularly in many parts of the developing world. The availability of combination drug therapy in resource rich countries led to a dramatic decline in the death rate and has transformed HIV infection into a chronic, but manageable disease. Current drug regimens are very effective in suppressing virus replication to undetectable levels. However, these drugs must be taken for the life of

Barry W. Hicks (ed.), *Methods in Molecular Biology, Viral Applications of Green Fluorescent Protein, vol. 515*

DOI: 10.1007/978-1-59745-559-6_11

the individual and over the long term are often accompanied by serious side effects. Studies employing structured interruptions of therapy revealed that viral load re-emerges to pretreatment levels in as short as 3–4 weeks *(1–3)*. The dynamics of T-cell and virus turnover during antiretroviral therapy revealed that cell death is rapid and newly divided naïve and memory T-cells become new targets for infection and thus provide a feedback loop that also results in the establishment of a small pool of latently infected cells. HIV-1 also infects cells derived from monocytes such as macrophages. These cells are present at mucosal surfaces, found in many tissues such as the brain, lung, spleen, and liver, and are not susceptible to the cytopathic effects induced by HIV-1. The potential for macrophages to serve as a site of persistent virus replication was recognized soon after the discovery that HIV infected this cell type, and more recently studies have tried to address the possibility that macrophages could also be a source of latent infection similar to that shown for T-cells *(4–6)*.

Macrophage–HIV in vitro culture systems can be difficult to establish and maintain. Some of the difficulties encountered include the different methods for isolation of monocytes, the type of culture dishes used for plating, plating density, the quality of the human and fetal sera, and the operation of the incubator. In addition, variability in the efficiency of infection between different donor cells can make the system intractable for certain types of studies. To overcome some of these limitations, we developed a green fluorescent protein (GFP)-tagged macrophage-tropic HIV-1 reporter virus *(4, 7)*. The enhanced GFP gene is under the control of the viral promoter and produced as an early protein that can be detected in macrophages about 48 h after infection. In cells that have a more rapid division cycle such as mitogen-activated T-cells or after transfection of cell lines with proviral plasmid DNA, GFP can be detected after about 24 h. With optimization of culture conditions, infection frequencies between 2 and 30% can be routinely obtained with different donor macrophages. After detachment from the culture dish, the number of infected GFP-positive cells can be determined by flow cytometry. In addition, combined with staining of macrophage surface molecules involved in immunity, the HIV-GFP reporter system provides a powerful tool to study the impact of HIV replication in this cell type. Furthermore, infected macrophage cultures can be sorted to provide pools of cells that are actively producing viral particles and those in which the viral promoter is inactive (GFP-negative), but have still integrated proviral DNA. The sorted GFP-negative pool can be purified to 99.9%. However, a few contaminating GFP + macrophages can be detected by microscopic examination. To remove these rare contaminants, laser ablation can be used *(4)*. The sort-ablated pools can then be cultured for viral reactivation studies or subjected to molecular assays.

2. Materials

2.1. Transfection of 293T Cells for Generation of Virus and Virion Quantitation

1. Human kidney epithelial cells transformed with the SV40 T-antigen (HEK293T) cells.
2. HIV-GFP-tagged reporter virus, pSF162R3 Nef+ *(7)*.
3. DMEM (medium only) and complete medium: DMEM supplemented with 1% penicillin/streptomycin (stock), 1% glutamine (stock), and 10% fetal bovine serum (FBS).
4. Trypsin–EDTA (Invitrogen, Carlsbad, CA).
5. DMRIE-C transfection reagent.
6. OPTI-MEM medium (Invitrogen).
7. 0.45 μm syringe filters (Millipore, low protein binding).
8. T-75 cm^2 culture flasks (Corning Inc., Corning, NY).
9. SlideFlask (Nalge Nuc International).

2.2. Isolation and Culture of Monocytes

1. Human leucocytes (40–50 mL, New York Blood Center).
2. 10× PBS, 1× PBS.
3. 46% Percoll (GE Healthcare, Piscataway, NJ) made by mixing 1 mL of 10× PBS, 11.5 mL of Percoll, and 12.5 mL of 1× PBS.
4. Ficoll-Paque™ Plus (GE Healthcare).
5. Macrophage medium: RPMI 1640 supplemented with 1% penicillin/streptomycin (stock), 1% glutamine (stock), and 20% FBS.
6. Hank's Balanced Salt Solution (HBSS; Invitrogen).
7. Human AB serum (Cambrex, Rockland, ME).
8. DMSO (Sigma, St. Louis, MO) or CryoStor™ CS-10 cryopreservation solution (BioLife Solutions, VWR).
9. T-25 cm^2 culture flasks (Corning).
10. Conical tubes: 15 and 50 mL (Falcon).

2.3. Flow Cytometry Analyses of Infected Macrophages

1. PBS-EDTA: PBS containing 5 mM EDTA (Sigma); PBS-EDTA-NaN_3: PBS with 5 mM EDTA and 10 mM sodium azide (Sigma).
2. Formaldehyde (Tousimis, Rockville, MD).
3. Accutase™ and Accumax™ (Chemicon International).
4. 5-mL round bottom tubes (Falcon).
5. Teflon cell scrapers (Fisherbrand).
6. Propidium iodide, 1 μg/mL (Sigma).
7. 5 mL round bottom polypropylene tubes with 35 mm nylon mesh cap (BD Biosciences, Palo Alto, CA).

2.4. Laser Ablation or Capture of Infected Macrophages

1. PALM Microlaser (Zeiss, Bernied, Germany).
2. PALM® duplex dish (#1440-0550) and adhesive caps microfuge tubes (#1440-0250) (Zeiss).

3. Methods

A few different methods can be used to prepare human monocyte-derived macrophages (MDM). After isolation by Ficoll gradient centrifugation, peripheral blood mononuclear cells (PBMCs; lymphocytes) can be directly plated into culture dishes and the monocyte fraction will adhere while the nonadherent cells will largely remain floating. The time allowed for adherence is typically 2–3 h to overnight and then the medium is changed to remove the nonadherent cells. This may be followed by a couple of washing steps to remove any remaining nonadherent cells. As T-cells and monocytes can form cell clusters that provide complementary growth and stimulatory factors, contamination of 14-day-old macrophage cultures with T-cells can be detected by flow cytometric analyses *(7)*. This fact may or may not be problematic depending on the type of experimental question being asked. One solution to restrict analyses to the macrophage fraction is to immunostain for a macrophage-specific surface molecule such as CD14 *(7)*. However, for experiments requiring purer cultures of macrophages, the PBMC fraction is further separated using a 46% Percoll gradient. Alternatively, magnetic bead selection for the subset of monocytes expressing CD14+ can also be used. No matter what technique is utilized, there is some variability among donors in the efficiency of plating and differentiation that is likely due to genetic variation.

3.1. Transfection of HEK293T Cells for Generation of Virus and Virion Quantitation

1. Thaw a low passage aliquot of HEK293T cells in complete medium. Pass the cells at least two to three times before plating for transfection.
2. For a T-75 flask, plate 2×10^6 cells in the late afternoon in a total volume of 8 mL of complete medium. Allow cells to grow overnight until they are about 70% confluent.
3. Follow the manufacturer's instructions for use of the DMRIEC lipid reagent. Use 15 μg of proviral DNA with 15 μg of DMRIE-C lipid reagent in a total volume of 3 mL OPTI-MEM. Incubate at room temperature for 40 min (*see* **Note 1**).
4. Wash cells once with 4 mL of OPTI-MEM and replace with 3 mL of OPTI-MEM.
5. Add DNA/lipid to cells which will bring the total volume to 6 mL. Incubate at 37 °C, in a 5% CO_2 incubator for 6 h.

6. Add 5 mL of complete medium with 20% FBS (*see* **Note 2**).
7. After 3 days harvest viral supernatants and centrifuge at 500 × *g* to pellet any cells present. Filter the supernatant through a 0.45-μm filter attached to a 30-cc syringe.
8. Aliquot into freezing vials and store at −70 °C (*see* **Note 3**).

3.2. Isolation of Monocytes and Differentiation into Macrophages

1. Aliquot 10 mL of blood into a 50-mL conical tube. Dilute with 30 mL of HBSS and mix by pipeting up and down until homogeneous. Underlay 10 mL of Ficoll. To do this, place the pipet at the bottom of the tube and slowly but continuously expel the Ficoll. Centrifuge with the brake off for 30 min, 25 °C at 450 × *g* (*see* **Note 4**).
2. Remove 15 mL of the top layer containing the platelets with a 10-mL pipet and discard. Carefully collect the slightly yellowish PBMC fraction located at the interface into a 50-mL conical tube using a 10-mL pipet. Depending on the yield of cells, the interface will usually be a two or more millimeter thick fluffy layer.
3. To remove any remaining platelets, wash the PBMCs with 30-mL of HBSS twice. Spin at 25 °C, 450 × *g* for 5 min to collect the cells. Typically, PBMC collected from two interfaces can be combined into one tube.
4. Resuspend PBMC isolated from one leukopak in a total final volume of 10 mL of HBSS. Place 7 mL of 46% Percoll into two 15 mL conical tubes. Carefully and slowly layer 5 mL of the washed PBMCs on top of the Percoll. Holding the Percoll tube at a 45° angle is useful for minimizing mixing of the two layers. Spin with the brake off at 25 °C, 450 × *g* for 30 min.
5. Collect the tight whitish interface into a 50-mL conical tube containing 30 mL of HBSS. Spin at 25 °C, 450 × *g* for 5 min to collect the monocyte-enriched fraction.
6. The cells can be counted and frozen at 10^7/vial in macrophage medium containing 10% DMSO (*see* **Note 5**).
7. If plating freshly isolated monocytes seed 6 × 10^6 per T-25 cm^2 flask in a volume of 8 mL of macrophage medium containing 10% human AB serum. **Table 1** shows volumes for other culture dish formats.
8. Allow for differentiation to occur for 3 days and then change the medium to macrophage medium only, without human serum.

3.3. Infection of Macrophages

1. The day before infection (day 6–9 postdifferentiation), feed the MDM monolayers with fresh macrophage medium (*see* **Note 6**).
2. Rapidly thaw the required number of vials of virus in a 37 °C water bath. Virus should be diluted into infection medium so that 2 mL is added per T-25 cm^2 flask (*see* **Note 7**).

Table 1
Plating guidelines for Percoll purified monocytes in different culture dishes

Culture dish	Fresh monocytes	Frozen monocytes	Volume of medium	Infection volume
T-25 cm^2	6×10^6	1×10^7	8 mL	2 mL
Flaskette	3×10^6	5×10^6	3 mL	1 mL
24-well plate	2×10^5	5×10^5	300 µL	300 µL
96-well plate	3×10^4	7×10^4	100 µL	100 µL

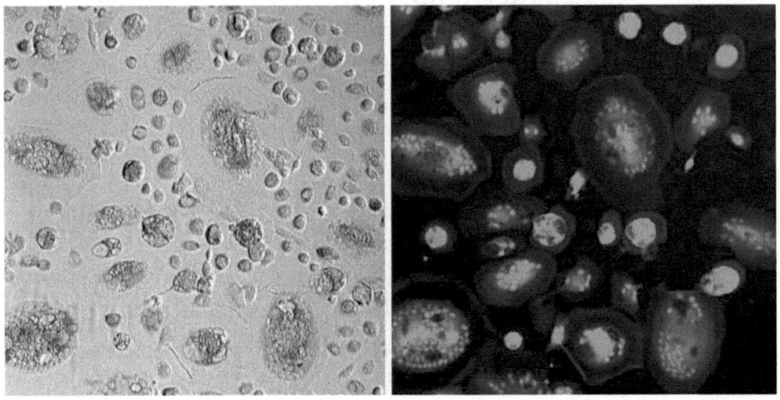

Fig. 1. Human MDM infected with the HIV-1 GFP-tagged reporter virus pSF162R3 Nef+. Phase contrast and fluorescent images of long-term macrophage monolayers in T-25 flasks at 42 days postinfection are shown.

3. Remove the macrophage medium and replace with the infection medium containing HIV-1. Allow viral entry and infection to occur 5 h to overnight.
4. Remove the inoculum and rinse with 5 mL of PBS. Feed MDM every 3–4 days with 8 mL of macrophage medium.
5. GFP-expressing MDM can be seen by fluorescent microscopy at 48 h postinfection (**Fig. 1**).

3.4. Flow Cytometry Analyses of Infected Macrophages

1. Remove medium from MDM monolayer and rinse with 4 mL of PBS. Add 2 mL of Accutase and incubate at 37 °C for 30 min. Forcefully tap the T-25 flask to dislodge the cells. To detach any remaining MDM, use a cell scraper to gently remove the remaining cells. Collect the cells in PBS-EDTA (*see* **Note 8**).
2. Resuspend the MDM in PBS-EDTA-NaN_3 and aliquot about 5×10^4 to 1×10^5 cells into the required number of 5 mL round bottom tubes. Pellet cells at 500 × *g* for 5 min. Resuspend

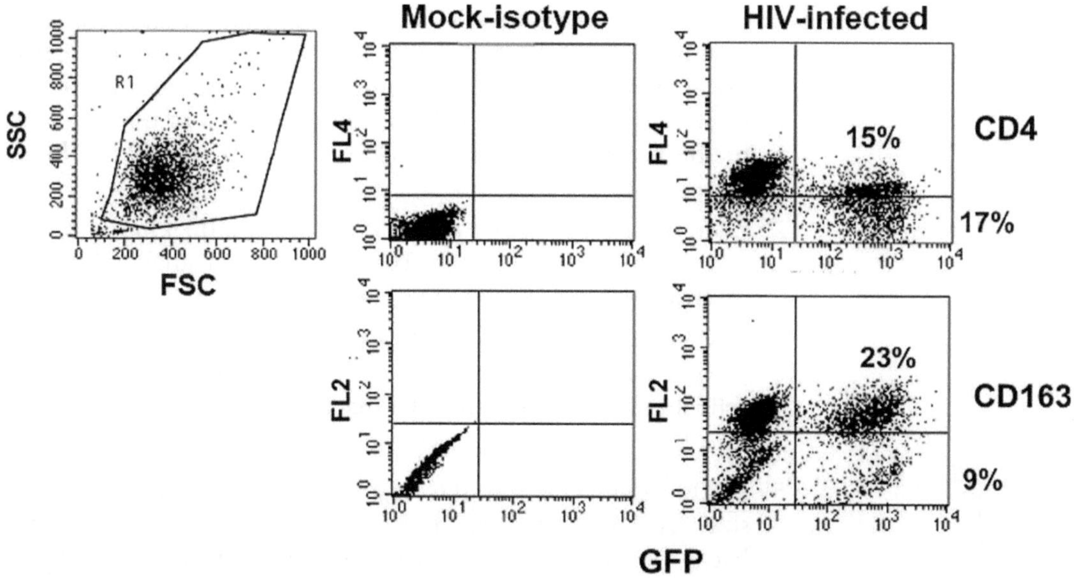

Fig. 2. Flow cytometric analyses of MDM infected with the GFP-tagged pSF162R3 Nef+ reporter virus at 8 days postinfection. MDM were stained with antibodies against CD4-allophycocyanin (APC) (eBiosciences) and CD163-phycoerythrin (PE) (BD Biosciences, Palo Alto, CA). The numbers shown represent the percentage of GFP + MDM in each quadrant as a fraction of the total cell number. A total of 11,070 events were collected. The HIV-1 Nef protein has been shown to be responsible for the majority of the activity responsible for downregulating the CD4 receptor on a variety of cell types. Evidence of this function in macrophages is revealed here. As seen in the *top rightmost* panel, over 50% of the GFP positive MDM have no CD4 on their cell surface. In contrast, CD163 is expressed on the majority of GFP + MDM.

the cells in PBS-EDTA-NaN_3 containing 2% human serum for 10 min at room temperature to block Fc receptors.

3. Add the desired fluorescently conjugated antibodies to the cells, mix by flicking the tube, and incubate in the dark for 20 min. Wash twice in PBS-EDTA-NaN_3 and resuspend in 200–300 μL of PBS-1% formaldehyde and store in the dark at 4 °C.
4. Analyze the MDM by flow cytometry. The number of events collected will depend on the extent of infection with the HIV-GFP reporter virus and whether Nef wildtype or Nef-defective mutant reporter viruses are used *(7)*. With the culture conditions described in this protocol, infection levels up to 30% can be obtained with some donor macrophages. Typically 10,000–100,000 events may need to be collected as seen in **Fig. 2**.

3.5. Cell Sorting of HIV-EGFP-Infected Macrophages

1. Remove medium from MDM monolayer and rinse with 4 mL of PBS. Be sure to include a mock-infected control flask that is required to set up the cell collection gates on the flow

cytometer. Add 2 mL of Accutase and incubate at 37 °C for 30 min. Forcefully tap the T-25 flask to dislodge the cells. To detach any remaining MDM, use a cell scraper to gently remove the remaining cells. Collect the cells in macrophage medium containing 5 mM EDTA and keep on ice (*see* **Notes 9** and **10**).

2. Concentrate the cells by centrifugation at 500 × *g* for 5 min and resuspend at about 1 × 10^6 cells/mL in a small volume of macrophage medium containing 50% Accumax.
3. The MDM suspension should also be passed through a 35-mm nylon mesh attached to a 5-mL conical tube to prevent large cell clumps from clogging the machine.
4. Add 5–10 μL of propidium iodide to stain for dead cells.
5. A 100-μm nozzle is used to accommodate the MDM of varied sizes. A neutral density filter should be used to decrease the background autofluorescence that is typical of differentiated macrophages. To obtain the highest purity, the sort single 1 mode is used.
6. Sorted cells can be collected in macrophage medium.

3.6. Laser Ablation or Capture of Infected Macrophages

1. Seed MDM obtained from cell sorting into the center of the PALM dish in 400–500 μL of macrophage medium. Allow the cells to adhere for 3 h and then bring up the volume to 2 mL (*see* **Notes 11** and **12**).
2. Before laser capture or ablation, remove the majority of the medium from the MDM culture dish. Keep a tube with medium nearby to rewet the surface since the plate will dry out during the work period.
3. If you are unable to focus with the 20× objective, tape the dish to the inner circle of the stage and perform three to four laser cuts on an unwanted portion of the dish. This will soften and stretch the membrane and then it should be easier to focus. PALM recommends scoring the dish before plating of your cells. However, for macrophage culture in general and particularly for long-term maintenance, sterility is very important that I would suggest not scoring before plating.
4. The optimal energy and focus settings for cutting and catapulting must be determined empirically. Start by testing a region of the sample that is not critical to the experiment to find the optimal settings (*see* **Note 13**; **Fig. 3**).
5. The 20× objective is best for laser capture or ablation as the entire circumference of the MDM can be seen.
6. Monitor the efficiency of capture by examining the microfuge lid for cells (*see* **Note 14**).

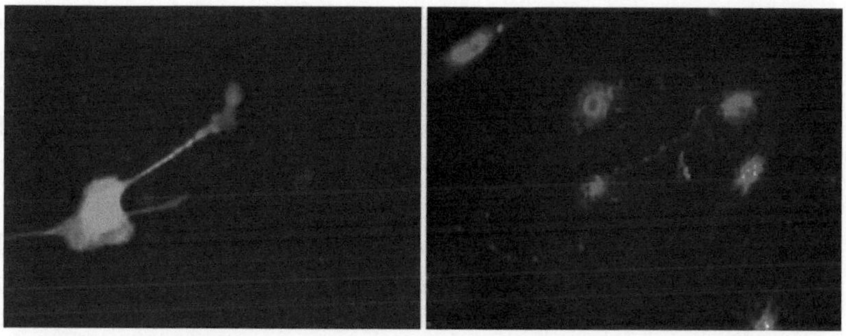

Fig. 3. Use of laser ablation to generate pure cultures of GFP-negative MDM. After sorting for high purity of MDM infected with the HIV-1 GFP-tagged reporter virus pSF162R3 Nef+ into GFP+ and GFP− pools, the GFP− pool nevertheless contains a few contaminating GFP+ cells. To remove these cells, the sorted GFP− pool was plated onto PALM dishes as described in **Subheading 11.3**. The marking tool was used to outline the circumference of the GFP+ cell to be eliminated and then it was catapulted from the surface of the membrane. Thorough washing and aspiration of the remaining cell monolayer ensured removal of the catapulted cells and membrane.

Notes

1. The DNA used for transfection should be of high purity as obtained using the Qiagen Endotoxin free kit or other similar DNA extraction kits.
2. If desired, the next day after transfection, remove the medium that contains residual DNA and lipid reagent and replace with 11–12 mL of fresh DMEM complete medium. In order to avoid contamination with plasmid DNA in downstream molecular applications, viral supernatants are often treated with DNase (50–100 U/mL).
3. The amount of viral particles can be quantified using the Alliance HIV-1 p24 ELISA Kit (Perkin Elmer). It is necessary to dilute and test the viral stocks out to at least 1:10,000 in order to obtain results that fall within the range of the standard curve. The kit provides 96-well plates that are precoated with a mouse monoclonal antibody against HIV-1 p24. The standards and controls are prepared as indicated in the manufacturer's protocol and included on the plate to provide the standard curve from which the concentrations of the unknown viral supernatants will be determined. After incubation with the primary biotinylated polyclonal anti-p24 and secondary streptavidin-HRP antibodies, a colorimetric HRP substrate is added and the plate read at 490 nm on a plate reader.
4. An important factor for obtaining high-level infection is to test individual lots of FBS and human sera for optimum monocyte differentiation and growth. This has to be done

empirically by noting the efficiency of adherence (strong), the rate of differentiation (rapid), and the rate of acidification of the medium (slow). Once a good lot of each type of serum is found, it is recommended to buy several bottles from the favorable lot. Also be sure that your incubator is free of contamination and that the humidifier works properly.

5. Cryopreservation agents such as CryoStor™ CS10 that provide improved cell recovery after freeze–thaw over DMSO have become widely available and their use to store purified monocytes should be investigated. Although the viability based on trypan blue staining of monocytes after freeze–thaw in DMSO/medium is greater than 95–98%, a large proportion of the cells do not maintain the ability to adhere and differentiate into macrophages. In this regard, trypan blue staining is not very sensitive at detecting preapoptotic cells.
6. For infection, typically 0.2–1 μg of viral p24 per T-25 flask is used.
7. A small population of macrophages is permissive to HIV-1 infection as soon as 1 day after plating. However, 7–10 days postdifferentiation is the most efficient window for infection of MDM generated using the Percoll gradient method. The volume of infection medium must be adjusted according to the size of the culture dish used (*see* **Table 1**).
8. Do not use pipetman and pipet tips to resuspend MDM as this will cause the cells to be sheared. Use 1–5 mL plastic pipets to resuspend MDM in all of the following procedures.
9. For sorting of HIV-infected cells, a biohazard level 3 cell (BSL3) sorting facility is required.
10. After incubation in EDTA or Accutase, the MDM should roundup and become spherical. The viability of MDM is not significantly affected by prolonged incubation in Accutase. However, more than 30 min incubation in 5 mM EDTA is not recommended.
11. Monocytes can be differentiated into macrophages directly on the PALM dishes. However, the cost of the dishes and the difficulty of observing cell morphology of unstained cells on the membrane surface will render this approach less attractive.
12. MDM are placed onto the center of the PALM dish to allow ready laser capture or ablation. Cells that are near to the edge of the plate are difficult to access. In addition, by keeping the cells confined to a limited area, one can be sure that all of the cells of interest have been captured or ablated.
13. To avoid the body of the microfuge tube from hitting the PALM dish when it is loaded over the sample to be captured, it is best to separate it from the adhesive lid.

14. Nucleic acids for use in PCR-based methods can be isolated using a variety of kits that have been optimized for cells obtained by laser capture.

Acknowledgments

The author thanks Peter Lopez and Hao Zhang for their expertise in flow sorting, Carlos A. Pardo for use of the PALM Microlaser, and Suzanne Gartner for sharing her love and knowledge of monocyte–macrophage culturing and for making these studies possible. This work was supported by grants from The Campbell Foundation (A. Brown), National Institutes of Health R01 AI057007 (S. Gartner), and K08 DA16160 (C. Pardo).

References

1. Chun, T. W., R. T. J. Davey, D. Engel, H. C. Lane, and A. S. Fauci. (1999). Re-emergence of HIV after stopping therapy. *Nature* 40:874–875.
2. Davey, R. T. J., N. Bhat, C. Yoder, T. W. Chun, J. A. Metcalf, R. Dewar, V. Natarajan, R. A. Lempicki, J. W. Adelsberger, K. D. Miller, J. A. Kovacs, M. A. Polis, R. E. Walker, J. Falloon, H. Masur, D. Gee, M. Baseler, D. S. Dimitrov, A. S. Fauci, and H. C. Lane. (1999). HIV-1 and T cell dynamics after interruption of highly active antiretroviral therapy (HAART) in patients with a history of sustained viral suppression. *Proc. Natl. Acad. Sci. USA* 96:15109–15114.
3. Harrigan, P. R., M. Whaley, and J. S. G. Montaner. (1999). Rate of HIV-1 RNA rebound upon stopping antiretroviral therapy. *AIDS* 13:F59–F62.
4. Brown, A., H. Zhang, P. Lopez, C. P. Pardo, and S. Gartner. (2006). In vitro modeling of the HIV-macrophage reservoir. *J. Leukoc. Biol.* 80:1127–1135.
5. Honda, Y., L. Rogers, K. Nakata, B. Y. Zhao, R. Pine, Y. Nakai, K. Kurosu, W. N. Rom, and M. Weiden. (1998). Type I interferon induces inhibitory 16-kD CCAAT/enhancer binding protein (C/EBP)beta, repressing the HIV-1 long terminal repeat in macrophages: pulmonary tuberculosis alters C/EBP expression, enhancing HIV-1 replication. *J. Exp. Med.* 188:1255–1265.
6. Hoshino, Y., K. Nakata, S. Hoshino, Y. Honda, D. B. Tse, T. Shioda, W. N. Rom, and M. Weiden. (2002). Maximal HIV-1 replication in alveolar macrophages during tuberculosis requires both lympho-cytes contact and cytokines. *J. Exp. Med.* 195:495–505.
7. Brown, A., S. Gartner, T. Kawano, N. Benoit, and C. Cheng-Mayer. (2005). HLA-A2 down-regulation on primary human macrophages infected with an M-tropic EGFP-tagged HIV-1 reporter virus. *J. Leukoc. Biol.* 78:675–685.

Chapter 12

GFP-Lentiviral Vectors Targeting for NeuroAIDS

Yuanan Lu

Summary

Human immunodeficiency virus type 1 (HIV-1) is the causative agent of acquired immunodeficiency syndrome (AIDS). HIV-1 can infect human brain macrophages and microglial cells, causing HIV-associated dementia, or neuroAIDS, an increasingly common disorder of the central nervous system (CNS) that affects 20% of HIV-1-infected individuals. Current treatments for neuroAIDS are hampered by the poor efficiency of many antiretroviral drugs to cross the blood-brain barrier (BBB). Circulating blood monocytes and their derived macrophages are known to migrate across the BBB and enter the CNS under normal physiologic conditions and certain circumstances; some of these cells can subsequently mature into long-lived tissue-resident brain macrophages and microglia. Thus, the natural homing/migratory properties of blood monocyte-derived macrophages (MDM) can be potentially utilized as an effective genetic tool for delivering anti-HIV-1 genes to the CNS in a noninvasive and nonsurgical manner. To test and establish this macrophage-based gene therapy for the CNS, we have constructed and generated high-titered defective lentiviral vectors (DLV) expressing enhanced green fluorescent protein (GFP) as a reporter and optimized protocols for the isolation and long-term cultivation of primary MDM from humans and mice. We have demonstrated that primary cultures of human and mouse MDM can be efficiently modified in vitro using GFP-DLV vectors without apparently adverse effects on cellular biological properties. We have also shown that primary mouse MDM can enter the brain. The efficiency of CNS uptake of these cells can be enhanced through the use of bradykinin or a hypertonic mannitol solution for transient disruption of the BBB. These experimental methods and findings lay the initial groundwork for future in vivo studies on the ability of GFP-DLV-modified blood MDM to introduce anti-HIV-1 and neuroprotective genes into the CNS.

Key words: Defective lentiviral vector (DLV), Monocyte-derived macrophages (MDM), Enhanced green fluorescent protein (GFP), Gene transfection, Cell transduction, MDM transmigration, Blood–brain barrier (BBB).

1. Introduction

In the USA, highly active antiretroviral therapy (HAART) has substantially reduced the overall morbidity and mortality associated with human immunodeficiency virus type 1 (HIV-1) infection *(1)*.

Barry W. Hicks (ed.), *Methods in Molecular Biology, Viral Applications of Green Fluorescent Protein, vol. 515*

DOI: 10.1007/978-1-59745-559-6_12

However, HIV-1-associated neurocognitive impairments remain high. In fact, the prevalence of HIV-1 dementia has increased during this post-HAART era *(2)*. It has further been suggested that HAART may be altering the presentation and/or pathogenesis of HIV-1 dementia, resulting in a more slowly progressive, chronic form of the disease *(3–5)*. These observations highlight the need for new, adjunct therapeutic approaches aimed specifically at ameliorating or preventing HIV-1 dementia. One potential approach to treating HIV-1 infection within the central nervous system (CNS) is the genetic modification of specific cells in the brain, making them resistant to HIV-1 infection and able to secrete anti-HIV-1 molecules that would reduce the spread of HIV-1 to other susceptible brain cells.

Monocytes and macrophages are readily available cell types, with natural abilities to penetrate otherwise inaccessible end organs such as the brain *(6–12)*. As a result, it has been suggested that it may be possible to use genetically modified monocyte-derived macrophages (MDM) to carry exogenous neuroprotective and anti-HIV-1 genes across the blood-brain barrier (BBB) for purposes of neuroAIDS gene therapy. There is now growing interest in testing the natural homing/migratory properties of hematopoietic stem cells, including MDM, for use as novel cellular vehicles for CNS gene therapy. The evaluation of this novel approach to CNS gene therapy will require considerable preclinical analysis, but a critical first step lies in the development of efficient methods for isolation and cultivation of primary monocytes/macrophages, as well as ex vivo gene transfer into these cells.

Although both viral vectors and nonviral delivery systems can be used for genetic modification of monocytes and macrophages, HIV-1-based lentiviral vectors have been widely tested as the most effective method for stable transfection of DNA into these hematopoietic cells. HIV-1-based vectors are attractive gene delivery tools for the ex vivo transduction of primary MDM because of their ability to efficiently transduce target cells at both dividing and nondividing phases, as well as their capacity to establish long-lasting transgene expression due to chromosomal integration of the proviral DNA. Present reports describe the development of efficient protocols for the isolation of human and mouse monocyte/macrophages, in vitro primary cultivation, and ex vivo transduction of these cells using GFP-lentiviral vectors. Current studies also show that the efficiency of gene transfer can be enhanced through (i) short-term in vitro culture of cells prior to exposure to viral vectors and (ii) increased multiplicity of infection. These transduced cells showed stable, long-term green fluorescent protein (GFP) expression and exhibited no apparent difference from untransduced MDM in terms of cell morphology and growth kinetics *(13, 14)*. Finally, experimental methods were also established to assess the transmigration of modified mouse macrophages across the BBB into the brain. The efficiency of macrophage transmigration to the CNS can be effectively enhanced through temporary disruption of the BBB *(15)*.

2. Materials

2.1. Isolation and In Vitro Cultivation of Primary MDM

1. BD vacutainers with K_2EDTA (BD Biosciences, Franklin Lakes, NJ).
2. Dulbecco's phosphate buffered saline (DPBS) (Cellgro, Mediatech, Herndon, VA).
3. Sequester-Sol anticoagulant solution (13.8% K_2EDTA) (Cambridge Diagnostic Products Inc., Fort Lauderdale, FL).
4. Heat-inactivated fetal bovine serum (FBS) (Hyclone, Logan, UT); heat-inactivated human serum (Gemini Bio-Products, West Sacramento, CA); heat-inactivated equine serum (Hyclone).
5. Lympholyte-Mammal (Cedarlane Laboratories Ltd., Hornby, Ont.) and Ficoll-Paque Plus gradient solution (GE Healthcare Bio-Sciences Corp., Piscataway, NJ). Store at 4 °C.
6. MDM growth medium: RPMI-1640 medium with l-glutamine (Mediatech) containing 10% heat-inactivated human serum and 20% heat-inactivated define FBS for human MDM, or 20% heat-inactivated equine serum and 10% FBS for mouse MDM, also with 100 U/mL penicillin, 100 g/mL streptomycin sulfate (Sigma) (*see* **Note 1**).
7. Wash buffer: 0.2 μM SFCA sterile-filtered (Nalgene Nunc, Rochester, NY) and 0.5% bovine serum albumin (EMD Chemicals, Gibbstown, NJ) in DPBS.
8. Recommended media (RM): 0.2 μM SFCA sterile-filtered phosphate buffered saline (Mediatech) containing 2% FBS and 2 mM EDTA (Promega, Madison, WI). Store at 4 °C.
9. SpinSep Negative Selection Mouse Monocyte antibody cocktail (StemCell Technologies, Vancouver, BC, Canada) for Murine cells, SpinSep Dense Particles, SpinSep Density Medium. Store at 4 °C.
10. 15- and 50-mL polypropylene centrifuge tubes (VWR, West Chester, PA).
11. T-12.5 cm^2 tissue culture flasks (BD Biosciences), 12- and 96-well tissue culture plate (Corning Incorporated, Corning, NY), and 125-mL plastic flasks with plain bottom (Nalgene Nunc).

12. Human CD14 monoclonal antibody conjugated with R-phycoerytherin, mouse CD11b, and PE-F4/80 antibodies (Caltag Laboratories, Burlingame, CA).
13. 3.0 mL syringe and 22G needles (BD Biosciences).
14. Analog Orbital Shaker (OS-500, VWR) and CO_2 incubators.
15. Highly purified recombinant human macrophage colony stimulating factor (MCSF).

2.2. Production of Defective Lentiviral Vector Expressing GFP (GFP-DLV) and Vector Concentration

1. The backbone of transfer plasmid (pHR-CMV-hB7-GFP) was kindly provided by Dr. V. Planelles at the University of Utah. This construct is modified to generate pDLV-GFP to coexpress a TNF-α receptor fusion protein (TNFR-Fc) and a reporter gene GFP linked by an IRES element (**Fig. 1**).
2. Packaging plasmid, pCMVΔR8.2Δvpr (kindly provided by Dr. V. Planelles), is a helper-free, second generation lentiviral packaging plasmid that expresses HIV-1 genes except *vpr* and *env*.
3. Modified envelope plasmid, pCMV-VSV-G (kindly provided by Dr. V. Planelles), is a plasmid that expresses vesicular stomatitis

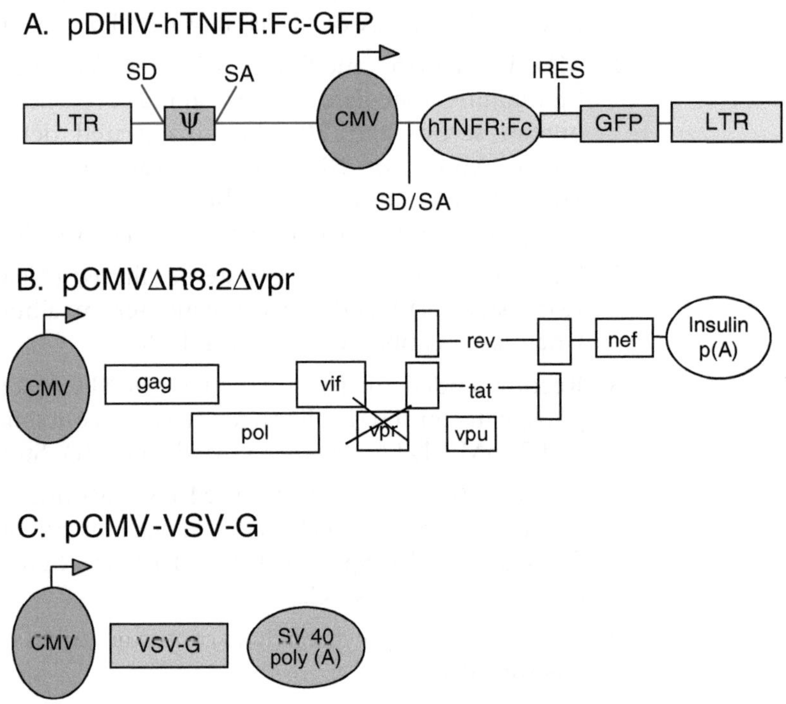

Fig. 1. Schematic illustration of the design and components of the defective lentiviral vector (DLV-GFP): **A** HIV-based transfer vector for TNFR-Fc constructs, **B** the lentiviral packaging construct, and **C** the envelope construct encoding the vesicular stomatitis virus (VSV) glycoprotein (G).

virus glycoprotein (VSV-G). All plasmid DNA used for transfection is cloned in *E. coli* strain DH5α and is extracted using HiSpeed Plasmid Midi Kit (Qiagen, Valencia, CA).

4. Dulbecco's modified Eagle's medium (DMEM) (Sigma) for 392T cells containing 10% FBS define, 100 U/mL penicillin, 100 μg/mL streptomycin sulfate, and 4.0 mM l-glutamine. Store at 4 °C.
5. Trypsin–EDTA solution (Sigma): 10 mL of 10× trypsin stock diluted with 90 mL of EDTA solution. EDTA solution: 8.0 g NaCl, 0.2 g KH_2PO_4, 0.2 g KCl, 1.15 g Na_2HPO_4, 0.2 g EDTA di-sodium salt, 0.01 g phenol red in 1 L. EDTA solution is autoclaved and stored at 4 °C.
6. Human embryonic kidney 293T (GenHunter Co., Nashville, TN) and T-cell line CEM (ATCC, Manassas, VA).
7. T-75 and T-25 cm² tissue culture flasks and 0.45 filters (Corning).
8. Modified 2× HEPES buffered saline (2× HBS) for transfection contains 50 mM HEPES (Sigma), 1.5 mM Na_2HPO_4, 280 mM NaCl, 20 mM KCl, and 0.6% dextran-2000 (Amersham Biosciences, Uppsala, Sweden), pH 7.12. Sterilize using a 0.2 μm filter. Keep at 4 °C (stable for 2 months) or store single use aliquots (100 μL/vial) at –80 °C.
9. 2.0 M $CaCl_2$ (Sigma) is prepared in ddH_2O and sterilized through 0.2 μm filters, and kept at 4 °C.
10. 50% (w/v) sucrose solution is prepared by dissolving 50 g of sucrose (Sigma) in 100 mL of DPBS buffer. Autoclave and store at 4 °C.
11. Hexadimethrine bromide (polybrene) (Sigma). 0.8 mg/mL solution (100×) is prepared in DMEM and sterilized by filtration. Aliquots of 1.0 mL/tube are stored at –30 °C.
12. Optimal L-90 Ultracentrifuge or comparable, centrifuge rotor SW28, and 25 × 89 mm ultra-clear centrifuge tubes (Beckman Coulter).
13. Sterile disposable Pasteur pipets (12.7 cm) (VWR).

2.3. GFP-DLV-Mediated Transduction of Primary Cultures of MDM

1. RPMI-1640 medium (Sigma).
2. 100× polybrene solution (0.8 mg/mL).
3. Concentrated GFP-DLV vector virus stocks (50–100 μL/vial).
4. Orbital shaker (Cole Parmer) and CO_2 incubator (VWR).
5. Primary cultures of mouse and human MDM at day 3 or more in vitro.

2.4. Stable Long-Term GFP Expression in Transductant MDM In Vitro

1. Nikon phase contrast inverted fluorescent microscope (Nikon 2000) with digital camera.
2. SuperScript II RT kit (GIBCO BRL) containing SuperScript II RT, first strand cDNA synthesis buffer (FSB 5×), and 0.1 mM DTT. Store at −30 °C.
3. PCR amplification reagents (Perkin-Elmer Corporation, Norwalk, CT) containing Teq-DNA, 10× PCR buffer, $MgCl_2$ (25 mM), and dNTPs (2.5 mM), nucleic acid free H_2O. Store at −30 °C.
4. TRIzol reagents (GIBCO BRL) for cellular RNA extraction and DNase (Ambion, Austin, TX).
5. 2% (w/v) agarose gel prepared in TBS and stored at 4 °C, gel loading buffer, 100-bp DNA ladders, and agarose gel electrophoresis apparatus (BioRad Laboratories).
6. Spectrophotometer and Coulter AD340 micro plate reader (Beckman), the Visionary gel documentation system (Bio-Rad).
7. Western blot assay for TNFR-Fc expression including goat anti-human IgG-Fc and rabbit anti-goat IgG conjugated with horseradish peroxidase (KPL, Gaithersburg, MD), prestained molecular weight markers (Invitrogen), metal-enhanced DAB substrate kit (Pierce, Rockford, IL). Store at −30 °C.
8. 2× SDS sample buffer; 30% acrylamide/bis solution and *N,N,N,N'*-tetramethylethylenediamine (TEMED) (Fisher Scientific); 10% (w/v) ammonium persulfate in water, freshly prepared or single-use aliquots immediately stored at −30 °C.
9. 10× running buffer (Fisher Scientific). Store at room temperature (rt) and dilute to 1× with ddH_2O before use. Transfer buffer is freshly prepared containing 25 mM Tris, 190 mM glycine, and 20% (v/v) methanol.
10. Tris-buffered saline with Tween-20 (EMD Bioscience, San Diego, CA) (TBS-T) is prepared as 10× stock and stored at rt. Dilute 100 mL with 900 mL ddH_2O prior to use.
11. Nitrocellulose membrane (0.2 μM) (Invitrogen).
12. ELISA for quantitative measurement of TNFR-FC production from transduced MDM including ELISA-enhanced 96-well plates (BD Biosciences), 10× coating solution (KPL). Store at 4 °C.
13. Coating antibody goat anti-human IgG Fc (KPL), recombinant human TNFRII protein (R&D Systems, Minneapolis, MN), detection antibody-biotin-labeled goat

anti-human IgG Fc (Rockland, Gilbertsville, PA), and peroxidase-labeled streptavidin (Rockland), substrate 1-Step Ultra TMB (Pierce).

14. Blocking buffer: 1% BSA (EMD) in PBS; stop buffer 2 M sulfuric acid. Store at 4 °C.

2.5. Modeling Transmigration of MDM Across the BBB to Enter the Brains

1. PKH26-GL (Sigma).
2. CD146 monoclonal antibody (Chemicon International Inc). Dilute to 1–10 μg/mL with PBS buffer. Store at 4 °C.
3. Cryostat microtome (Leica CM1900 or comparable).
4. Poly-L-lysine (0.1% w/v, Sigma) stored at rt. Dilute 1:10 with ddH2O prior to coating slides. Coated microscopic slides are kept in clean boxes and stored at rt.
5. Isoflurane (Abbott, UK). 2–4% isofluorene is in a mixture of 70% nitrous oxide and 30% oxygen and placed in a stereotactic frame (Stoelting Co., Wood dale, IL).
6. Hamilton syringe and other surgical syringes: 1/3 cc, 1 cc, and 60 cc and needles: 16G and 26G (BD Biosciences).
7. Avertin (1%, w/v) 2,2,2-tribromoethanol and 1% (v/v) *tert*-amyl alcohol.
8. Bradykinin (Sigma) is dissolved in DPBS at 1.0 g/L and mannitol (Sigma) dissolved in DPBS at 0.24 g/L, filtered through 0.2 μM filters and stored at 4 °C.
9. Perfusion saline solution containing 0.9% NaCl (w/v) stored at rt.
10. Rabbit anti-laminin (Sigma) stored at –80 °C. It is freshly diluted 1:1,000 in 1% nonfat milk in PBS prior to use and kept in 4 °C.
11. Rabbit anti-rat IgG and goat anti-rabbit IgG conjugated with FITC (Sigma), and rat anti-mouse MOMA-2 (New England Bolas). Store at –80 °C.
12. Isopentane (Sigma) and acetone (Fisher Scientific). Store at –80 °C.
13. 0.5% Triton-100 in PBS (Sigma) and 0.1% Sudan Blank B in 70% ethanol. Store at rt.
14. Phosphate buffered saline (PBS) solution (pH 7.4) and 0.1% v/v) nonfat milk in PBS.
15. Microscopic slides and mounting buffer (Fisher Diagnostics).
16. Sterile dissection items (Stoelting) – surgical scissors and forceps, carb-edged scissors, cotton swabs, surgical cotton string.
17. DAPI (Research Organics).

3. Methods

3.1. Isolation and Cultivation of Primary MDM from Human and Mice

3.1.1. Primary Culture of Human MDM

1. Collect 100–150 mL of blood from healthy donors by intravenous puncture into 12–15 BD Vacutainers ACD (Beckman Dickson). Pool samples into a 500-mL sterile serum bottle and dilute with DPBS to a ratio of 1:2 (blood:DPBS).
2. Slowly layer 25 mL of diluted blood over 20 mL of Ficoll-Paque Plus in a 50-mL polypropylene centrifuge tube stabilized at rt. The interface should be clearly visible.
3. Spin the blood cells in a centrifuge, with the break off, at 1,000 × *g* for 30 min. The upper layer is removed by slow aspiration. Peripheral blood mononuclear cells (PBMC) form a white layer at the interface. Carefully collect the interface layer and pool into a fresh 50-mL centrifuge tube.
4. Dilute the pooled PBMCs with DPBS using a 1:2 ratio and then centrifuge at rt for 15 min at 460 × *g*. Resuspend the cell pellet in 15–20 mL of DPBS, and centrifuge again at 220 × *g* for 8 min (*see* **Note 2**).
5. Resuspend the recovered PBMC pellet in 10 mL of RPMI-1640 medium containing no FBS. Count the cells using a hemocytometer, and seed the purified PBMC into T-12.5 cm^2 flasks at densities of about 1.5×10^7 cells/flask with 4 mL of RPMI growth medium, or 12-well plates at 3.5×10^6 cells/well with 1 mL/well of the medium. Incubate the cells at 37 °C with 5% CO_2 to allow blood monocyte attachment (*see* **Note 3**).
6. After 48 h of incubation, remove the supernatant containing nonadherent cells (*see* **Note 4**). Wash the attached cells representing MDM three times using DPBS. Add fresh RPMI-1640 growth medium. On days 4 and 7, wash the cultures again to completely remove any remaining nonadherent cells.
7. Observe cell growth and monolayer formation daily using a phase-contrast inverted microscope (**Fig. 2**).

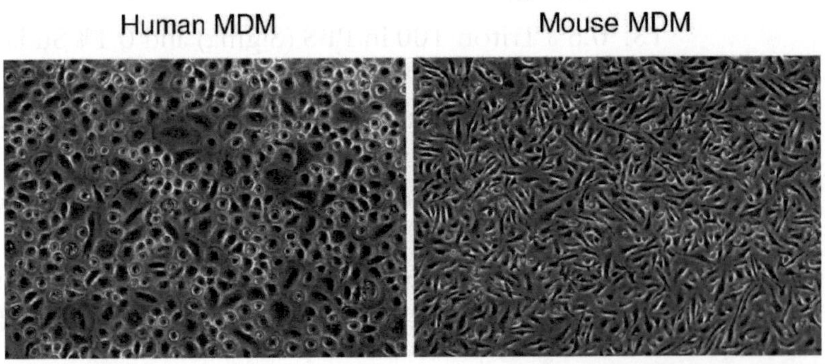

Fig. 2. Photomicrographs of primary cultures of monocyte-derived macrophages (MDM) from humans and laboratory mice in vitro at incubation days 12 and 6, respectively (original magnification: 100×).

8. To verify the purity of the attached cells, wash the monolayer cultures twice with DPBS and stain them for 1 h with anti-human CD14 monoclonal antibodies conjugated with R-phycoerytherin and diluted 1:100 in DPBS. After washing twice with DPBS, examine the cells under an inverted fluorescent microscope. Mouse isotype IgG2a conjugated with R-PE is used as a negative control (*see* **Note 5**).
9. All MDM cells are easily visualized as red cells (**Fig. 3**). Purity of human MDM preparations are quantified by counting the number of red cells relative to the total number of cells contained within the same microscopic field. An average cell purity is obtained by counting the cells from three randomly selected fields.

3.1.2. Primary Cultivation of Mouse MDM

1. Euthanize mice in a CO_2 chamber. Mice should show no reflection when its feet are pinched.
2. Immediately collect peripheral blood by cardiac puncture, using a 3.0-mL syringe with 22G needle, and pool samples from different mice in a BD K_2EDTA vacutainer.

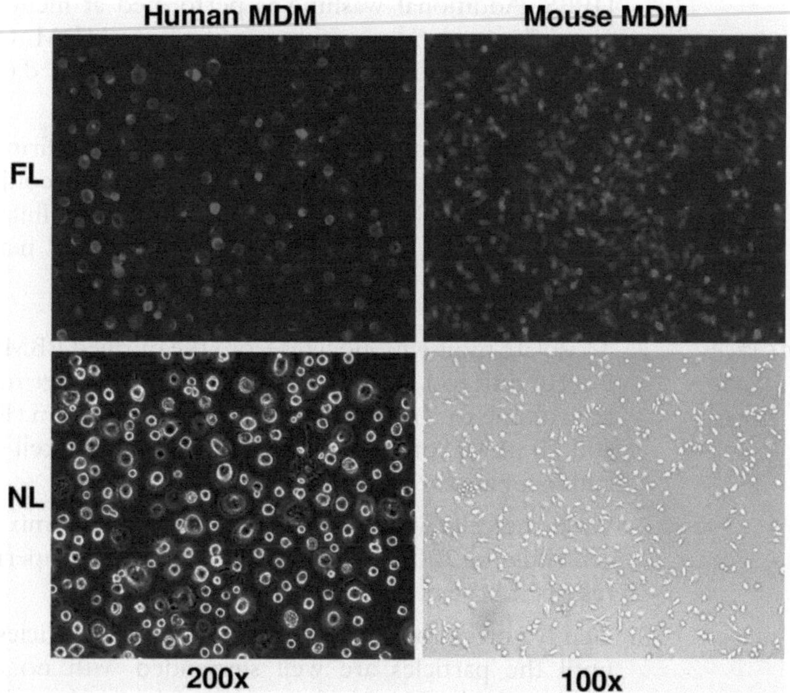

Fig. 3. Photomicrographs of primary cultures of human and mouse MDM. Human MDM at day 10 stained with mouse anti-human CD14 monoclonal antibody conjugated with R-phycoerytherin, and mouse MDM stained with rabbit anti-mouse CD11b monoclonal antibody conjugated with R-Phycoerytherin, showing the same fields under either normal light (NL) phase-contrast microscopy or fluorescent light (FL) microscopy. (Modified reproduction from **refs.** *13* and *14* with permission from John Wiley & Sons and Elsevier Science, respectively).

3. Dilute the pooled blood with DPBS at a ratio of 1:2 (blood:DPBS), then carefully load 6–8 mL of the diluted blood on top of 6 mL of lympholyte-M gradient medium in a 15-mL centrifuge tube. Centrifuge at rt for 30 min at 1,000 × *g*. Carefully collect the mouse PBMC, located at the interface as a white layer, and dilute 1:1 with DPBS. In a 15-mL tube, load 4–6 mL of cell suspension on top of 6 mL of a second gradient solution – Ficoll-Paque gradient. Centrifuge the tubes again at 820 × *g* for 20 min.
4. Collect the mouse PBMC from the interface and wash twice; first with DPBS at 460 × *g* for 15 min and again at 220 × *g* for 8 min. Resuspend the recovered PBMC in RPMI-1640 growth medium, count the cells, and dilute them to 1×10^7–2×10^7 cells/mL.
5. Seed the purified mouse PBMC into 96-well plates at ~1×10^6 cells/well in a total of 0.2 mL of RPMI-1640 growth medium supplemented with 20% equine serum, 10% FBS, 2 mM L-glutamine, 100 U/mL penicillin, 100 μg/mL streptomycin, and 2.5 μg/mL amphotericin B.
6. Following 48 h of cell attachment at 37 °C with 5% CO_2, remove nonadherent cells (*see* **Note 4**). Vigorously wash the attached cells representing mouse MDM three times with DPBS. Additional washing is performed at incubation days 4 and 7 to ensure pure cultures of mouse MDM. Cell attachment and monolayer formation are observed daily using a phase-contrast inverted microscope (**Fig. 2**).
7. The purity of primary mouse MDM is confirmed by the immunostaining of the cultured cells with PE-conjugated rat monoclonal antibodies directed against select cellular proteins, CD11b and F4/80 of mouse monocytes and macrophages (**Fig. 3**).

3.1.3. Suspension Cultures of Mouse MDM

1. To isolate mouse monocytes from the purified PBMC, transfer 2×10^7 cells to a 15-mL polypropylene centrifuge tube. Adjust the volume to 1 mL with recommended medium (RM). Then add 10 μL of SpinSep antibody cocktail to the cell suspension and incubate for 30 min on ice.
2. Wash the cells with 8 mL of RM, inverted to mix well, then centrifuge at 200 × *g* for 10 min. Aspirate the supernatant and resuspend the pelleted cells in 2 mL of RM.
3. Gently vortex the tube of Spin-Sep Dense Particles for ~30 s until the particles are well suspended with no observable clumps. Add 100 μL of dense particles to the cells and mix well. Incubate this mixture on ice for 20 min, with mixing every 5 min, or so.

4. Add 4.0 mL of Spin-Sep Density Medium per tube into 15 mL polypropylene centrifuge tubes. Dilute the cell-particle suspension to 6 mL with RM, and carefully layer it on top of the density medium. Centrifuge at 1,200 × *g* for 10 min with the centrifuge brake off.
5. Collect the enriched MDM cells from the interface layer and transfer to a 15-mL polypropylene centrifuge tube. Dilute the cells with 8 mL of RM, invert it to mix well, then centrifuge at 200 × *g* for 10 min. Discard the supernatant and resuspend the enriched cells in RPMI-1640 growth medium.
6. Count the cells with a hemocytometer and determine the purity of recovered cells by staining cells with PE-CD14 or PE-F4/80 antibody.
7. Pool the purified mouse monocytes/macrophages prepared from multiple tubes and seed 3×10^7 cells in a 125-mL plastic flask with 30 mL of growth medium (1×10^6 cells/mL).
8. Place the flask on an analog orbital shaker, which is set up in a 37 °C, 5% CO_2 incubator. Adjust the shaking speed to ensure no cell attachment. Change half of the growth medium every other day (*see* **Note 6**).
9. The viability of MDM culture in suspension can be calculated by staining the cells with trypan blue solution at selected cultivation days 1, 2, 4, and 7.

3.2. Production and Concentration of GFP-DLV Vector

1. High-titer GFP-DLV vectors are produced in 293T cells by co-transfection of transfer plasmid pDLV-GFP plasmid with pCMV-ΔR8.2ΔVpr and pCMV-VSV-G. Perform transfections with a dextran- and polybrene-enhanced calcium phosphate precipitation (*see* **Note 7**) *(16)*.
2. Harvest 293T cells during exponential growth phase and seed into T-75 cm^2 flasks 1 day prior to transfection (*see* **Note 8**).
3. Replace conditioned medium with fresh DMEM containing 10% FBS, 2–3 h prior to transfection (*see* **Note 9**).
4. To transfect 293T cells in the T-75 cm^2 flask, mix a total of 480 μL of plasmid DNA solution containing 12.5 μg DLV transfer plasmid, 12.5 μg pCMV-ΔR8.2ΔVpr, and 2.5 μg pCMV-VSV-G in nucleic acid free water, with 66.5 μL of 2.0 M $CaCl_2$. Following incubation on ice for 5 min, add 546.5 μL 2× HBS to the DNA mixture drop by drop, while simultaneously mixing the solution by vigorous agitation. Keep this transfection cocktail on ice for 20 min and briefly vortex every 5–10 min.
5. Add the cocktail to transfect 293T cells by dropwise addition to the medium, with gentle mixing. Incubate at 37 °C with 5% CO_2 for 8 h.

6. Carefully aspirate the transfection solution from the flask and add 10 mL fresh DMEM containing 5% FBS. Return the cultures to the 37 °C incubator with 5% CO_2 (*see* **Note 10**).
7. Collect cell culture medium from transfected cultures every 24 h for 5–8 days, centrifuge to remove cell debris, pool to a new tube, and store at −80 °C. The pooled culture supernatant can be directly used for vector concentration or for transduction (*see* **Note 11**).
8. For vector concentration, thaw frozen supernatant in a 25 °C water bath and spin for 30 min at 3,000 × *g*. Filter recovered supernatant through a 0.45-μm filter and centrifuge at 113,000 × *g* for 3 h using a SW28 rotor (*see* **Note 12**).
9. Following ultracentrifugation, completely aspirate the supernatant and resuspend the pellet in 100 μL/tube of DMEM containing no FBS. Pool the concentrated vector in a 1.5-mL tube, vortex briefly, and keep at 4 °C overnight.
10. Spin the vector solution in a bench top centrifuge at full speed for 30 s. Prepare aliquots of 50–100 μL/vial and store at −80 °C for transduction.

3.3. GFP-DLV Transduction of Primary Cultures of MDM

1. To transduce primary cultures of mouse and human MDM, remove the cell medium from primary cultures at selected cultivation times (3–7 days). Gently wash MDM cultures three times with DPBS to remove all cellular debris and residual serum.
2. Initiate transduction by adding 0.1 mL of vector stock to a T-25 cm^2 flask in a total of 0.2–0.3 mL inoculum, or 50 μL/well of a 96-well plate, in the presence of 8 μg/mL polybrene. Incubate for 1 h with gentle shaking every 15–20 min, then rinse the transduced cells twice with DPBS to remove residual vectors. Feed the transduced cells with RPMI-1640 growth medium.
3. Using an inverted phase-contrast fluorescent microscope, search for transduced cells by visual inspection for expression of the reporter gene (GFP). Determine the efficiency of vector-mediated transduction of primary MDM at day 3 posttransduction by calculating the percentage of GFP-positive cells within the transduced cell population (**Fig. 4**).
4. Transduction of mouse MDM suspension cultures is done similarly. First, wash harvested MDM twice with DPBS and then resuspend the cells with concentrated vector stock at an MOI of more than 20, in the presence of 8 μg/mL polybrene. Following 1 h incubation at 37 °C with gentle tapping every 15–20 min, transfer transduced cells directly into a 120-mL culture flask and grow in RPMI-1640 growth medium. Transduction efficiency is determined by seeding

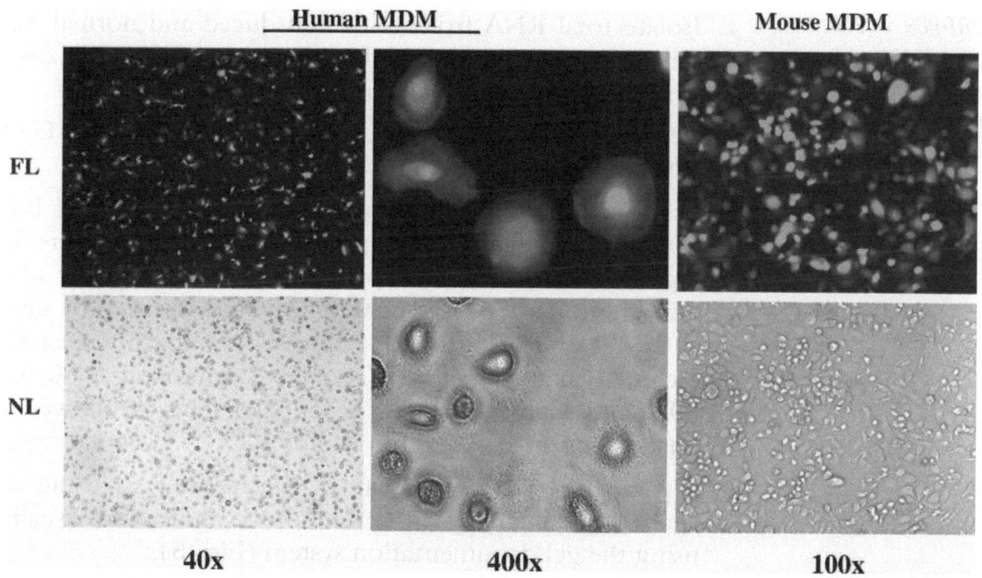

Fig. 4. Photomicrographs of DLV-GFP-mediated transduction of primary cultures of human and mouse monocyte-derived macrophages (MDM), showing stable expression of GFP in transduced cells at posttransduction days 7 (human) and 5 (mouse). *FL* = fluorescent light and *NL* = normal light. (Modified reproduction from **refs.** *13* and *14* with permission from John Wiley & Sons and Elsevier Science, respectively).

a small portion of the transduced cells in a few wells of a 96-well plate and then determining the percentage of GFP-positive cells within the transduced cell population at incubation day 3.

5. Under the described experimental conditions, transduction efficiencies of primary human and mouse MDM ranges between 20–30% and 40–80%, respectively.
6. Stable expression of the target transgene (sTNFR-Fc) is quantified by Western blot and ELISA using samples collected at selected incubation days.

3.4. Stable Long-Term GFP Expression in Transductant MDM In Vitro

1. Primary human and mouse MDM cultures transduced with GFP vectors are stable in vitro for over 7 and 2 weeks, respectively. There is no apparent difference between transduced and normal control MDM cells under phase-contrast microscopy.
2. Identify successful transduction of primary MDM cultures by one or more of the following methods: (a) fluorescent microscopic examination of transduced cells for GFP expression since affected cells are greenish, making them readily differentiated from untransduced cells (**Fig. 4**); (b) RT-PCR detection of the expression of the transgenes using PCR-specific primers for the respective genes; (c) Western blot assay; or (d) ELISA to detect the transgene products in culture media of transduced MDM.

3.4.1. RT-PCR Detection of GFP Expression in Transduced MDM

1. Isolate total RNA from both transduced and normal MDM cells at varying in vitro cultivation times using TRIzol reagents and quantify by spectrophotometery.
2. Remove residual DNA by treating RNA with DNase (1 U/1 μg RNA) at 37 °C for 30 min.
3. Perform RT-PCR amplification using SuperScript II RT kit (or comparable kits) with a pair of GFP-specific primers flanking, for example, a 373-bp fragment. Following the reverse transcription at 58 °C for 30 min, PCR is initiated by heating at 95 °C for 15 min, followed by 35 amplification cycles that include denaturing at 95 °C for 30 s, annealing at 60 °C for 30 s, and extension at 72 °C for 30 s. This is followed by a final extension at 72 °C for 7 min.
4. Separate the PCR products by electrophoresis using a 2% agarose gel, staining with ethidium bromide, and visualizing using the gel documentation system (**Fig. 5**).

3.4.2. ELISA Detection of Transgene Expression in Transduced MDM (See Note 13)

1. Coat an ELISA-enhanced 96-well plate with 2 μg/mL (1:500 dilution) of goat anti-human IgG Fc, 100 μL/well, in coating solution (10× stock diluted to 1× in ddH_2O). These plates are placed at 4 °C overnight.
2. Tap the plates out to remove the antibody solution, then block with 1% BSA in PBS, using 300 μL/well, for 30 min at rt. Note also that this BSA blocking buffer is used to dilute antigen, detection antibody, and streptavidin HRP.

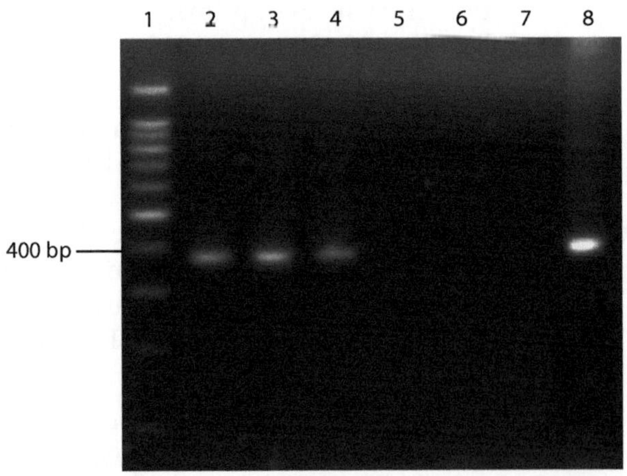

Fig. 5. RT-PCR detection of stable expression of GFP in DLV-GFP transduced human monocyte-derived macrophages (MDM) at selected posttransduction days. RT-PCR amplification was conducted with the use of GFP-specific primers flanking a 373-bp fragment and PCR products were subjected to 2% agarose gel electrophoresis. Lanes: *1* = 100 bp DNA ladder; *2* = transduced cells at day 5; *3* = transduced cells at day 39; *4* = transduced cells infected with HIV-1 at day 25; *5* = normal MDM at day 25; *6* = negative control (water) for rt; *7* = negative control (water) for PCR; *8* = positive control, pDLV-GFP. (Reproduced from **ref.** *13* with permission from John Wiley & Sons).

3. Tap the plates out again to remove BSA blocking buffer, and then wash five times with the TBS-T Buffer (0.05% Tween-20 in TBS), 300 µL/well. Incubate the fourth wash in the plate for 5–10 min. Tap out the excess buffer after the fifth wash.
4. Apply the antigen to the plate in duplicate or triplicate. For standards, use recombinant human TNF RII/TNFRSF1B/Fc Chimera ranging from 125 to 8,000 pg/mL. In addition, prepare "blank" (BSA blocking buffer) and "no-antigen control" (BSA blocking buffer). Dilute test samples ranging from 1:100 to 1:5,000. At least two different dilutions per sample should be used. Incubate these for 1 h at rt on an orbital shaker.
5. Wash the plates five times as described in **step 3**.
6. Apply 200 ng/mL of detection antibody – Rockland's biotin-labeled goat anti-human IgG Fc, 1:10,000 dilution. Use 100 µL/well and incubate for 1.5 h at rt on an orbital shaker. Do not apply antibody to "blank" wells; use BSA blocking buffer instead.
7. Wash the plates five times as in **step 3**.
8. Apply 100 µL/well of Rockland's peroxidase-labeled streptavidin at 100 ng/mL and incubate it for 1 h at rt on an orbital shaker. Protect the plates from light, and do not apply streptavidin HRP to "blank" wells; use BSA blocking buffer instead.
9. Wash the plates five times as in **step 3**.
10. Add 100 µL/well of 1-Step Ultra TMB and incubate at rt for 5–15 min. Keep the plates covered and monitor them for color development every couple of minutes.
11. When sufficient color development is observed, gently shake plates to mix the wells, then stop the reaction by adding 100 µL/well of 2 M sulfuric acid.
12. Read the absorbance of the plates at 450 nm using Beckman Coulter AD340 or comparable spectrophotometer.

3.4.3. Detection of Transgene Expression in Transduced MDM by Western Blot

1. To determine the stable expression of TNFR-Fc in transduced MDM by Western blot, perform SDS-PAGE using either medium from primary MDM transduced with GFP-DLV or brain tissues homogenates prepared from experimental mice infused with transduced MDM and clarified by centrifugation.
2. Transfer separated proteins onto a NC-membrane, and block the NC membrane with 5% (w/v) nonfat milk in PBS for 1 h at rt on a shaker.

3. Discard the blocking buffer, and wash the membrane three times with TBS-T, 5 min/wash with 30 mL TBS-T. Expose to 1:1,000 diluted goat-anti-human IgG-Fc antibody for 1 h at rt on a shaker, then remove the primary antibody.
4. Wash the membrane again three times and soak in freshly prepared secondary antibody (rabbit-anti-goat IgG HPR; 1:5,000 dilution with TBS-T) for 1 h at rt on a shaker.
5. Discard the secondary antibody and wash the membrane three times with TBS-T. Expose to 3,3-diaminobenzidine tetrahydrochloride (DAB) substrate for color development.
6. Stop the color development by washing the membrane three times with tap water at the time when the specific band appeared clearly (**Fig. 6**).

3.5. Modeling Transmigration of GFP-DLV Transduced Monocytes Across the BBB to Enter Mouse Brains

1. Anesthetized experimental mice (40–50 g) with 2–4% isoflurorene in a mixture of 70% nitrous oxide and 30% oxygen using a stereotactic frame. Test animals by toe pinch reflex periodically during the procedure to ensure a proper surgical plane.
2. Administer 200 μL of bradykinin (1.0 g/L) or 200 mannitol (2.4×10^{-4} g/L) intravenously into the tail vein using a 1.0-mL syringe with 25 ½ G needles.
3. Shave the incision site on the mouse's neck and then disinfect well with 70% ethanol before beginning the surgical procedure.
4. Make one incision just above the right clavicle using sterile scissors. Push connective and adipose tissues aside using curb-edged scissors. Locate the neck artery by its location between the sternohyoid, sternomastoid, and omohyoid muscles, and medial to, and below, the jugular vein.
5. Expose the right CCA, and infuse ~50 μL of primary MDM ($>1 \times 10^6$ cells) either transduced with GFP-DLV or stained with PKH26 into the CCA of recipient mouse using a Hamilton syringe. A sterile cotton swab is used to press the injection site to prevent any possible bleeding (*see* **Note 14**).

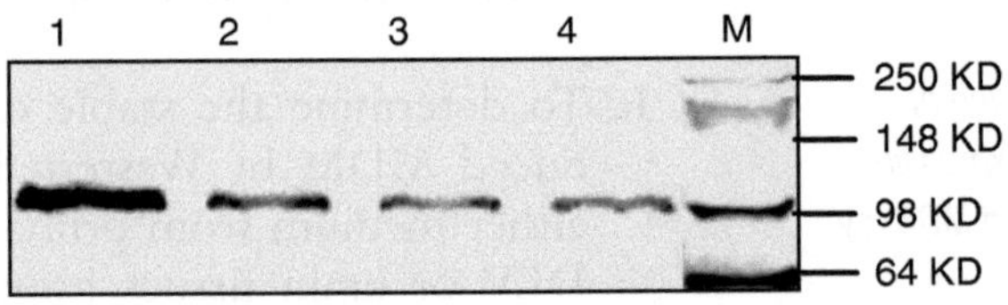

Fig. 6. Western blot detection of stable expression of transgene (TNFR-Fc) in primary cultures of human and mouse MDM transduced with DLV-GFP. Media from transduced cells mixed with 2× SDS sample buffer at 1:1 and then loaded on 12% separating SDS-PAGE. Separated proteins were transferred onto a NC membrane, incubated with goat anti-human IgG-Fc, and followed with rabbit anti-goat IgG-HRP conjugate. The membrane was exposed to 3,3-diaminobenzidine tetrahydrochloride (DAB) substrate for color development. Lanes: *1* = CHME-5, *2* = BV-2, *3* = human MDM, *4* = mouse MDM, and *M* = protein MW marker.

6. Pull together the connective and adipose tissues, and seal the incision using surgical cotton string. Keep experimental mice on a 37 °C warm trap until they awaken and then return them to their cages (*see* **Note 15**).
7. At predetermined intervals, anesthetize experimental mice by intraperitoneal injection with avertin [1% (w/v) 2,2,2-tribromoethanol and 1% (v/v) *tert*-amyl alcohol] at a dose of 250 mg/kg body weight.
8. Test animals for 5 min by toe pinch reflex until no reaction is observed. Cut the skin at the thoracic area just below the rib cage and remove the sternum to expose the mouse's heart. Insert a 26-G needle into the aorta through the left ventricle while the right auricle is snipped to allow blood and saline buffer to leave the body during perfusion.
9. Perfuse until ~20–50 mL/mouse of saline solution is delivered at rt via gravitational force. By this time, the fluid should become clear, and the animal will be dead.
10. Open the top of the skull using sharp surgical scissors and forceps. Remove the brain tissue and freeze it rapidly in cooled isopentane (20–30 s). The brain is then packed in tinfoil and kept at −30 °C until cryosection.

3.5.1. Detection of Transmigrated MDMs in the Brain

1. Prepare serial coronal sections (20μm thickness) from the caudal diencephalon (CD) and rostral mesencephalon (RM) using a cryostat microtome.
2. Transfer sections onto poly-l-lysine-coated microscopic slides and allow them to adhere for 3 min at rt, and then put them into the section boxes kept at −30 °C for future use.
3. Wash the slides gently five times for 10 min with PBS, and then aspirate dry from one corner.
4. To confirm the transmigration of infused cells across the BBB, stain sections with CD146 mouse monoclonal antibody targeting murine endothelial cell marker and follow with Alexa Fluo 647 goat anti-mouse IgG.
5. You can reduce tissue autofluorescence by incubating the sections with 0.1% Sudan Black B in 70% ethanol for 10 min, followed by washing three times for 3 min each with 70% ethanol.
6. Place a coverslip over the sections with mounting buffer. Infused MDM cells can be identified and photographed using a fluorescent microscope (excitation wavelength 546 nm) (**Fig. 7**).

3.5.2. Immunochemistry Staining for Brain Macrophages

1. Fix brain sections with cold acetone for 15 min and wash with PBS for 5 min.

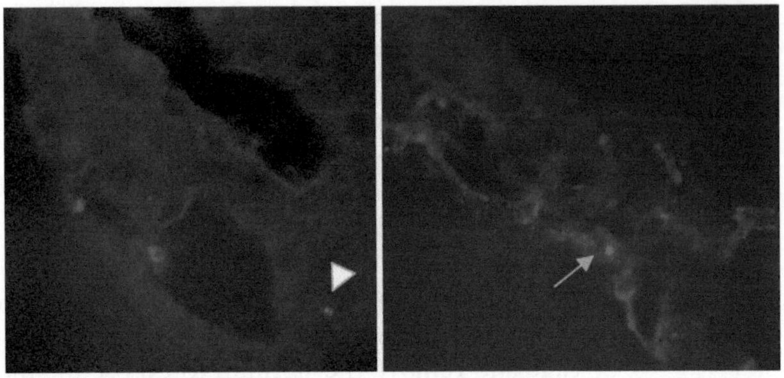

Fig. 7. Fluorescence micrographs of brain sections from experimental mice infused with PKH26-labeled MDM stained with mouse monoclonal antibody CD146 (endothelial cell marker), followed by Alexa Fluor 647 goat anti-mouse IgG (H + L) with (*left*), or without (*right*) perfusion. PKH26-positive cells were detected in the brain (*arrowhead*) and trapped in blood vessels (*arrow*).

2. Block sections with PBS containing 0.2% BSA for 30 min and wash with PBS containing 0.5% Triton for 20 min before incubation with primary antibody.
3. Apply diluted MOMA-2, a rat anti-mouse macrophage/monocyte monoclonal antibody, and incubate for 1 h at rt. Gently wash the brain sections three times with PBS buffer containing 0.3% Triton X-100.
4. Apply the secondary antibody (rabbit anti-rat IgG) for 1 h at rt, and follow with a detection antibody such as goat anti-rabbit IgG-FITC conjugate for 1 h at rt. Then wash the sections three times with PBS.
5. Add 300μL DAPI solution to each section for 10 min at rt to stain the DNA and identify the nuclei.
6. To reduce tissue autofluorescence, the sections can be incubated with 0.1% Sudan Black B in 70% ethanol for 10 min.
7. Wash the slides three times with 70% ethanol and coverslip with mounting buffer. Brain macrophages can be identified and photographed using a fluorescence microscope.

Notes

1. Human serum needs to be pretested. There are some batch-to-batch variations in terms of their support for primary cell culture growth.
2. It is important to check the purity and number of recovered PBMCs following each washing step. Centrifuge speed and

spin times for a particular centrifuge may be subjected to minor adjustments to ensure maximum recovery of purified PBMC, with fewer platelets at each wash.

3. Cell suspensions should be mixed very well prior to each counting, and counted at least —two to three times. We have found that the initial cell seeding density is very critical for producing a good monolayer of primary MDM cultures, following culture washings to remove nonadherent cells.
4. After seeding cells in culture flasks or plates, mix the cells by gentle shaking to ensure a more even distribution of cells. The cultures should be incubated at 37 °C for 48 h without touching. For the first wash, it is necessary to collect all cell suspensions together, and reseed them in fresh flasks and plates, to such a density that is three- to fivefold more than the initial seeding. We have found that good preparations of primary MDM cultures occur during the second seeding in some cases.
5. We have developed this immunostaining protocol for primary human MDM cultured in flasks and plates. Since these cells are still viable and can be cultured for other tests, all reagents and buffers should be sterile. The washing and staining processes should be performed very gently. Following immunostaining and checking, the cells are washed again with RPMI containing no FBS. Then growth medium is added and the cultures are returned to 37 °C incubator with 5% CO_2.
6. Viability and differentiation of primary mouse MDM suspension cultures can vary, and depend largely on the initial seeding density, shaking speed, medium changing, and use of 1,000 U/mL of highly purified recombinant human macrophage colony stimulating factor (MCSF).
7. It is important to add polybrene to the DNA solution after $CaCl_2$ solution is added to ensure no unwanted DNA precipitation occurs.
8. We routinely maintain 293T cells in DMEM with 10% FBS at 37 °C in a humidified 5% CO_2 incubator. Under these conditions, we pass cells every 3–4 days. Cells at logarithmic growth phase are split 24 h before transfection and seeded to $\sim 2.6 \pm 0.2 \times 10^6$ cells per T-75 cm^2 flask to allow the formation of ~70% cell monolayer the following day. The highest titer is obtained when the cell monolayer is transfected at around 70% confluency.
9. Some protocols skip this medium change prior to adding transfection cocktail. We found that this medium change is helpful for increasing transfection efficiency.

10. Since transfected 293T cells at this point can easily come off, handle these cell cultures very carefully to avoid cell detachment.
11. It is important to centrifuge collected medium immediately at 4,000 rpm for 20 min after each collection to remove cellular debris, since broken down debris can be toxic or interfere with subsequent cell transduction.
12. Deposit 0.3 mL/tube of 50% (w/v) sucrose in PBS at the bottom of the centrifuge tubes using a sterile Pasteur pipet before ultracentrifugation. This protocol can be adapted to other uses in vector and virus concentration.
13. Bring all buffers, plates, and TMB to rt before use. Use ELISA sealing film for all incubation steps.
14. To ensure cell infusion, blood vessels need to be stabilized by positioning a pair of artery forceps under the vessel during the injection using a microsyringe with 28G needle.
15. We have found that keeping the experimental mice on the 37 °C warm trap is helpful for animal recovery. No antibiotic treatment is needed for the infused animals if the inoculation procedure is conducted aseptically. However, Trimethoprim sulfamethoxazole (TMS) at a dose of 30 mg/kg body weight is helpful for recovery of any mice sick from the surgical procedure.

Acknowledgments

This study was supported by grants from the National Institutes of Health (S11NS043499 and G12RR003061) and by the Hawaii Community Foundation.

References

1. Palella, F.J., Delaney, K.M., Moorman, A.C., Loveless, M.Q., Fuhrer, J., Satten, J.A., Aschman, D.J., and Holmberg, S.D. (1998) Declining morbidity and mortality among patients with advanced human immunodeficiency virus infection. *N. Engl. J. Med.* 338:853–60.
2. Masliah, E., DeTeresa, R.M., Mallory, M.E., and Hansen, L.A. (2000) Changes in pathological findings at autopsy in AIDS cases for the last 15 years. *AIDS.* 14:69–74.
3. Portegies, P., Solod, L., Cinque, P., et al. (2004) Guidelines for the diagnosis and management of neurological complications of HIV infection. *Eur. J. Neurol.* 11:297–304.
4. Dore, G.J., McDonald, A., Li, Y., Kaldor, J.M., Brew, B.J., et al. (2003) Marked improvement in survival following AIDS dementia complex in the era of highly active antiretroviral therapy. *AIDS.* 17:1539–45.
5. Neuenburg, J.K., Brodt, H.R., Herndier, B.G., Bickel, M, Bacchetti, P., Price, R.W., Grant, R.M., and Schlote, W. (2002) HIV-related neuropathology, 1985 to 1999: rising prevalence of HIV encephalopathy in the era of highly active antiretroviral therapy. *J. Acquir. Immune Defic. Syndr.* 31:171–7.
6. Rutherford, M.S., Witsell, A., and Schook, L.B. (1993) Mechanisms generating functionally

heterogeneous macrophages: chaos revisited. *J. Leukoc. Biol.* 53(5):602–18.

7. Pereira, C.F., and Nottet, H.S.L.M. (2000) The blood-brain barrier in HIV-associated dementia. *NeuroAids.* 3:1–8.
8. Zhu, T.F. (2000) HIV-1 genotypes in peripheral blood monocytes. *J. Leukoc. Biol.* 68:338–44.
9. Priller, J., Flugel, A., Wehner, T., Boentert, M., Haas, C.A., Prinz, M., et-al. (2001) Targeting gene-modified hematopoietic cells to the central nervous system: use of green fluorescent protein uncovers microglial engraftment. *Nat. Med.* 7:1356–61.
10. Ross, J.A., and Auger, M.J. (2002) The biology of the macrophage. In: Burke, B. and Lewis, C.E. (eds), The Macrophage. Oxford University Press. Oxford, New York. Pp. 1–72.
11. Fischer-Smith, T., and Rappaport, J. (2005) Evolving paradigms in the pathogenesis of HIV-1-associated dementia. *Expert Rev. Mol. Med.* 27:1–26.
12. Eugenin, E.A., Osiecki, K., Lopez, L., Goldstein, H., Calderon, T.M., and Berman, J.W. (2006) CCL2/monocyte chemoattractant protein-1 mediates enhanced transmigration of human immunodeficiency virus (HIV)-infected leukocytes across the blood-brain barrier: a potential mechanism of HIV-CNS invasion and NeuroAIDS. *J. Neurosci.* 26:1098–106.
13. Zeng, L.B., Planelles, Y., Sui, Z., Gartner, S., Maggirwar, S.B., Dewhurst, S., Ye, L.B., Nerurkar, V.R., Yanagihara, R., and Lu, Y. (2006a) Human immunodeficiency virus type 1 (HIV-1) based defective lentiviral vectors efficiently transduce human monocyte-derived macrophages and suppress replication of wild-type HIV-1. *J. Gene Med.* 8:18–28.
14. Zeng, L.B., Yang, S.M., Wu, C.X., Ye, L.B., and Lu, Y. (2006b) Effective transduction of primary mouse blood- and bone marrow-derived monocytes/macrophages by HIV-1-based defective lentiviral vectors. *J. Virol. Methods.* 134:66–73.
15. Wu, J.M., Yang, S.M., Luo, H.Y., Nguyen, M., Zeng, L.B., and Lu, Y. (2006) Quantitative evaluation of monocyte transmigration into the brain following chemical opening of the blood-brain barrier in mice. *Brain Res.* 1098:79–85.
16. Wu, C.X., and Lu, Y., and (2007) Addition of high molecular weight dextran in calcium phosphate-mediated transfection significantly improves gene transfer efficiency. *Cell. Mol. Biol.* 53(4):67–74.

Chapter 13

Construction of Cell-Type-Specific Promoter Lentiviruses for Optically Guiding Electrophysiological Recordings and for Targeted Gene Delivery

Aaron M. Jasnow, Donald G. Rainnie, Kimberly A. Maguschak, Jasmeer P. Chhatwal, and Kerry J. Ressler

Summary

It is often advantageous to identify and alter gene expression of specific cell populations within the brain. Currently, it is not possible to a priori identify specific cell types within the brain of rats for electrophysiological recordings, nor is it possible to routinely alter gene expression in specific cell types within the CNS of a variety of species. Here, we describe a general method for the relatively rapid screening of specific promoter activity in cell culture, in acute brain slice preparations, and in vivo. As an example, we describe the examination of an ~3 kb promoter region of the neuroactive peptide cholecystokinin (CCK) compared to the ubiquitous cytomegalovirus (CMV) promoter. We find a high degree of cell-type specificity in vivo using lentiviral approaches in rats and mice.

Key words: Lentivirus, Electrophysiology, CCK, Cholecystokinin, Behavior, Neuroscience.

1. Introduction

The potential for using lentiviral vectors in experimental therapeutics is very promising, owing largely to their ability to integrate into both dividing and postmitotic cells, and to achieve stable, long-lasting transgene expression *(1)*. Additionally, in comparison to adeno-associated-based and sindbis-based vectors, lentiviral vectors allow for the packaging of larger inserts, opening the possibility for the use of larger promoter lengths to achieve gene expression in selected populations of cells *(2–4)*. Achieving restricted expression would be beneficial to

Barry W. Hicks (ed.), *Methods in Molecular Biology, Viral Applications of Green Fluorescent Protein, vol. 515*

DOI: 10.1007/978-1-59745-559-6_13

both basic science studies of different cell types interacting in complex biologic systems, and to the development of therapeutic gene-delivery systems that are active only within specific types of cells. Despite these potential benefits, the process of determining minimal promoter lengths to achieve biologically appropriate expression is made difficult by the inability to identify a priori enhancer and silencer regions within putative promoter regions, and by the possibility that regions of the 5′ untranslated region (5′ UTR) of a given gene may have an important bearing on its expression. In this context, the screening of putative promoters of varying lengths has largely depended on the selective expression of transfected promoter–reporter constructs in cultured cells, based on the notion that selectivity in vitro may indicate selectivity in vivo. However, this may not always be the case.

Here, we outline an alternative method of screening promoters for cell-type-specific expression in vitro and in vivo. As this method makes use of viral vectors to assess appropriate promoter expression, it offers several advantages over transgenic approaches, including lower cost, more rapid throughput and most importantly, the ability to use the resulting vectors in a variety of different species. We assess cell-type-specific expression using in vitro, acute brain slice electrophysiology and in situ hybridization methods.

2. Materials

2.1. Plasmid Design and Construction

1. Vector NTI Advance™ 10 (free for academic use from Invitrogen Corp., Carlsbad, CA), or equivalent sequence analysis program.
2. Species-specific genomic DNA or BAC DNA for amplifying the promoter region of interest.
3. Antisense and sense primers designed to amplify the promoter region of interest, with custom restriction sites incorporated into the original primers used for subcloning the promoter region in the lentivirus.
4. Any of the following TOPO® cloning vectors with electroporation or chemically competent cells (Invitrogen Corp.): pCR®2.1-TOPO®, pCR®II-TOPO®, pCR®-blunt II-TOPO®, pCR®4-TOPO®, or equivalent cloning vectors (*see* **Note 1**).
5. Phusion™ High Fidelity PCR Kit (Finnzymes cat.no. F-553S or F-553L).
6. dNTPs (dATP, dTTP, dGTP, dCTP, 100 mM each or 2.5 mM mix).
7. 0.2 mL PCR tubes.

8. Thermal cycler.
9. Restriction enzymes (e.g., *Bam*HI, *Eco*RI, *Cla*I) for restriction mapping and subcloning.
10. A Lentivirus packaging construct. We use pCMV-GFP-dNhe (a kind gift from Inder Verma, Salk Institute, La Jolla, CA). Other backbones are now available commercially.
11. Ampicillin (at 100 mg/mL) and kanamycin (at 100 mg/mL).
12. 100-mm round Petri dishes.
13. Miller LB Broth.
14. Miller LB Agar plates: dissolve 37 g of LB Agar (amount may differ, see instructions on bottle) in 1 L of H_2O, autoclave, and allow to cool to ~50 °C. When cooled to 50 °C, add appropriate antibiotic (ampicillin and/or kanamycin) and pour ~15 mL into each 100-mm culture plate, cover and allow to cool. Store at 4 °C protected from light.
15. Gas burner.
16. Bacterial cell spreader.
17. Miniprep kit (Qiaprep Spin miniprep kit, Qiagen cat no. 27106 or equivalent) and Gel Extraction Kit (Zymoclean Gel DNA Recovery Kit™ Zymo Research Corp cat. no. D4002, or equivalent).
18. Vacuum manifold (Qiagen, QIAvac 24 plus Vacuum Manifold, 19413). Not essential, but recommended.
19. Antarctic Phosphatase (NEB cat. no. M0289S).
20. Agarose gel electrophoresis equipment.
21. Electroporator.
22. ElectroTen-Blue® electroporation competent cells (Stratagene cat. no. 200159).
23. T4 DNA ligase and buffer (NEB cat. no. M0202T).
24. Incubator set at 37 °C.
25. Shaking incubator.
26. X-Gal reagent (Gold BioTechnology cat. no. X4281C): dissolve 40 mg of X-Gal in 1 mL of dimethylformamide (Fisher Scientific cat. no. D119-1).
27. 14-mL BD Falcon round bottom culture tubes (Fisher Scientific cat. no. 14-959-45).
28. Sterile glycerol.

2.2. Assessment of Promoter Activity In Vitro

1. Tissue culture capabilities.
2. 293T Human embryonic kidney (HEK) cells (Invitrogen cat. no. R700-07) or equivalent.

3. 1× PBS (Invitrogen cat. no. 14025-092).
4. Trypsin–EDTA.
5. 100× Penicillin/streptomycin.
6. 24-well tissue culture plates.
7. Complete DMEM: Dulbecco's modified Eagle's medium (DMEM) (BioWhittaker cat. no. 12-741F) with 10% fetal bovine serum (FBS) (Gibco/Invitrogen cat no. 26140-079) and 1× penicillin/streptomycin.
8. Transfection reagents: Lipofectamine or calcium phosphate (*see* ref. *6* for calcium phosphate reagent setup).
9. Fluorescent microscope.

2.3. Virus Production and Concentration and Titration

Tissue culture capabilities for working with lentiviral vectors (*see* **Note 2**).

1. Lentivirus packaging construct (*see* **Subheading 2.1**, **item 10**).
2. Lentiviral packaging vectors: pVSVG, pΔ8.9 (a kind gift from Inder Verma, Salk Institute or pMDL, pRev, and pVSVG (Invitrogen cat. no. K4975-00).
3. Endotoxin-free maxipreps (Endo-free Maxiprep kit, Qiagen cat no. 12632, or equivalent); resuspend DNA at 1 μg/μL.
4. Terrific broth (TFB): dissolve 25 g of LB Broth (amount may differ, see instructions on bottle) in 1 L of H_2O and add 5 g peptone, 5 g yeast extract, and 10 mL glycerol. Autoclave and allow to cool.
5. Avanti J-26 Series Centrifuge or equivalent (Beckman Coulter).
6. JLA-10.500 centrifuge rotor or equivalent (Beckman Coulter cat. no. 369681).
7. 500-mL centrifuge bottles or equivalent (Beckman Coulter cat. no. 361691).
8. See materials and reagents **#1–5** and **#7–9** from **Subheading 2.2**.
9. 150 mm tissue culture dishes.
10. 0.45 μm filter units, 500 mL capacity (Corning cat. no. 430773).
11. Conical bottom ultracentrifuge tubes (Beckman cat. no. 358126).
12. Round bottom ultracentrifuge tubes (Beckman cat. no. 326819).
13. Beckman SW28 rotor (or equivalent).
14. Beckman ultracentrifuge (or equivalent).

2.4. Virus Injections and In Vivo Assessment of Promoter Activity

1. Stereotaxic instrument (David Kopf Model 962 Dual Ultra Precise Small Animal Stereotaxic, or equivalent).
2. Microinfusion pump; Stereotaxically mounted (e.g., World Precision Instruments UMP 2) or stand alone (e.g., Harvard Apparatus PHD 2000) (*see* **Note 3**).
3. World Precision Instruments Nanofil 10 μL syringe (WPI cat. no. NANOFIL) with 35G beveled Nanofil needle (WPI cat. no. NF35BV-2), or equivalent Hamilton syringe (*see* **Note 3**).
4. Sterile filtered 1% bovine serum albumin (BSA). BSA has been found by some to decrease molecular interactions that might lead the virus particles to "stick" to the injection apparatus.
5. 100% ethanol.
6. Wound clips or sutures.
7. Clippers for trimming hair prior to surgery.
8. Betadine or equivalent 10% providone-iodine solution topical antiseptic.

3. Methods

3.1. General Considerations

One of the most important considerations in the design of cell-type-specific promoter lentiviral vectors is the size and region of the promoter used to drive cell-specific expression patterns. Therefore, it may be advantageous to design several different size promoters keeping in mind that lentiviruses can accept inserts of up to ~8 kb and still successfully package virus particles. Another important consideration is to incorporate the region of the promoter containing the ATG transcription start sequence downstream of the stretch of the promoter to be tested. This promoter region will be inserted into the lentivirus packaging construct such that the ATG sequence of the original promoter remains in frame with the ATG transcription start sequence of the green fluorescent protein (GFP) sequence in the packaging construct. Incorporating the original ATG translation start site ensures that splice junctions near the ATG and possible cell-type control over translation initiation are likely to be conserved within the new construct.

One method for designing different size promoter regions is to examine regions of conservation within the promoter using the UCSC Genome Bioinformatics website's Genome Browser and incorporate regions of high conservation among different species into the region of the promoter to be cloned into the lentiviral vector. There may be several regions of high conservation along different points of the promoter. For instance, there may be regions of conservation at both 2.5 and 7 kb upstream of the ATG sequence. One would then design two sets of primers to generate separate mini promoters of 2.5 and 7 kb incorporating the regions of conservation. Keep in mind, this may not always be possible, and may not always result in cell-type-specific expression patterns.

3.2. Plasmid Design and Construction

1. Evaluate your cloning strategy to assure that the ATG transcription start sequence within the promoter of interest will be in frame with the ATG sequence of the GFP coding sequence when cloned into the lentiviral packaging construct.
2. Design primers taking into account the above considerations and include custom restriction sites that will allow you to easily ligate your promoter into the lentiviral paging construct. In the case of the pCMV-GFP-dNhe lentiviral construct, the restriction sites will be 5′-*Cla*I site on the sense primer and a 5′-*Bam*HI site on the antisense primer (*see* **Note 4**).
3. Excise the CMV promoter from the pCMV-GFP-dNhe lentiviral construct using a *Bam*HI/*Cla*I digest.
4. Following the digest, treat the reaction mix with Antarctic phosphatase to remove 5′ phosphate groups from the DNA and prevent self-ligation.
5. Gel-purify the 6.8 kb virus backbone fragment minus the CMV promoter. When using a spin column kit for purification, elute the DNA in ddH_2O. Store at –20 °C until ready for use (*see* **step 15** below).
6. Meanwhile, set up a PCR reaction using the Phusion™ High Fidelity PCR Kit (or any equivalent kit) to amplify the promoter region of the gene of interest by following the instructions provided with the PCR kit (*see* **Note 1**).
7. Verify the size of your amplified fragments by running the reaction(s) on a 0.5% agarose gel and then gel purify the PCR fragment(s) of the correct size. When using a spin column kit for purification, elute the DNA in autoclaved ddH_2O.
8. Ligate the purified promoter PCR product into a TOPO or equivalent vector following the manufacturer's instructions (*see* **Note 5**).
9. Transform electrocompetent or chemically competent bacterial cells using the manufacturer's instructions.
10. Warm two plates per transformation in a 37 °C incubator while your cells are shaking. If you plan to use blue/white screening to identify positive transformations, spread 40 μL of X-Gal (40 mg/mL) on each plate and let warm for 15–30 min before plating.

11. Plate 50 and 150 μL of each culture onto LB plates containing the appropriate antibiotic. Place the plates upside down in a 37 °C incubator and incubate overnight.
12. If you are using blue/white screening and TOPO vectors from Invitrogen, pick ten white colonies using a sterile pipette tip and inoculate 5 mL of LB in a 14-mL round bottom falcon tube. If you are using another manufacturer's vector, you may need to pick blue colonies. If you are not using blue/white screening, pick ten colonies and inoculate 5 mL of LB. In both cases, shake at 260 rpm overnight in a 37 °C shaking incubator.
13. Miniprep 1 mL of overnight culture from each colony and analyze using appropriate restriction enzymes. Send positive clones off for sequencing to confirm restriction digest analysis. Be sure to make bacterial stocks (1 mL of bacterial culture + 10% sterile glycerol) for each colony, and store in –80 °C.
14. Excise the promoter region from the TOPO clone with a *Bam*HI/*Cla*I digest. Run the digest on a 0.5% agarose gel and gel-purify the promoter fragment. Elute the fragment in ddH_2O and store in –20°C until use.
15. To ligate the cell-type-specific promoter region into the prepared lentiviral packaging construct from **step 5**, set up a reaction in a 10-μL volume and use up to 7 μL of insert DNA and 1 μL of packaging construct DNA, plus 1 μL of T4 DNA ligase and 1 μL T4 DNA ligase buffer (including ATP). Bring up to 10 μL with ddH_2O if you do not use 7 μL of insert DNA. Incubate reaction at 16 °C for 4 h. Larger reaction volumes may also be used.
16. Transform the ligated product into ElectroTen® Blue cells according to the manufacturer's instructions.
17. Plate onto LB agar plates (50 and 150 μL) containing 100 mg/mL ampicillin and 100 mg/mL kanamycin. ElectroTen® Blue cells are kanamycin resistant and the pCMV-GFP-dNhe lentiviral packaging construct is ampicillin resistant. Therefore, using both antibiotics selects for the appropriate constructs. Incubate plates upside down overnight in a 37 °C incubator.
18. Pick 10–12 colonies from the plates and inoculate into 5 mL of LB containing both ampicillin and kanamycin. Shake overnight in a shacking incubator at 260 rpm.
19. Miniprep 1 mL of overnight cultures and use appropriate restriction enzymes to determine positive clones. Send positive clones off for sequencing to confirm. Remember to make a bacterial stock (1 mL of bacterial culture + 10% sterile glycerol) and store at –80 °C.

3.3. Assessment of Minimal Promoter Activity In Vitro

Before production of active viral particles, the viral constructs should be transiently transfected into HEK 293T cells to determine if the cell-type-specific promoters will provide for transgene expression and determine their level of expression compared with the CMV promoter.

1. Seed 1 × 10^5 293T HEK cells in a 24-well plate with 500 μL of Complete DMEM. Grow cells to 90–95% confluent on the day of transfection.
2. Transiently transfect cells with the cell-type-specific promoter packaging construct using your method of choice (e.g., lipofectamine or calcium phosphate).
3. Also transfect control cultures with the pCMV-GFP-dNhe construct.
4. Incubate cells in a 37 °C incubator for 18–48 h then examine cultures for GFP expression and compare to expression from cells utilizing the CMV promoter (**Fig. 1**).
5. Be aware that in vitro GFP expression using a cell-type-specific promoter may not necessarily confer appropriate or robust expression in vivo (*see* **Note 9**).

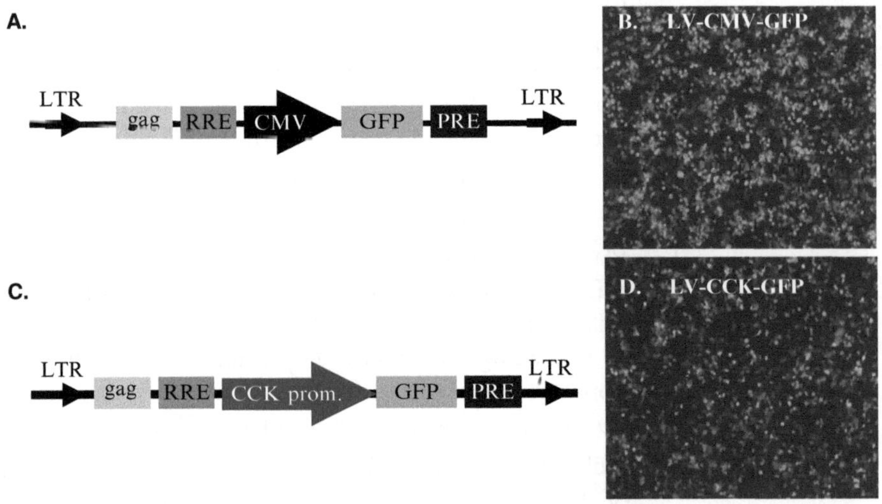

Fig. 1. GFP expression vectors. Here we describe the analysis of an ~3 kb promoter region of the neuroactive peptide cholecystokinin (CCK) compared to the ubiquitous cytomegalovirus (CMV) promoter. Before production of active viral particles, the viral constructs for CMV-GFP **(A)** and CCK-GFP **(C)** were transiently transfected into HEK293 cells to determine if the CCK promoter would provide for transgene expression, and determine its level of activity compared to the CMV promoter. Examination of cells 48 h posttransfection indicated that the CCK promoter was indeed able to generate detectable expression of GFP in HEK293 cells **(D)**. The level of expression attained with the CCK promoter appeared to be quite high, although still far less than the expression of the CMV promoter-containing vector **(B, D)**. This figure is adapted from Chhatwal et al. *(9)*.

3.4. Virus Production, Concentration, and Titration

1. Inoculate 5 mL of TFB with stock cells of the HEK-screened construct and grow at 37 °C overnight using appropriate antibiotic.
2. Using appropriate antibiotic, inoculate 500–1,000 mL of TFB with ~1 mL of the overnight culture from **step 1**.
3. Pour cultures into 500-mL or 1-L centrifuge bottles, balance precisely, and spin for 10 min in the Beckman or Sorvall centrifuge at the rate of 7,000 rpm at 4 °C (*see* **Note 6**).
4. Prepare plasmid DNA using Endotoxin-free Maxiprep kit (follow instructions with the kit) and resuspend DNA at 1 μg/μL (*see* **Note 7**).
5. For viral production, our plasmid lentiviral transfer vector (LV-CCK-GFP) encoding GFP under the regulation of the CCK promoter and the lentiviral envelope and packaging vectors (pΔ8.9, pVSVG) were transfected into the HEK 293T cells using the calcium phosphate method. Cells were observed and supernatant was harvested on days 2, 4, and 6 following transfection, filtered, and then concentrated. Viral pellets were resuspended in PBS with 1% BSA, aliquoted, and stored at –80 °C (*see* **Note 2**).

3.5. In Vivo Virus Injections

1. Anesthetize animal following approved animal care guidelines.
2. Shave the top of the head where the incision will be made.
3. Place the animal in the stereotaxic frame and scrub the shaved area with Betadine.
4. Preload the syringe with the 1% BSA for a few minutes and expel before loading the virus into the syringe. Be sure to load slightly more virus than needed for a typical injection (usually 2 μL) (*see* **Note 8**).
5. After determining the location of bregma and appropriate coordinates for the injection, drill a small hole through the skull to dura. Verify that the needle is not clogged by expelling a small volume of liquid, and slowly lower the syringe to the appropriate depth below dura (lower the syringe to 0.1 mm below the desired depth and then raise it back to up to the appropriate depth).
6. Start the infusion pump. The typical volume used in our experiments is 2 μL injected at a rate of 0.2 μL/min. After the injection is finished, leave the needle in place for an additional 5 min to allow for diffusion of the virus, then slowly raise the needle.
7. Flush the syringe with ddH_2O between infusions to eliminate any debris and reduce the possibility of clogging.

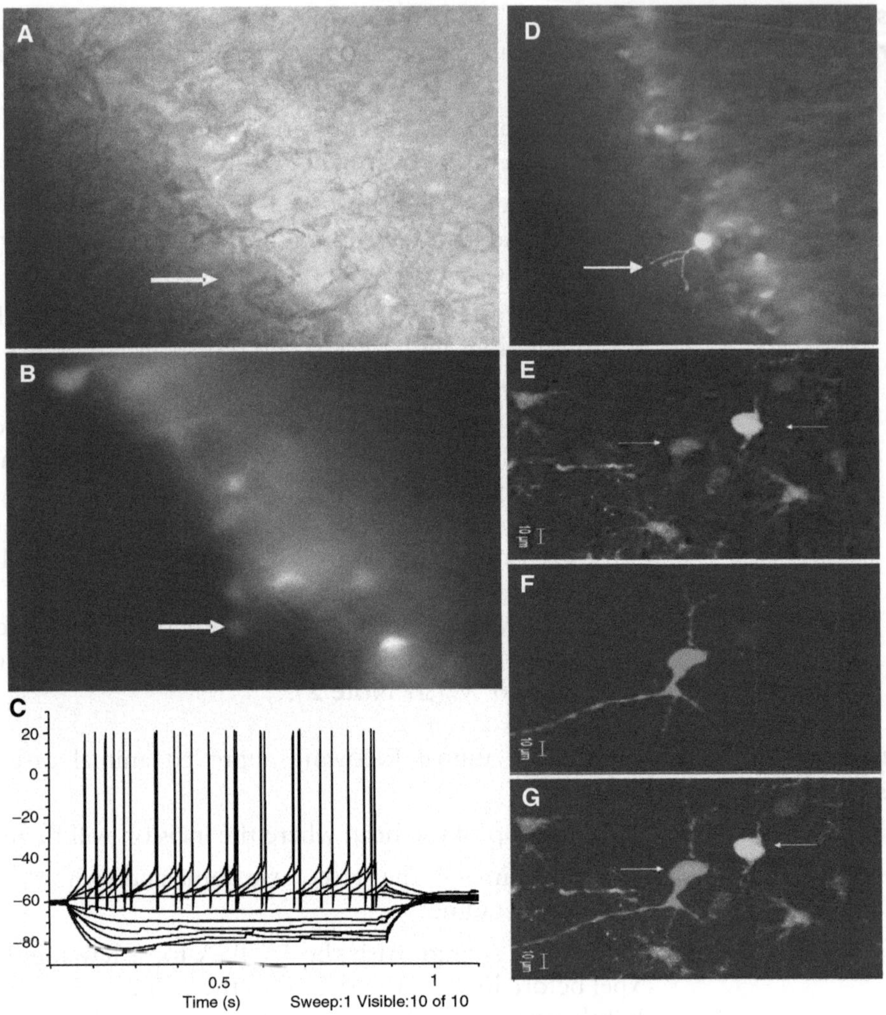

Fig. 2. Using GFP-expressing lentivirus to record from rat interneurons. Acute brain slices were prepared from juvenile rats 12 days postinfusion of CCK-GFP into the CA1 region of the hippocampus **(A–D)**. GFP fluorescence **(B)** was first used to identify a putative CCK interneuron within the slice (*arrow*), which was then located using normal brightfield illumination (DIC, **a**). The fluorescently identified cell was then patched and electrophysiologic recordings obtained (example trace shown in **C**). During the recording process, the cell was passively filled with 0.3% biocytin. Slices were then fixed overnight, resectioned at 75 μm, and visualized using a streptavidin-Texas Red conjugate. This neuron exhibited a voltage-dependent depolarizing sag in the voltage response to hyperpolarizing current injection (**C**). The amplitude and rate of onset of this rectification became more pronounced with increasing hyperpolarization, which is indicative of activation of hyperpolarization-activated nonselective cation channels. In the depolarizing direction, this neuron exhibited a rhythmic pattern of action potential firing, and a pronounced fast after-hyperpolarizing potential (fAHP) following each spike, which are properties similar to those previously reported for hippocampal CCK interneurons. The same CCK-GFP virus was also used to label rat BLA neurons, and to identify them for electrophysiologic recording **(E–G)**. Confocal images of CCK neurons are shown at 60× magnification with *arrows* denoting fluorescing neurons. One of these putative CCK neurons (**E–G**, *left arrow*) was patched and recorded. During recording, this cell was filled with a biocytin and visualized (**F**, **G**) using the same post hoc technique as used in the hippocampus. This figure is adapted from Chhatwal et al. *(9)*.

8. Suture the incision and administer appropriate pain medication. Animals should recover for at least 2 weeks before assessment of in vivo GFP expression, as well as behavioral and/or

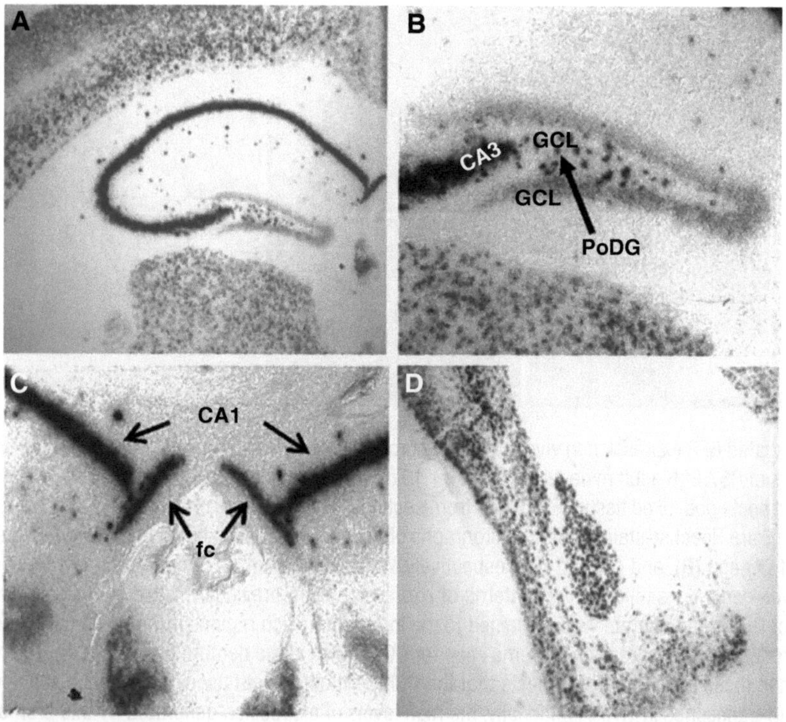

Fig. 3. Endogenous CCK mRNA expression. CCK mRNA was examined using in situ hybridization, demonstrating high levels of expression within the hippocampal formation (low power **A**, high power **B**, **C**) and within the basolateral complex of the amygdala **(D)**. Intense CCK mRNA expression is present within the CA3 subfield and within the interneuron rich region (polymorph layer of the dentate gyrus, PoDG) separating the granule cell layers of the dentate gyrus (GCL; **A**, **B**). Notably, the GCL lacks CCK mRNA + cell bodies. CCK mRNA is highly expressed in the CA1 and fasciola cinereum (fc; **A**, **C**). This figure is adapted from Chhatwal et al. *(9)*.

electrophysiological experiments **(Fig. 2)**, to allow for sufficient viral transduction and transgene expression.

9. Clean the syringe with 100% ethanol and ddH_2O before storing.
10. The cell-type specificity of the virus should be examined using in situ hybridization **(Fig. 3)** combined with in vivo virus injections **(Fig. 4)**. If possible, dual-labeled fluorescence in situ hybridization for GFP and the gene of interest should be used in order to demonstrate co-localization of viral-induced GFP expression and endogenous gene expression. Immunocytochemistry may also be used to verify specificity before proceeding with electrophysiological and/or behavioral experiments. The plasmid packaging construct may also be used to generate transgenic mouse lines **(Fig. 5)**.

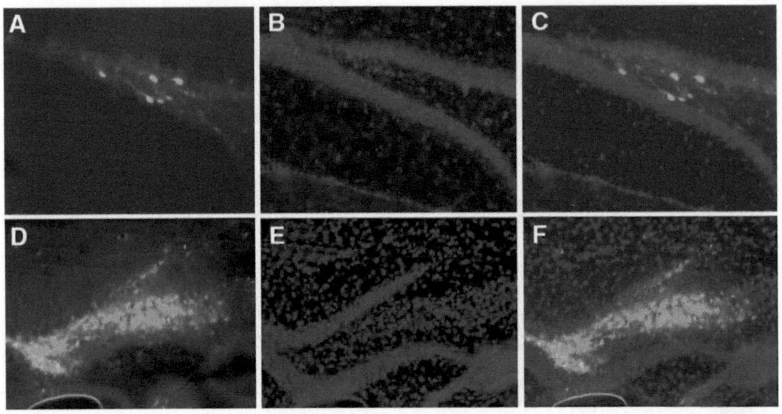

Fig. 4. Virally mediated GFP expression in vivo to examine promoter specificity. Virus encoding CCK-GFP was injected into the dentate gyrus of C57/BL6 adult mice (2 μL/side over 10 min). Animals were sacrificed 10 days later and fluorescence was assessed on sectioned, fixed tissue. Examples from sections showing sparse **(A–C)** and dense expression **(D–F)** are shown above. **B**, **E** are Hoechst-stained photomicrographs of the same sections shown in **(A)** and **(D)**, respectively. **c** and **f** are overlays of **(A)** and **(B)**, and **(D)** and **(E)**, respectively. We observed that GFP expression in various subfields of the hippocampus was generally similar to the patterns of expression seen previously. More specifically, we observed that the expression of GFP in the dentate was contained to the interneuron-rich regions in-between the pyramidal cell layers of the dentate, reflecting the expression patterns seen for CCK mRNA in the dentate gyrus **(Fig. 3)**. The high level of GFP expression seen in these experiments suggests that the CCK promoter length used here may be suitable to drive selective expression of transgenes in vivo. Additionally, the high levels of expression generated by this promoter may make it useful in the generation of transgenic mouse strains. This figure is adapted from Chhatwal et al. *(9)*.

Notes

1. If you are using Taq polymerase, any of the T-A cloning vectors will be sufficient (e.g., pCR2.1, pCRII, pCR4); however, if you are using a proofreading polymerase, you will need to either add A-overhangs onto the PCR product or use a blunt cloning vector (e.g., pCRII-blunt-TOPO or equivalent).
2. Working with lentiviral vectors requires Bio-safety Level II containment. Although current lentiviral vectors have been constructed to be self-inactivating (SIN) as a safety feature, some residual transcription activity has been reported *(5)*. Some institutions may require replication competence assays when using lentiviral vectors. Several chapters in this volume offer detailed descriptions of viral production and titering, and an excellent protocol describing lentivirus production and concentration is available *(6)*.
3. Numerous infusion methods are available for in vivo lentiviral injections; however, some consideration of the needs of your experiment is required before proceeding. The most common injection method is to use a Hamilton syringe attached to an infusion pump that is connected to a stereotaxically mounted

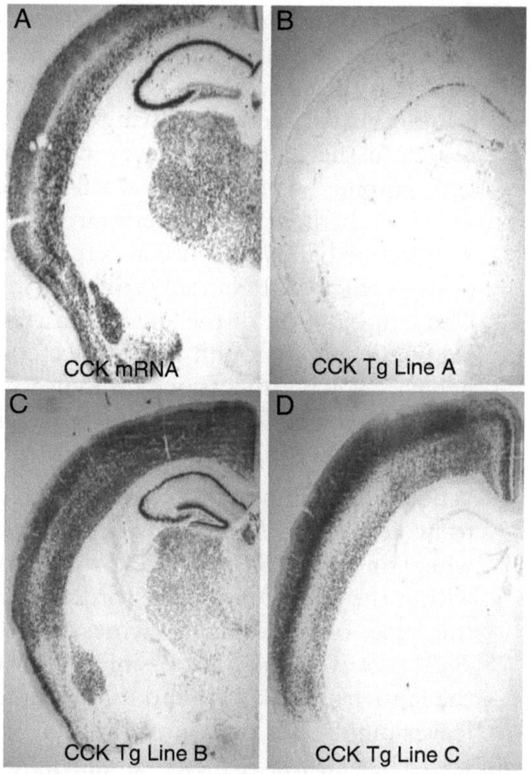

Fig. 5. Transgenic production with cell-type-specific viral vectors to examine promoter specificity. While the aforementioned viral vectors can be used to induce transgene expression in restricted regions, in many cases widespread transgene expression is desirable. To assess the suitability of using promoter lengths identified in the aforementioned manner for the production of transgenic mice, we subcloned the coding sequence for Cre-IRES-DsRed into the CCK-GFP vector. This new plasmid (CCID) was then linearized and microinjected into C57/BL6 embryos. These injections yielded three founders carrying the CCID gene, and these were subsequently mated with RosaLacZ reporter mice to assess transgene expression. The patterns of Cre-mediated LacZ expression suggest that the CCK promoter was capable of driving Cre expression in the transgenic construct. The expression of LacZ varied widely among the three founder lines, although ubiquitous, nonselective expression was not observed in any of the lines. The variability of expression suggests that there may be significant silencing of transgene expression depending on the insertion site of the recombinant sequence, a commonly observed phenomenon in neural systems (this was especially true of the DsRed portion of the gene, which appeared to be completely silenced in all three transgenic lines). Animals from CCID line B showed the widest expression of LacZ, which was essentially identical to endogenous CCK gene expression (**Fig. 4**A, C), suggesting that they may be suitable for use in generating CCK-selective knockouts when crossed with animals containing floxed transgenes. This figure is adapted from Chhatwal et al. *(9)*.

infusion needle or glass-pulled pipette via polyethylene (PE) tubing. This method generally works well; however, the connection between the Hamilton syringe and injection needle or glass pipette is frequently unreliable because air bubbles trapped within the system that often store the energy of the pressure injection and cause erratic infusions. Backfilling the syringe and PE tubing with mineral oil is generally a better method, but we find that although it works for most injections, it is still often unreliable when infusing lentiviral preparations.

The type of injection apparatus used is often determined by the demands of the experiment. For instance, a glass-pulled pipette is the preferred method when conducting electrophysiological experiments following lentiviral infusions in order to reduce tissue damage. However, the viscosity of the lentivirus combined with the very small diameter tip often results in inconsistent injections. A compromise that appears to work well in our hands is to use a stereotaxically mounted infusion pump with a WPI Nanofil syringe with a 35G beveled needle. The combination of the small diameter tip size reduces damage yet is large enough to produce consistent injections. For behavioral experiments, we often use a Hamilton syringe with a 26-G needle. This increases tissue damage but allows for very consistent injections because of the larger tip diameter.

4. When determining your cloning strategy, it may be necessary to avoid using a *Bam*HI restriction site on your primers (i.e., when there are *Bam*HI sites within the promoter region or within the TOPO vector) in order to avoid partial digests. In this case, one may use enzymes with compatible ends (e.g., *Bgl*II), or if this is not possible, blunting the *Bam*HI site on the lentivirus backbone and using a unique restriction site that leaves blunt-ends on your promoter.
5. For large inserts (>1 kb), allow the ligation to incubate for 30 min at room temperature instead of the recommended 5 min.
6. If necessary, the pellets can be stored frozen indefinitely at –20 or –70 °C.
7. In our hands, Endotoxin-free Maxiprep kits do not always provide consistent plasmid DNA preparations. In addition, they are expensive for each preparation. We isolate our plasmid DNA using large-scale alkaline lysis and purify it using cesium chloride/ethidium bromide gradients. We then remove the ethidium bromide using *n*-butanol saturated with H_2O and remove CsCl by dialysis. This method provides milligram quantities of very pure plasmid DNA, is cheaper, and does not require much more actual bench time over the maxiprep kits. Excellent protocols are available in *Molecular Cloning (7)* as well as *Current Protocols in Molecular Biology (8)*.
8. Keep virus on ice at all times when not loaded into the syringe. If you are using the WPI Nanofil with the 35G needle for injection, because of the viscosity of the virus, it is easier to fill the syringe with the 26G needle, then draw the virus into the glass portion of the syringe, and then change to the 35G needle.
9. Examining minimal promoter expression in vitro using HEK cells is a first step at determining level and specificity of promoter activity. GFP or other gene expression driven by cell-type-specific

promoters, in general, will express more poorly and/or more slowly than vectors using a CMV promoter. As noted above, this does not necessarily confer appropriate or robust expression in vivo. In other words, poor and/or slow expression in vitro using HEK cells does not indicate poor or inappropriate expression in vivo, nor does it indicate robust or appropriate expression.

10. An additional in vitro method for determining appropriate cell-type expression is primary cell culture from an area containing the cell type of interest. Appropriate cell-type expression can then be examined using dual-fluorescent in situ hybridization or double-labeled immunocytochemistry. This method is generally more difficult and time consuming than using HEK cells; however, it provides a better indication of appropriate and specific promoter activity. Of course, the virus should always also be examined for cell-type-specific expression in vivo using dual-labeled immunocytochemistry and/or dual fluorescent in situ hybridization if possible before proceeding with experiments.

References

1. Wong, L. F., Goodhead, L., Prat, C., Mitrophanous, K. A., Kingsman, S. M., and Mazarakis, N. D. (2006) Lentivirus-mediated gene transfer to the central nervous system: therapeutic and research applications. *Hum. Gene Ther.* 17. 1–9.
2. Ehrengruber, M. U., Hennou, S., Bueler, H., Naim, H. Y., Deglon, N., and Lundstrom, K. (2001) Gene transfer into neurons from hippocampal slices: comparison of recombinant Semliki Forest Virus, adenovirus, adeno-associated virus, lentivirus, and measles virus. *Mol. Cell. Neurosci.* 17. 855–871.
3. Fleming, J., Ginn, S. L., Weinberger, R. P., Trahair, T. N., Smythe, J. A., and Alexander, I. E. (2001) Adeno-associated virus and lentivirus vectors mediate efficient and sustained transduction of cultured mouse and human dorsal root ganglia sensory neurons. *Hum. Gene Ther.* 12. 77–86.
4. Zufferey, R., Dull, T., Mandel, R. J., Bukovsky, A., Quiroz, D., Naldini, L., and Trono, D. (1998) Self-inactivating lentivirus vector for safe and efficient in vivo gene delivery. *J. Virol.* 72. 9873–9880.
5. Logan, A.C., Haasm D.L., Kafri, T., and Kohn, D.B. (2004) Integrating self-inactivating lentiviral vectors produce full-length genomic transcripts competent for encapsidation and integration. *J. Virol.* 78. 8421–8436.
6. Tiscornia, G., Singer, O., and Verma, I.M. (2006) Production and purification of lentiviral vectors. *Nat. Protoc.* 1. 241–245.
7. Sambrook, J., and Russell, W. D. (2001) Plasmids and Their Usefulness in Molecular Cloning, in *Molecular Cloning: A Laboratory Manual* (Argentine, J., ed), Cold SPring Spring Harbor Laboratory Press, Cold Spring Harbor, New York, pp, . 1.65--1.68.
8. Heilig, J. S., Elbing, K. L., and Brent, R. (1998) Large-Scale Preparation of Plasmid DNA, in *Current Protocols in Molecular Biology* (Ausubel, F. M., Brent, R., Kingston, R. E., Moore, D. D., Seidman, J. G., Smith, J. A., and Struhl, K., eds), John Wiley & Sons, New York, pp, . 1.7.1--1.7.15.
9. Chhatwal, J.P., Hammack, S.E., Jasnow, A.M., Rainnie, D.G., and Ressler, K.J. (2007) Identification of cell-type-specific promoters within the brain using lentiviral vectors. *Gene Ther.* 14(7). 575–583.

Chapter 14

Selective Protein Expression Within the CNS Using Hybrid Lentivirus

Fredrik Gussing, Ingrid van Marion, Igor Ralets, and Cecilia Lundberg

Summary

Lentiviral vectors offer many advantages within the central nervous system. They are capable of transfecting neurons and can be readily manipulated for selective expression. This chapter describes the production and use of lentiviruses to deliver GFP to the central nervous system.

Key words: Cell specific, GFP, Lentivirus, Brain, Expression.

1. Introduction

The ability to express GFP in cells of the nervous system is very useful for a number of research applications. We have chosen to work with lentiviral vectors as they are able to transfect all cell types, including postmitotic neurons. Thus, lentivirus can be used to express GFP in the brain. Furthermore, they are fairly easy to produce and can easily be modified to carry different promoters. We have previously shown that it is feasible to use cellular promoters to drive expression of GFP in specific cell types in the brain *(1, 2)*. Here we describe in detail how to construct your own vector, and how to produce and concentrate the vector. Moreover, we describe the basics of stereotaxic surgery for intracranial injections of the vectors as well as the essential methods for postmortem analysis.

Barry W. Hicks (ed.), *Methods in Molecular Biology, Viral Applications of Green Fluorescent Protein, vol. 515*

DOI: 10.1007/978-1-59745-559-6_14

2. Materials

2.1. Cloning of Cell-Specific Promoters

1. We use the second generation of HIV-1-based lentiviral packaging system which includes three components:
   a. Vector construct: pHR™cppt *PROMOTER GENE* WPRE SIN (*see* **Note 1**) (Fig. 1).
   b. Packaging construct: pBRΔ8.91.
   c. Envelope plasmid: pMDG (expresses the vesicular stomatitis virus g protein, VSV-G).
2. PCR reagents including AccuPrime™ *Pfx* DNA Polymerase (Invitrogen, Carlsbad, CA, USA).
3. One Shot® Stbl3™ Chemically Competent Cells (Invitrogen).
4. DYEnamic ET Terminator Cycle Sequencing Kit (GE Healthcare, Uppsala, Sweden) is used to check that the sequence of the final vector is correct.

2.2. Virus Transfection and Production

1. We use human 293T cells. This cell line is immortalized and derived from embryonic kidney and is easy to culture in large quantities. The culture media for this cell line is standard DMEM culture media supplemented with 10% fetal bovine serum and antibiotics.
2. Standard culture media: 500 mL DMEM (Dulbecco's modified Eagle's medium), 5 mL L-glutamine (200 mM), 5 mL penicillin (10,000 U/mL)/streptomycin (10,000 μg/mL) sterile filtered through a 0.2-μm filter containing 50 mL of heat-inactivated fetal bovine serum (10% final).

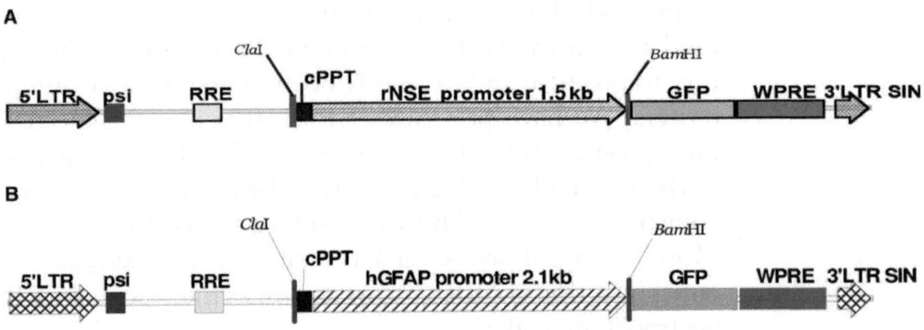

Fig. 1. The vector constructs with rNSE (**A**) and hGFAP (**B**) as cell-specific promoters. The promoter region was inserted into the pHR™-vector between the *ClaI* and *BamHI* restriction sites. Included in these vectors are the reporter gene GFP and psi-sequence that is necessary for packaging viral genomic RNA with the reporter into virions. The Rev response element (*RRE*) stabilizes the viral RNA and allows for nuclear export. To enhance the nuclear translocation and transduction efficiency in vivo, a central polypurine tract (*cPPT*) is inserted before the promoter of choice. To further increase the expression of the transgene, a woodchuck hepatitis posttranscriptional regulatory element (*WPRE*) is introduced, which stabilizes the transgenic nonspliced mRNA and may be needed when using a weaker promoter. Finally, most of the original U3 region of the 3′ LTR is deleted so that it becomes self-inactivating (*SIN*), lacking active viral promoter, after the process of reverse transcription.

3. Virus production media: Standard culture medium containing 2.5 mL HEPES (50 mM).
4. 2× HeBS: 16.36 g NaCl (0.28 M final), 11.9 g HEPES (0.05 M final), 0.213 g Na_2PO_4 anhydrous (1.5 mM final), 800 mL H_2O (distilled), adjust pH to 7.0 with NaOH, add H_2O (distilled) to a final volume of 1 L.

2.3. Determination of Transducing Units of Viral Vector

1. Trypsin.
2. HBSS (1×): 50 mL Hanks Balanced Salt Solution (10×), 50 mL H_2O (distilled), 2.5 mL 7.5% $NaHCO_3$.
3. Proteinase K.
4. Lysis buffer for genomic DNA: 50 mM Tris-HCl pH 8.2, 100 mM NaCl, 5 mM EDTA, 0.5% SDS.
5. RT-PCR Reagents: Taqman Universal Mastermix (Applied Biosystems, Foster City, CA, USA), LV2 forward and reverse primers, LV2 Probe.

2.4. Intracerebral Injections

1. Shield all solutions from light.
2. Anesthesia solution: 10 mL of Fentanyl (Meda AB, Sweden) 50 μg/mL, 0.5 mL Domitor (Orion Pharma, Finland); mix and use 6.2 mL/kg.
3. Antidote: Antisedan (Orion Pharma, Finland), 5 mg/mL. Dilute 1:15 in dH_2O and use 0.9 mL/kg.
4. Analgesia: Temgesic (Shering-Plough Europe, Belgium), 0.3 mg/mL. Dilute tenfold in saline and use 0.9 mL/kg.

2.5. Postmortem Analysis

1. Phosphate buffer: 0.2 M phosphate buffer (PB), pH 7.4, made by mixing 21.8 g of Na_2HPO_4, 6.4 g NaH_2PO_4, and 1 L distilled water.
2. Perfusion fixative: 0.5× PB containing 4% (w/v) paraformaldehyde. Make fresh for each experiment. Prepare 8% paraformaldehyde in water. Heat to 60–65 °C while stirring. Add a few drops of 1N NaOH until solution is clear and continue to stir to dissolve. Cool the solution, filter, adjust volume and add equal volume of 0.2M PB to get final 0.1M PB, set pH to 7.4.
3. Isotonic saline (0.9% NaCl).
4. 25% (w/v) sucrose in 0.1 M PB.
5. Potassium–phosphate buffered saline (KPBS) 6× stock solution 0.12 M 6.86 g KH_2PO_4, 40.08 g K_2HPO_4, 126 g NaCl, add dH_2O to 2.5 L and adjust pH to 7.0–7.4.
6. Antifreeze soultion: 5.02 g $NaH_2PO_4{\bullet}H_2O$, 17.44 g $Na_2HPO_4{\cdot}2H_2O$, 1,280 mL dH_2O, 960 mL ethylene glycol, 960 mL glycerol.
7. Blocking buffer: KPBS containing 0.25% Triton X-100 and either 5% normal donkey serum (NDS) (fluorescent staining) or 5% normal rabbit serum (NRS) (peroxidase staining).

8. Washing buffer: KPBS with 2% NDS/NRS and 0.25% Triton X-100.
9. Chicken anti-GFP antibody (cat# AB 16901, Chemicon (Millipore), Billerica, MA, USA).
10. Donkey anti-chicken antibodies conjugated with Cy2 (cat# 703-225-155, JacksonLabs, Bar Harbor, ME, USA).
11. PVA-DABCO *(5)*.
12. Biotinylated rabbit anti-chicken antibodies (cat# G289A, Promega, Madison, WI, USA).
13. Avidin–biotin–peroxidase complex (Vectastain Elite ABC Kit PK-100; Vector Labs, Burlingame, CA, USA).
14. 3,3-Diaminobenzidine (DAB) (0.5 mg/mL; Sigma, St. Lous, MO, USA).
15. DPX (BDH, Poole, UK)

3. Methods

3.1. Cloning of Cell-Specific Promoters

As cell-specific promoters, we have used the rat neurospecific enolase (rNSE) and the human glial fibrillary acidic protein (hGFAP) *(1, 2)* (**Fig. 2**). These were made using standard molecular biology techniques (*see* **Notes 2** and **3**).

Our choice of cell-specific promoters can of course easily be exchanged with your favorite promoter, the only restriction being the actual size limitation of ~5–6 kb. Exceeding this size will

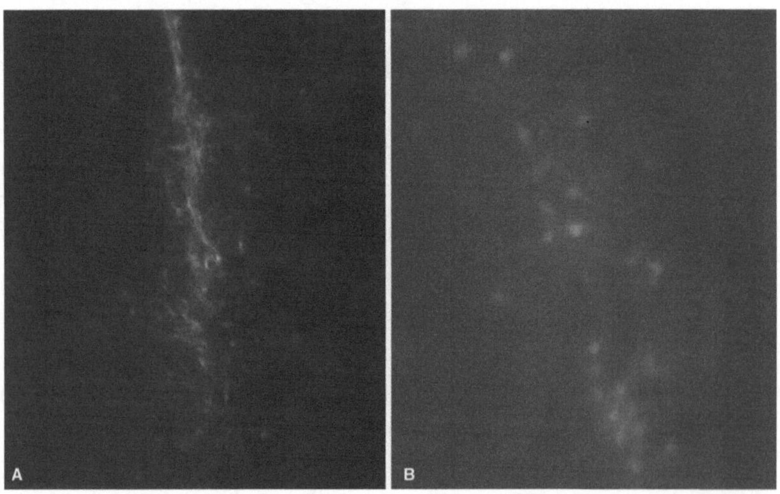

Fig. 2. Photomicrographs of cells in the striatum transduced with lentiviral vectors. In (**A**), the human GFAP promoter drives expression in astrocytes. The green cells are clustered around the injection tract overlapping with the reactive astrocytes that form the scar at the site of injection *(1, 2)*. In (**B**), the rat NSE promoter was used to express GFP specifically in neurons. The GFP fills out the whole cell and projections can be traced to target nuclei of the striatal output neurons, for example, the globus pallidus *(1)*.

limit your options of transgenes and eventually decrease titer as you approach the maximal cloning capacity of the vector. Today, despite the growing number of bioinformatic tools, there are no easy ways to determine where the tissue- and cell-specific regulatory domains are for the expression of a certain gene. Start with a literature search to see whether others have defined the regulatory sequence through in vitro experiments or in transgenic mice. If not, you will have to construct and evaluate a few vectors with varying length of the upstream sequence in order to get the right specificity. As starting material for the cloning, you can either use genomic DNA or quite easily order a BAC clone containing the entire promoter and gene of interest from, for example, http://bacpac.chori.org/home.htm.

If a PCR-based cloning is chosen, we believe that it is necessary to use a proofreading polymerase in the PCR reaction such as the AccuPrime™ *Pfx* DNA Polymerase (Invitrogen) or equivalent (*see* **Note 4**).

3.2. Virus Transfection and Production

1. In the evening of the first day, split a confluent T75 flask of 293T cells (1:10) into ten 10-cm culture dishes. Each dish should contain roughly 1 million producer cells. Add 10 mL of virus production media to each dish.
2. In the evening of the second day, for each plate, make a transfection solution containing 15 μg packaging construct plasmid, 5 μg envelope construct plasmid, and 20 μg transfer construct plasmid, diluted with 1.25 mM HEPES to a total volume of 250 μL. This can also be made as a 2.5-mL "master solution" for all ten plates.
3. Add 250 μL of 0.5 M $CaCl_2$ to each 250 μL of transfection solution and mix thoroughly.
4. Mix the plasmid/$CaCl_2$ solution with an equal amount of HeBS (*see* **Note 5**).
5. Leave the resulting mixture undisturbed for 25 min in order for a small precipitate to form (*see* **Note 6**).
6. After incubation, slowly add 1 mL of precipitate to each 10-cm Petri dish with 293T cells, while gently shaking the culture dish to distribute the mix evenly. It should be virtually impossible to detect a precipitate using the microscope at this stage.
7. In the morning of the third day, after gentle but thorough stirring of the dishes, replace the media containing the precipitate with 10 mL of prewarmed fresh producing media (*see* **Note 7**). It may be beneficial to add 1 mM sodium butyrate to the culture media in order to enhance vector production.
8. On the fourth day, harvest the culture media, which now contains viral vector particles, and centrifuge at 2,500 rpm for 10 min at 4 °C to remove cell debris. Store the supernatant overnight at 4 °C, until the second batch of supernatant is harvested.

9. Add 10 mL of fresh prewarmed virus production media to the 10-cm Petri dishes and incubate for an additional 24 h to facilitate ongoing viral vector production.
10. On the fifth day, the cells will begin to show the toxic effects of overexpressing VSV-G and HIV proteins. Collect the culture media from the 10-cm Petri dishes, and centrifuge at 2,500 rpm for 10 min at 4 °C. The producer cells can now be disregarded.
11. Pool the supernatant with the supernatant that was collected the previous day. Ten plates of producer cells will result in ~200 mL of vector suspension (*see* **Note 8**).
12. To concentrate the vector, filter the supernatant through a 0.45-μm filter and carry out a double round of ultracentrifugation at 25,000 rpm at 4 °C for 90 min (*see* **Note 9**).
13. Discard the supernatant and resuspend the pellet in 200 μL of medium (culture medium without penicillin/streptomycin). Repeatedly vortex the resuspended pellet, that is, a few seconds every 5–10 min, for the next hour to completely dissolve the pellet. Remove any remaining insoluble material with a brief spin.
14. Prepare aliquots of 10 μL and store at −80 °C.

3.3. Determination of Transducing Units of Viral Vector

The concentration of the vector should now be ~10^8–10^9 transducing units (TU/mL). To estimate the concentration of TU, titration should be performed. In the case of GFP, one may use the following method:

1. Seed 100,000 293T cells in each well of a 6-well dish (9.5 cm^2 of growth surface per well) using normal culture medium. After 2–3 h, add a serial dilution of virus stock (1, 0.3, 0.1, 0.03, 0.01, and 0.0 μL) to transduce the cells. Gently swirl the plates to make sure the vector particles are evenly spread over the cells.
2. After 48 h, manually count the GFP-positive cells in each well in relation to GFP-negative cells using a fluorescence microscope and calculate the percentage of GFP-positive cells.
3. Use the dilution that results in ~10–20% of GFP-positive cells for the calculation of the concentration of functional transducing units (TU/mL) since the risk of two copies in one cell is low here and thus you will not risk under estimating your titer (*see* **Note 10**).

The titration method as described above is sufficient when using GFP as a transgene and a ubiquitous promoter such as CMV. An alternative method for titration that can be used when a cell-specific promoter or a nonvisual transgene is present is based on real-time PCR. The technique we use in our laboratory is described in more detail by Sastry et al., and gives a titer value that is dependent on viral integration *(3)*. The method as it has been carried out in our laboratory is as follows:

1. Seed 50,000 293T cells in each well of a 6-well dish using standard culture medium.
2. After 16–18 h, add a serial dilution of virus stock (1, 0.3, 0.1, 0.03, 0.01, and 0.0 μL) to transduce the cells. Gently swirl the plates to make sure the vector particles are evenly spread over the cells.
3. Remove the media from the wells after 72 h and add 200 μL of trypsin to each well. Incubate for 5 min at 37 °C and transfer the cells to 1.5 mL tubes using 700 μL HBSS.
4. Harvest the cells by centrifuging at 1,000 × *g* for 5 min and discard the supernatants.
5. Extract the DNA for real-time PCR analysis as follows: add 18 μL lysis buffer and 2 μL of proteinase K (10 mg/mL) and incubate each tube at 55 °C for 30 min.
6. Subsequently, add 180 μL of H_2O and boil samples for 10 min.
7. In our laboratory, real-time PCR analysis is carried out using Taqman Universal Mastermix (Applied Biosystems) and an ABI Prism 7700 Sequence Detection System (Applied Biosystems). The reaction contains the LV2 primers and probe as listed in **Table 1**. The probe holds the fluorescent dye "Fam" in its 5′-end and the quencher "Tamra" in its 3′ end.
8. Set up PCR reactions in triplicate including a known control, containing the following: 12.5 μL Taqman Universal Mastermix (2×), 0.5 μL Probe (0.2 μM final concentration), 0.8 μL forward primer (0.32 μM final concentration), 0.8 μL reverse primer (0.32 μM final concentration), 8.4 μL H_2O (MilliQ), and 2 μL cell extract. Run a PCR program consisting of the following: 50 °C for 2 min, 95 °C for 10 min followed by 40 cycles of 95 °C for 15 s, and 60 °C for 1 min.

Table 1
LV2 primer and probe sequences

Oligo	Sequence	GC%	T_m
LV2 FP	5′-ACTTGAAAGCGAAAGGGAAAC-3′	42.9	55.9
LV2 RP	5′-CACCCATCTCTCTCCTTCTAGCC-3′	56.5	64.2
LV2 Probe	5TM-Fam-TGCTGACGCAACCCCCACTGGT-Tamra-3′	66.7	69.6

9. After obtaining the PCR data, plot the average C_T value against the logarithm of the volume of added virus stock. Perform regression analysis and use the resulting equations to calculate the relative concentration ratio (integrating units) between the produced virus and the known control.

3.4. Intracerebral Injections

Because of the complex structure of the brain, precise delivery of viral vector is important, especially if the aim of the study is to manipulate a specific subset of cells in a given nucleus. Therefore, the method of choice for intracerebral delivery of the GFP-expressing viral vectors is the use of stereotaxic surgery.

1. We use either injections with fentanyl/Domitor®, Orion Phama, Finland (0.3 mg/kg body weight) or isoflurane/N_2O (400 cc/min) during the whole surgical procedure to anesthetize the animals.
2. Place the rat into a stereotaxic frame (Kopf Instruments, Tujunga, CA, USA). Expose the skull, find the appropriate place for your injection using coordinates from a rat brain atlas, for example, Paxinos and Watson, and drill holes *(4)* (*see* **Note 11**).
3. Inject the viral vector using a Hamilton syringe (87943, 22s gauge, Hamilton Europe) (*see* **Note 12**). Injections should be performed at 0.5 μL/30 s. Leave the syringe in place for 2–3 min after the final injection to normalize the pressure in the brain parenchyma and reduce back flow of the viral vector. Retract the syringe slowly.
4. Close the wound using wound clips and administer analgesia and an anesthetic antidote (*see* **Note 13**).

3.5. Postmortem Analyses

Visualization of the GFP-expressing cells can be done either by analysis of the GFP fluorescence or by immunohistochemical approaches. The peak of transgene expression takes some time, typically 1 week after injections, thus we routinely wait at least 3 weeks before we analyze the material. Perfusion is used to preserve native tissue structure and allow long-term storage of the sectioned brain. Note that paraformaldehyde used as fixative in this method has retarding effect on GFP fluorescence.

1. Anesthetize the rat by injecting 1 mL of pentobarbital (60 mg/kg body weight, IP). Test the corneal reflex and pain reflexes to make sure that the animal is deeply anesthetized.
2. Fix the animal to a corkboard and introduce a syringe attached to tubing into the ascending aorta. Make sure to clamp the descending aorta and to cut the right atrium open. This ensures perfusion of the upper body only.
3. Pump ~50–60 mL of isotonic saline through the heart at flow rate of 50 mL/min until fluid flowing from the right atrium becomes clear of blood.

4. Switch the pump to ice-cold fixative (*see* **Note 14**) and run 250–300 mL through the system at a flow rate of 50 mL/min.
5. Excise the brain using bone rongeurs to open the skull and a spatula to gently remove the brain. Be careful to remove the dura mater prior to moving the brain.
6. Post-fix the brain in 4% paraformaldehyde/0.1 M PB at 4 °C for 3–4 h and subsequently replace the fixative for a phosphate buffered 25% sucrose solution, ensuring the amount of sucrose solution is sufficient for the brain to float. Incubate at 4 °C.
7. When brain has sunk to the bottom of the vial (usually it takes 16–20 h), it can be cut on a freezing microtome.
8. Section the brain into 40 μm thick sections on a freezing microtome through the injected area and transfer the sections into antifreeze solution (*see* **Note 15**).
9. Rinse sections thoroughly in KPBS. If peroxidase is used at later stages, treat sections with 3% H_2O_2 and 10% methanol for 15 min to quench endogenous peroxidase activity and rinse 3× 10 min in KPBS.
10. To reduce unspecific background, incubate sections for 1 h at room temperature in blocking buffer (KPBS with 5% normal donkey serum (NDS) (fluorescent staining) or 5% normal rabbit serum (NRS) (peroxidase staining) and 0.25% Triton X-100).
11. Incubate sections overnight in 1:1,000 chicken anti-GFP antibody in blocking buffer.
12. Wash sections 3× 10 min in washing buffer (KPBS with 2% NDS/NRS and 0.25% Triton X-100).

The following steps depend on the type of staining undertaken.

3.5.1. Fluorescent Staining (*see* **Note 16**)

1. Incubate sections with 1:400 donkey anti-chicken antibodies conjugated to Cy2 in washing buffer for 1 h.
2. Wash 3× 10 min in washing buffer and mount on chrome-alum-coated slides.
3. Leave slides standing for several minutes to dry and coverslip using PVA-DABCO *(5)* (*see* **Note 17**).
4. Analyze the results using an epifluorescence or confocal microscope. Make sure to store the slides shielded from light at 4 °C.

3.5.2. Peroxidase Staining

1. Incubate sections with 1:200 biotinylated rabbit anti-chicken antibodies in washing buffer for 1 h.

2. Rinse 3× 10 min with KPBS and incubate with avidin–biotin–peroxidase complex for 1 h.
3. Rinse 3× 5 min with KPBS and visualize the reaction using 0.5 mg/mL DAB as chromogen. The reaction time should be monitored by eye and the reaction must be stopped when sections turn brown.
4. Mount sections on chrome-alum-coated slides and dehydrate in ascending alcohol concentrations (3× 5 min in 95%, 3× 5 min in 100%) and clear in xylene (3× 15 min).
5. Coverslip using DPX.
6. Analyze the results in a light microscope.

Notes

1. When using nonubiquitous promoters, we suggest that you have cPPT, WPRE, and SIN in your lentiviral backbone *(6–9)*. These modifications will increase the transduction and expression levels of your lentivirus and thus aid your gene expression level from the sometimes weaker cell-specific promoters.
2. When transforming the vector construct, use Stbl3™ chemically competent cells. This will reduce the frequency of unwanted homologous recombination of long terminal repeats.
3. It may be worthwhile to switch to the Gateway® Technology (Invitrogen). We have now done so enabling us to easily introduce and/or exchange promoter as well as gene for a specific experiment.
4. The addition of 0.8 μg/μL BSA (final concentration) to the PCR reaction might help when doing long sequences. Moreover, do not forget to increase the elongation time by 1 min/kb for sequences over 1 kb in size.
5. The pH of the HeBS solution is very important and good mixing in a slow manner is a crucial step in virus production. To facilitate this, the HeBS solution can be put in a 50-mL tube, which is then continuously vortexed on maximum speed. During the vortexing plasmid/$CaCl_2$ is added slowly, one drop at a time.
6. Do not let the mixture stand for more than 30 min.
7. Use a microscope to observe the precipitate in the Petri dishes. It is visible as a lot of small grains covering the empty spaces between the producer cells. There should be a clearance visible in the area directly surrounding the cells as the cells should have taken up the $CaPO_4$/DNA precipitate. Furthermore, sinc GFP is the transgene, it will be possible to detect green cells using a fluorescence microscope.

8. The concentration of the vector at this stage is estimated to be 10^5 TU/mL, which is too low for in vivo injections into the brain. Concentrating vectors can easily be done using ultracentrifugation.
9. In our laboratory, the ultracentrifugation is carried out in a Beckman L-70 Ultracentrifuge with rotor type SW28 using six plastic centrifuge tubes (Beckman). After the first round of centrifugation, the supernatant is discarded and pellets collected and resuspended in 1–2 mL of medium. The resuspended pellets are pooled, transferred to a single tube suitable for use with rotor-type SW41, and a second round of ultracentrifugation was carried out using rotor-type SW41.
10. An alternative method to determine the functional titer is to perform a FACS analysis instead of determination of amount of GFP-positive cells using a fluorescence microscope.
11. We are using the dural surface to define the zero point for the DV coordinate. We find it to be less variable than the skull of adult rats.
12. You may equip the syringe with a glass capillary tip *(10)* for a less traumatic and more precise injection.
13. We keep the rats separated from noninjected animals for 48 h to minimize the risk of vector cross-contamination. The half-life of the vector in vivo is about 1 h *(11)*.
14. Formaldehyde is toxic. Wear gloves while handling it and avoid inhaling powder/fumes. Work in a well-ventilated room and preferably use a ventilated sink.
15. We cut 40 μm sections since it allows for detailed morphological analysis and stereological quantification. Therefore, to improve penetration of antibodies into the sections, all steps prior to mounting are done on free-floating sections.
16. During all of the following steps, slides should be shielded from light.
17. Fresh preparation of PVA-DABCO should be used if possible. Even frozen aliquots must not be used after 4 weeks of storage to avoid formation of microbubbles *(5)*.

References

1. Jakobsson, J., Ericson, C., Jansson, M., Bjork, E., and Lundberg, C. (2003) Targeted transgene expression in rat brain using lentiviral vectors. *J. Neurosci. Res.* 73, 876–85.
2. Jakobsson, J., Georgievska, B., Ericson, C., and Lundberg, C. (2004) Lesion-dependent regulation of transgene expression in the rat brain using a human glial fibrillary acidic protein-lentiviral vector. *Eur. J. Neurosci.* 19, 761–5.
3. Sastry, L., Johnson, T., Hobson, M. J., Smucker, B. and Cornetta, K. (2002) Titering lentiviral vectors: comparison of DNA, RNA and marker expression methods. *Gene Ther.* 9, 1155–62.
4. Paxinos, G., and Watson, C. (1996) The rat brain in stereotaxic coordinates. Academic Press Inc., USA.
5. Peterson, D. (2004) *Quantitative metods in neuroscience* (Evans, S., Jansen, A. M., and

Nyengaard, J. R., Eds.), pp. 85–114, Oxford University Press, Oxford.

6. Follenzi, A., Ailles, L. E., Bakovic, S., Geuna, M., and Naldini, L. (2000) Gene transfer by lentiviral vectors is limited by nuclear translocation and rescued by HIV-1 pol sequences. *Nat. Genet.* 25, 217–22.
7. Zufferey, R., Nagy, D., Mandel, R. J., Naldini, L., and Trono, D. (1997) Multiply attenuated lentiviral vector achieves efficient gene delivery in vivo. *Nat. Biotechnol.* 15, 871–75.
8. Zufferey, R., Dull, T., Mandel, R. J., Bukovsky, A., Quiroz, D., Naldini, L., and Trono, D. (1998) Self-inactivating lentivirus vector for safe and efficient in vivo gene delivery. *J. Virol.* 72, 9873–80.
9. Zufferey, R., Donello, J. E., Trono, D., and Hope, T. J. (1999) Woodchuck hepatitis virus posttranscriptional regulatory element enhances expression of transgenes delivered by retroviral vectors. *J. Virol.* 73, 2886–92.
10. Nikkhah, G., Olsson, M., Eberhard, J., Bentlage, C., Cunningham, M. G., and Bjorklund, A. (1994) A microtransplantation approach for cell suspension grafting in the rat parkinson model: a detailed account of the methodology. *Neuroscience* 63, 57–72.
11. Croyle, M. A., Callahan, S. M., Auricchio, A., Schumer, G., Linse, K. D., Wilson, J. M., Brunner, L. J., and Kobinger, G. P. (2004) PEGylation of a vesicular stomatitis virus G pseudotyped lentivirus vector prevents inactivation in serum. *J. Virol.* 78, 912–21.

Chapter 15

Construction and Analysis of Alphaherpesviruses Expressing Green Fluorescent Protein

Bruce W. Banfield and Gregory A. Bird

Summary

Over the past 10 years, the use of fluorescently tagged herpesviruses has evolved from relative obscurity into a common component in the arsenal of many molecular herpesvirology laboratories. In this chapter we provide methods for construction and analysis of recombinant alphaherpesviruses using conventional co-transfection and homologous recombination procedures. In recent years many herpesviruses have been cloned into bacterial artificial chromosomes (BACs), which has facilitated their manipulation by sophisticated bacterial molecular genetic techniques [Messerle, M., Crnkovic, I., Hammerschmidt, W., Ziegler, H., and Koszinowski, U. H. (1997) *Proc Natl Acad Sci USA* **94,** 14759–63; Smith, G. A., and Enquist, L. W. (2000) *Proc Natl Acad Sci USA* **97,** 4873–8; Tischer, B. K., von Einem, J., Kaufer, B., and Osterrieder, N. (2006) *Biotechniques* **40,** 191–7]. These technological breakthroughs have allowed for the genetic analysis of virus gene products, including those that are essential for virus replication, with unprecedented ease. The main caveat to this approach is that one requires their virus strain of interest cloned into a BAC. If the virus strain under consideration has not been introduced into a BAC, it is far from trivial to do so. While comparatively antiquated, the procedures provided in this article can be used with any strain. Here we focus on pseudorabies virus (PRV), a swine pathogen, which is the alphaherpesvirus most amenable to genetic manipulation using this transfection-based approach.

Key words: Alphaherpesvirus, Pseudorabies virus, Green fluorescent protein, Homologous recombination, Plaque purification.

1. Introduction

The Herpesviridae is a very large family of viruses that infect and cause disease in a wide variety of animals ranging from oysters to elephants. The herpesviruses have been divided into three subfamilies: alpha, beta, and gamma based on host range, tissue tropism, and length of replicative cycle. The alphaherpesviruses

Barry W. Hicks (ed.), *Methods in Molecular Biology, Viral Applications of Green Fluorescent Protein, vol. 515*

DOI: 10.1007/978-1-59745-559-6_15

are often referred to as neurotropic because of their propensity to establish lifelong latent infections in neurons of the peripheral nervous system. Once a latent infection has been established, viruses of this group are capable of reactivating from infected neurons resulting in the production and shedding of infectious virus from the host. Key to both initial infection of neurons and reactivation of infection from these cells is the ability of the alphaherpesviruses to travel directionally within the nervous system. Moreover, spread of infection between neurons is restricted to those that are in synaptic contact. It is these properties of the alphaherpesviruses that has led to their development and extensive use as tracers of neuroanatomical circuitry. In this regard, the use of viruses expressing fluorescent proteins has provided significant contributions.

Included in the alphaherpesvirus subfamily are the common human pathogens, HSV-1, HSV-2, and varicella-zoster virus, and, most pertinent to this article, the swine pathogen pseudorabies virus (PRV). The PRV genome comprises linear double-stranded DNA that is 143,461 base pairs in length and codes for over 70 proteins *(1)*. Until recently, with the advent of virus genomes that can be manipulated in bacteria, the genome of PRV was the easiest to manipulate allowing for straightforward construction of recombinant viruses. The relative ease with which the PRV genome can be manipulated was a contributing factor to the early development of PRV as a tool for neuroanatomical analyses. The introduction of the *Escherichia coli* β-galactosidase gene and subsequently the gene for enhanced green fluorescent protein (EGFP) provided reporter systems that obviated the need for lengthy and tedious immunocytochemical methodologies for identifying virally infected cells *(2–4)*. Below, we describe procedures for construction of recombinant PRV strains that are outlined in **Fig. 1**.

To create recombinant viruses, one must construct a plasmid containing the DNA sequence to be recombined into the virus genome, flanked by homologous virus genome sequences that

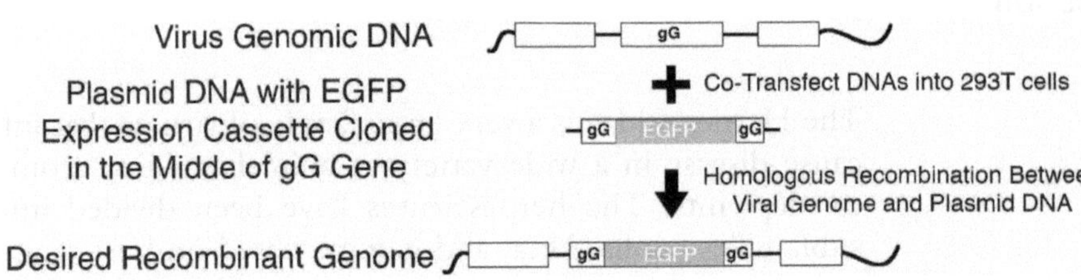

Fig. 1. Scheme for homologous recombination of an EGFP expression cassette with the PRV gG gene. The gG locus was chosen as a site for insertion of heterologous genes because elimination of gG has minimal effects on the ability of the virus to replicate and spread in the rodent central nervous system. Purified virus DNA and plasmid DNA are cotransfected into 293T cells. Approximately 3–4 days after cotransfection, when cells have lysed, the desired recombinant virus should be present in the culture supernatant at a frequency between 0.1 and 0.5%.

direct the recombination to the desired locus. This "transfer plasmid" is then co-transfected along with infectious virus DNA into permissive cells. Homologous recombination between the plasmid DNA and the virus DNA will occur and recombinant viruses will be released into the cell supernatant along with a vast excess of parental nonrecombinant virus. Finally, the desired recombinant must then be isolated from the parental strain.

2. Materials

2.1. Purification of Plasmid DNA

1. Luria Broth (Sigma, St. Louis, MO).
2. Solution 1: 25 mM Tris-HCl pH 8.0, 10 mM EDTA, 15% sucrose.
3. Solution 2: 0.2 M NaOH, 1% SDS.
4. Solution 3: 10 M ammonium acetate.
5. RNase A: 1 mg/mL.
6. Isopropanol.
7. 70% ethanol.
8. Buffer saturated phenol:choloroform (1:1).

2.2. Linearization of Plasmid DNA

1. Restriction endonuclease and buffer (New England Biolabs, Ipswich, MA).
2. Agarose (Sigma, St. Louis, MO).
3. Ethidium bromide (Sigma, St. Louis, MO).
4. TAE buffer 50× stock solution (1 L): 242 g Tris base, 57.1 mL acetic acid, 100 mL 0.5 M EDTA (pH 8.0). Adjust pH to 8.5. 1× is 40 mM tris-acetate, 1 mM EDTA.

2.3. Preparation of Viral DNA

1. 1× HIRT Lysis buffer: 10 mM Tris pH 8.0, 10 mM EDTA, 0.6% SDS, 0.25 mg/mL proteinase K.
2. Ethanol.
3. Pasteur pipettes (sealed by holding fine end over flame).

2.4. Cell Culture, Co-Transfection, and Viral Isolation

1. 293T cells (ATCC# CRL-11268).
2. PK15 cells (ATCC# CCL-33).
3. Dulbecco's modified Eagle's medium (DMEM) (Invitrogen, Carlsbad, CA).
4. Fetal bovine serum (FBS) (Invitrogen).
5. Phosphate buffered saline (PBS) (Invitrogen).
6. Trypsin–EDTA (Invitrogen).

7. Penicillin/streptomycin (Invitrogen).
8. 2× HBS: 50 mM HEPES, 280 mM NaCl, 1.5 mM Na_2HPO_4. Adjust pH to exactly 6.95 with HCl and sterilize using 0.2 μm filter.
9. 2 M $CaCl_2$.
10. Calf thymus DNA (Invitrogen).
11. Tissue culture plastics including 6-well cluster dishes and 100 mm plates.
12. 2% Carboxymethylcellulose (Fisher Scientific, Pittsburgh, PA); 10 g in 500 mL water. Autoclave in a 1-L bottle including medium stirbar.
13. DMEM powder (Invitrogen). Prepare 500 mL of 2× solution.
14. DMEM/5% FBS/1% carboxymethylcellulose: Once the 2% carboxymethylcellulose has cooled, add 500 mL of 2× DMEM. Stir at room temperature for 4–6 h until mixed well. Add penicillin/streptomycin and 50 mL of FBS. Stir for an additional hour. Store solution at 4 °C.
15. Snap cap tubes (BD Falcon #2054).
16. Sonicator.

2.5. Microscopy

1. We use a Nikon TE200 inverted epifluorescence microscope equipped with an X-Cite UV light source and Chroma (Rockingham, VT) filter set #41017 for EGFP, but any comparable fluorescence microscope can be used.

3. Methods

3.1. Preparation of Plasmid DNA

1. At least 2 days prior to co-transfection, streak an LB-agar plate containing the appropriate antibiotic selection with *E. coli* from a glycerol stock harboring the EGFP transfer plasmid (*see* **Note 1**).
2. Incubate the plate overnight at 37 °C. The next day isolated colonies should be apparent. Plates containing isolated colonies can be stored in a sealed container at 4 °C up to 1 week.
3. One day before co-transfection, inoculate 5 mL of LB (with selection), in a 15-mL snap-cap tube, with a single isolated colony from the LB-agar plate. Be sure that the cap remains loose to allow adequate aeration. Shake the tube at 250 rpm overnight at 37 °C.

4. The next day, transfer 1.5 mL of the overnight culture to a 1.5-mL microfuge tube and spin the culture at 12,000 rpm in a microcentrifuge for 30 s to pellet the bacteria.
5. Discard the supernatant and resuspend the pellet in 200 μL of solution 1. Add 400 μL of solution 2 and immediately mix by gently inverting the capped tube five or six times. The solution should become viscous.
6. Incubate at room temperature for 5 min. Longer incubation at this stage should be avoided.
7. Add 250 μL of solution 3. Mix by inverting the tube five or six times. A precipitate resembling cottage cheese should form almost immediately.
8. Incubate the tube on ice for a minimum of 10 min then centrifuge at 12,000 rpm for 10 min and immediately transfer the supernatant, containing plasmid DNA, to a clean 1.5 mL microfuge tube.
9. Add 800 μL of isopropanol and incubate tube on ice for at least 10 min. Centrifuge tube at 12,000 rpm for 10 min. Plasmid DNA will be contained within a white pellet at the bottom of the tube.
10. Discard the supernatant and add 70% ethanol to the pellet. There is no need to resuspend the pellet at this point. Centrifuge at 12,000 rpm for 2 min and remove the supernatant. Ensure that all of the supernatant has been removed, but avoid letting the pellet dry out.
11. Resuspend the pellet in 50 μL of dH_2O.

3.2. Linearization of Transfer Plasmid DNA

Linearization of the transfer plasmid ensures that genetically unstable viruses that have acquired the EGFP expression cassette via a single cross-over event are not isolated.

1. To linearize the DNA, digest 25 μL (~5 μg) of plasmid DNA with the appropriate restriction endonuclease in a final volume of 50 μL.
2. The addition of 1 μL of a 10 mg/mL solution of RNase A will ensure the degradation of contaminating RNA.
3. Incubate the reaction at the optimal temperature for the restriction endonuclease being used. The time required for complete digestion will vary depending on the enzyme used. Consult the recommendations from the enzyme supplier to determine appropriate incubation time. For most enzymes, a 1-h incubation will usually suffice.
4. After the digest is complete, add 150 μL of dH_2O and 200 μL of a 1:1 ratio of buffer saturated phenol:chloroform. Mix vigorously by inverting the capped tube several times.
5. Centrifuge the tube at 12,000 rpm for 10 min. This will partition the plasmid DNA containing aqueous phase on top of

the lower organic phase. The phases are often separated by a thin white layer of denatured protein at the interface.

6. Carefully transfer 150 μL of the upper aqueous layer to a new 1.5 mL microfuge tube. Be sure not to transfer any of the denatured protein or organic phase.
7. Add 20 μL of 3 M ammonium acetate and 500 μL of 100% ethanol. Mix the capped tube by inverting it several times, and incubate at −20 °C for at least 30 min.
8. Centrifuge at 12,000 rpm for 10 min. Immediately discard the supernatant and add 500 μL of 70% ethanol to the pellet.
9. Centrifuge at 12,000 rpm for 2 min. Remove and discard the supernatant. Ensure that all of the supernatant has been removed before proceeding. Resuspend the linearized plasmid DNA in 50 μL of dH_2O.

3.3. Preparation of Viral DNA

1. At least 2 days prior to co-transfection, infect a confluent 100-mm dish of PK15 cells with the parental virus strain at an MOI of 10 PFU/cell. Allow infection to proceed for a minimum of 16 h.
2. At this point cells should display robust cytopathic effects including margination of chromatin and significant cell rounding. Carefully remove the medium.
3. Add 10 mL of PBS to the plate and scrape the adherent cells into the PBS.
4. Transfer the cells to a 15-mL conical tube and centrifuge at 1,000 rpm for 10 min to pellet the cells.
5. Carefully remove supernatant and add 2 mL of 1× HIRT lysis buffer to cell pellet and mix gently. The solution should immediately become viscous.
6. Incubate tube overnight at 37 °C to ensure adequate digestion of proteins.
7. On the next day, extract three times with phenol:chloroform as described in **Subheading 3.2** (*see* **Note 2**).
8. Next, extract the recovered aqueous phase three times with chloroform. After the last chloroform extraction, aliquot the upper aqueous phase into three 5-mL snap cap tubes.
9. Precipitate the DNA by adding 2 vol. of ice-cold isopropanol. Mix by inversion until DNA "ghosts" appear.
10. Collect the DNA by spooling onto a sealed Pasteur pipette and allow it to air dry. The dried DNA can be stored on the end of the pipette inside a microfuge tube at −20 °C (*see* **Note 3**).

11. Resuspend one aliquot of the DNA in 200 μL of TE. Heat the tube at 37 °C for 30 min to help to dissolve the DNA. Determine the DNA concentration using a spectrophotometer.

3.4. Preparation of Cells and Co-Transfection

1. The day before co-transfection plate 293T cells on 100-mm dishes so that they will be at 50–70% confluence the next day.
2. Four hours prior to transfection, aspirate the growth medium from the cells and replace it with 9 mL of fresh growth medium (*see* **Note 4**).
3. To prepare the DNA for transfection, you will require two 5-mL snap cap tubes. Assemble contents of tubes as outlined below (*see* **Note 5**):

Tube A: 62.5 μL of 2 M $CaCl_2$, 20 μg of total DNA (5 μg viral DNA + 5 μg linearized plasmid DNA + 10 μg calf thymus DNA as carrier), H_2O up to 500 μL.

Tube B: 500 μL 2× HBS.

4. Add the contents of tube B to tube A dropwise while vortexing tube A at low speed (*see* **Note 6**).
5. Once the solutions are mixed together, incubate at room temperature for 20–40 min.
6. Add the DNA mixture dropwise into the 293T cell culture medium and onto the cells while gently rocking the plate back and forth.
7. Return the cells to the incubator and incubate overnight.
8. Early the next morning, remove the medium and replace it with 10 mL of fresh medium (*see* **Note 7**).
9. When all the cells show overt signs of infection, divide the medium into 1 mL aliquots in microcentrifuge tubes, then store at −80 °C.

3.5. Isolation of Recombinant Viruses

1. Recombinant viruses are purified by picking an isolated plaque. One day prior to performing the plaque assay, plate PK15 cells into 6-well cluster dishes. The cells should be 80–90% confluent the next day. Splitting a confluent 100-mm dish of cells to 12 wells will usually achieve the appropriate cell density of ~8.5×10^5 cells/well.
2. When the cells are ready, thaw an aliquot of the virus-containing medium (described in **Subheading 3.4**) at 37 °C.
3. Once thawed, transfer the tube to ice and sonicate ten times using 1 s pulses in a cup horn sonicator (*see* **Note 8**).

4. Perform serial tenfold dilutions of virus by transferring 100 μL of virus preparation to 900 μL of DMEM/10% FBS. Mix the contents of the tube by gentle trituration and discard the pipette tip into a waste container containing 10% bleach. Using a new tip transfer 100 μL of this tenfold dilution to a new tube containing 900 μL of DMEM/10% FBS. Once again mix the contents of the tube and discard the used tip. Repeat this step until you have prepared six dilutions of virus (i.e., 10- to 1,000,000-fold dilutions).
5. Remove medium from all six of the wells of PK15 cells in a cluster plate, and add 100 μL of each viral serial dilution to separate wells.
6. Rock the dish gently to ensure that the cells do not dry out and return the dish to the incubator.
7. Rock the cluster dishes every 15 min for 1 h.
8. After 1 h, add 2 mL of DMEM/5% FBS/1% carboxymethylcellulose, prewarmed to 37 °C, to each well.
9. Return the 6-well cluster dish to the incubator. Twenty-four hours after inoculation, remove the 6-well cluster dish from the incubator and examine wells for plaques under an inverted epifluorescence microscope using the appropriate EGFP filter set (**Fig. 2**).
10. Choose a well having the fewest plaques but that contains at least two or three green plaques. This will reduce the number of contaminating viruses isolated.
11. Using either a 4× or 10× objective, center an isolated green plaque in the middle of the field. Remove the lid of the 6-well cluster dish.
12. Using a P20 set to 20 μL and a sterile tip, aspirate the green cells into the pipette tip (*see* **Note 9**).

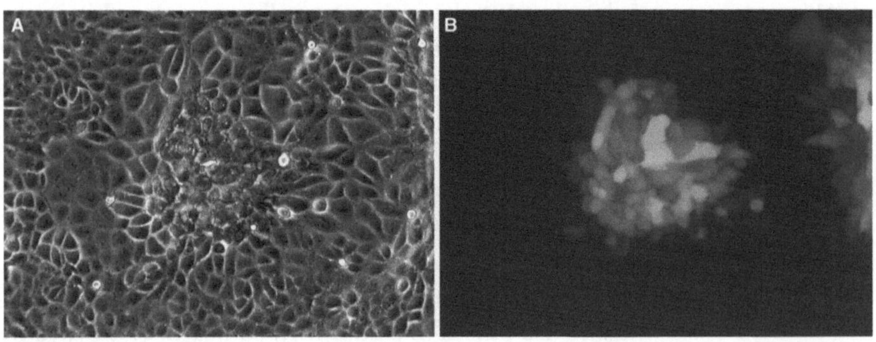

Fig. 2. Images of an isolated, EGFP-positive, PRV plaque at 17 h postinfection of a PK15 cell monolayer **(A)**. Phase contrast image of fluorescent plaque **(B)**.

13. Deposit the green cells into 500 μL of DMEM/10% FBS in a sterile microfuge tube.
14. If possible, pick three to five different plaques. The frequency with which green plaques are seen at this stage is variable, but typically ranges from 1/1000 to 50/1000.
15. Next, sonicate the medium containing a picked plaque as described previously.
16. Inoculate two wells of PK15 cells, one with 20 μL and the other with 5 μL of sonicated virus diluted into 100 μL of DMEM/10% FBS (*see* **Note 10**).
17. Once again, incubate and rock the dishes every 15 min for 1 h to ensure that the cells do not dry out. At the end of 1 h, add 2 mL of prewarmed DMEM/5% FBS/1% carboxymethylcellulose to each well.
18. The next day, examine the monolayers for green plaques and pick the most isolated ones as described above (*see* **Note 11**).

3.6. Analysis/Validation of Recombinant Viruses

It is critical to ensure that the expected recombination events have taken place. This is done by performing a Southern blot or by using PCR/DNA sequencing to analyze the 5' and 3' recombination sites (*see* **Note 12**). Protocols for these techniques can be found elsewhere *(5)*. It is also a good practice to perform single-step growth analysis on any recombinant strain to determine its growth characteristics in comparison to the parent strain. It is important to remember that the introduction of naked DNA into cells is inherently mutagenic and can lead to the introduction of unanticipated changes in the virus genome. These mutations can be deleterious, advantageous, or benign. Regardless, if studies of gene function are the goal of your experiments, it is critical that any recombinant virus is repaired back to the parental genotype to ensure that any phenotypes observed with the recombinant strain are due to the designed modification and not unanticipated mutations (*see* **Note 13**).

4 Notes

1. Not described in this chapter are protocols for the design of the transfer plasmids. These will vary depending on the application, the virus being used, and the site in the virus genome to receive the EGFP expression cassette. We routinely use an EGFP expression cassette isolated from pEGFP-N1 (Clontech). These sequences include the HCMV major immediate early promoter, EGFP coding sequences, and a SV40

polyadenylation signal. Homologous sequences required for recombination into the virus genome must be cloned upstream and downstream of the reporter cassette and these will vary based on the desired site of recombination. As a rule of thumb, we recommend at least 300 bp of homologous sequence on either side of the expression cassette. Decreasing the length of DNA available for recombination will result in a concomitant reduction in recombination frequency.

2. This must be done gently so as not to shear DNA. The use of a P1000 tip that has been cut to increase the size of the tip opening will aid recovery of the aqueous phase.
3. Viral DNA prepared in this manner is stable for upwards of 3 years. However, once the DNA is taken back into TE solution, it will only be stable for a few weeks, and it should be stored at 4 °C.
4. Generally speaking, low recombination frequency is a result of poor transfection efficiency of the 293T cells. When troubleshooting transfection efficiency, the first place to start is the cells. The 293T cells are very sensitive to over growth as defined by growth of the cells to 100% confluence. Be sure to split the cells before they reach 80% confluence. The cells should be plated for transfection the night before you plan to transfect. Cells that are allowed to grow for 2 or more days on the plate that is to be transfected will not take up DNA as well as cells that are seeded the night before transfection. Finally, transfection efficiency of 293T cells will decrease with increased number of passages. If the cells have been split more than 12 times, a new aliquot of cells should be thawed and expanded for transfection.
5. After determining the appropriate volumes, add the water to tube A first, then add other materials ensuring to place the pipette tip below the surface of the solution. Also, while somewhat counterintuitive, inclusion of more virus DNA in the co-transfection mixture does not increase the recovery of infectious virus. We have determined empirically that the use of 2–5 μg of virus DNA provides optimal results.
6. If the 2× HBS is added too rapidly to tube A, a poor quality precipitate will form and this will reduce the transfection efficiency. If the precipitate is allowed to sit too long, the size of the precipitate will become too large and the cells will not take up the DNA efficiently. We recommend incubating the DNA for 20–40 min and not longer. Additionally, if the viral DNA and/or plasmid DNA is not clean, this can lead to an increase in the size of the precipitate. Try an additional phenol/chloroform extraction of the viral and plasmid DNA.

7. You should start to see signs of overt infection that present as holes in the cell monolayer at 48 h posttransfection. Alternatively, for EGFP expressing strains, one can monitor the efficiency of the transfection by fluorescence microscopy with an inverted epifluorescence microscope, in which case infection can be detected at 24 h posttransfection. If the transfection has worked well, all the cells should display overt signs of infection by 96 h.
8. We use a setting of 80% using a Sonic Dismembrator Model 100 (Fisher Scientific). Sonication disrupts clumps of virus, liberates cell-associated virus, and aids in downstream purification.
9. With practice, one can visualize the collection of green cells through the microscope. Using fluorescent optics along with very dim transmitted light can allow visualization of cells and fluorescent cells. It is important to work quickly and use sterile technique to avoid microbial contamination of your isolated plaque. It is also important to collect the minimal amount of medium required to capture green cells. Doing so will reduce the amount of contaminating nonrecombinant virus.
10. If your cells are not yet ready at this stage, you can store the virus at −80 °C. The cells should be 80–90% confluent before inoculation.
11. Plaque purification is usually performed three times in order to ensure that the final recombinant virus isolated is pure. A good rule of thumb is to perform one more round of plaque purification past the point at which no contaminating plaques are observed. In order to expedite virus purification, ensure that you have 6-well cluster dishes of PK15 cells ready to infect everyday for 3 or 4 days.
12. In addition to Southern blot analysis to confirm proper insertion of the EGFP transgene, Western blot analysis can be used to confirm the knockout of a targeted gene. Additionally, when antisera is available to the protein products of the flanking genes, Western blot analysis can be used to confirm that insertion of the EGFP expression cassette does not alter the expression of adjacent genes.
13. Constructing viruses that express multiple fluorescent proteins is a great way to isolate virus that are knocked out for two or more genes. The main consideration when using a second fluorescent marker is that it does not contain sequence homology with the EGFP gene. The use of reporters that contain homologous sequences can lead to recombination between the reporters resulting in deletions or inversions in the viral genome and potential loss of reporter functionality.

One must also ensure that promoter and polyadenylation site sequences used for the different reporters do not contain significant homology. We have found that mRFP-1 has worked well for a second fluorescent marker in virus having EGFP *(6)*.

References

1. Klupp, B. G., Hengartner, C. J., Mettenleiter, T. C., and Enquist, L. W. (2004) Complete, annotated sequence of the pseudorabies virus genome. *J Virol* 78, 424–40.
2. Jons, A., and Mettenleiter, T. C. (1997) Green fluorescent protein expressed by recombinant pseudorabies virus as an in vivo marker for viral replication. *J Virol Methods* 66, 283–92.
3. Mettenleiter, T. C., and Rauh, I. (1990) A glycoprotein gX-&bgr;-galactosidase fusion gene as insertional marker for rapid identification of pseudorabies virus mutants. *J Virol Methods* 30, 55–65.
4. Smith, B. N., Banfield, B. W., Smeraski, C. A., Wilcox, C. L., Dudek, F. E., Enquist, L. W., and Pickard, G. E. (2000) Pseudorabies virus expressing enhanced green fluorescent protein: a tool for in vitro electrophysiological analysis of transsynaptically labeled neurons in identified central nervous system circuits. *Proc Natl Acad Sci USA* 97, 9264–9.
5. Sambrook, J., and Russell, D. W. (2001) Molecular Cloning, Cold Spring Harbor Laboratory Press, Cold Spring Harbor, New York.
6. Banfield, B. W., Kaufman, J. D., Randall, J. A., and Pickard, G. E. (2003) Development of pseudorabies virus strains expressing red fluorescent proteins: new tools for multisynaptic labeling applications. *J Virol* 77, 10106–12.

Chapter 16

The Use of Green Fluorescent Fusion Proteins to Monitor Herpes Simplex Virus Replication

Travis J. Taylor and David M. Knipe

Summary

The localization pattern of the seven herpes simplex virus (HSV) DNA replication proteins is dependent upon the status of viral DNA synthesis in the infected cell. Normally, the replication proteins accumulate within replication compartments, which expand as viral DNA synthesis increases. If viral replication is blocked, either by the addition of drugs or a genetic lesion, prereplicative sites are observed. Observing the distribution of a GFP-tagged HSV replication protein can monitor the progression of viral replication. Here, we demonstrate the use of an ICP8-GFP fusion protein to observe the status of HSV replication in cultured cells by the formation of viral replication compartments.

Key words: Herpes simplex virus, Replication compartments, Prereplicative sites, ICP8, Intranuclear localization.

1. Introduction

Viral processes are regulated not only temporally, but also spatially in the infected cell. There are multiple lines of evidence that demonstrate that viral DNA synthesis, transcription, capsid assembly, and DNA packaging do not occur diffusely within the nucleus but within specialized intranuclear structures called replication compartments. Replication compartments were initially defined by the presence of ICP8 *(1)*, an essential replication protein that binds single-stranded DNA *(2, 3)*, as detected by immunofluorescence, which suggested that DNA synthesis occurs within these structures. As expected, the six other viral replication proteins,

Barry W. Hicks (ed.), *Methods in Molecular Biology, Viral Applications of Green Fluorescent Protein, vol. 515*

DOI: 10.1007/978-1-59745-559-6_16

including the origin-binding protein (U_L9) *(4)*, the helicase-primase complex ($U_L5/U_L8/U_L52$) *(4–6)*, and the polymerase holoenzyme (U_L30/U_L42) *(2, 7)*, are also found within replication compartments. Cellular protein involved in cellular DNA synthesis, homologous and nonhomologous recombination, and chromatin remodeling are also found within replication compartments *(8, 9)*.

When viral DNA synthesis is blocked, either because of genetic defects in U_L30 or U_L42 or because of treatment with compounds that specifically block viral DNA synthesis, replication compartments do not form. Instead, structures called prereplicative sites are observed *(1, 5, 10)*. These structures are believed to be precursors to replication compartments because they are present in cells in the absence of DNA synthesis and represent an early intranuclear accumulation of viral proteins. Two populations of prereplicative sites are observed in infected cells *(11)*, which differ depending upon the status of cellular DNA synthesis. In cells actively synthesizing DNA, as shown by BrdU incorporation, ICP8 is distributed to numerous punctate structures, called S-phase-dependent sites, which co-localize with sites of cellular DNA synthesis *(11)*. In cells that are not actively synthesizing DNA, as shown by the lack of specific BrdU incorporation, ICP8 is distributed to only a few intranuclear structures, called S-phase-independent sites, that form adjacent to cellular ND sites *(11–14)*.

Prereplicative sites and replication compartments form at nonrandom sites within the nucleus that is dictated by preexisting nuclear architecture *(15)*. Assembly of these sites is initially influenced by the preexisting nuclear architecture. For example, in HSV-1infected binucleate CV-1 cells, mirror image patterns of replication compartments are observed in the sister nuclei *(15)*.

This technology can facilitate drug discovery screens by rapidly addressing the state of viral DNA synthesis using automated microscopy *(16)*. The use of ICP8-GFP can also determine where the potential block occurs in the viral life cycle because the assay can distinguish pre- and post-DNA replication events. For instance, if there is no or low fluorescence, the block occurs at entry, or an immediate-early event, or at the level of early gene expression. If there is a GFP signal, prereplicative sites or replication compartments will be observed depending upon the status of viral DNA synthesis. If there are replication compartments, the block occurs in a post-DNA synthesis event, possibly virion maturation or release. The 8GFP virus was used successfully recently to screen a compound library to identify potential HSV inhibitors *(16)*.

2. Materials

2.1. Cell Culture

1. Dulbecco's modified Eagle's medium (Mediatech, Inc) supplemented with 5% heat-inactivated fetal bovine serum and 5% calf serum (DMEM + 10% FBS), 2 mM glutamine, penicillin (100 U/mL), and streptomycin (100 mg/mL). For the ICP8-complementing V827 cell line, G418 was added to the culture medium at a final concentration of 400 μg/mL (*see* **Note 1**).
2. African green monkey kidney cells (Vero; ATCC). The ICP8-complementing V827 cell line *(17)* was derived from low passage Vero cells (*see* **Note 2**).
3. Versene solution (0.2 g/L EDTA; Gibco).
4. Trypsin solution (0.25% in 1 mM EDTA; Gibco).

2.2. Immunofluorescence and Live Cell Imaging

1. Time lapse medium: CO_2-independent medium (Gibco) plus 1% heat-inactivated FBS.
2. 12 mm round, 18 mm round, and 22 × 22 mm square glass coverslips (Sigma).
3. Viral stock dilution medium: PBS supplemented with 1% heat-inactivated FBS.
4. HSV replication inhibitor: phosphonoacetic acid (PAA) at 40 mg/mL (100× stock solution) in water.
5. Primary antibodies: anti-ICP8 39S conformation-specific *(18, 19)*, anti-ICP4 58S *(18)*, and anti-UL42 5H11D6 *(20)* monoclonal antibodies diluted in PBS (Gibco).
6. Secondary antibody: rhodamine-conjugated goat anti-mouse antibody (Cappel) diluted in PBS (Gibco).
7. Cell fixation solution: 3.7% formaldehyde (Sigma) in PBS (Gibco).
8. Cell permeabilization solution: acetone (Sigma) prechilled to −20 °C (*see* **Note 3**).
9. Coverslip mounting: glycerol gelatin (Sigma) containing the anti-bleaching agent p-phenyldiamine (Sigma) at 1.3 mg/mL.
10. A Chen coverslip staining rack to hold coverslips during wash, fix, and permeabilization steps.
11. A humid chamber for the incubation steps constructed from a Petri dish with a dampened paper towel and a layer of Parafilm on the bottom.

12. Microscope for short-term live cell analysis: An Axioplan 2 microscope (Zeiss) controlled with OpenLab software (Improvision) equipped with an Orca CCD camera (Hamamatsu Photonics) and a Zeiss 63× Apochromat objective.
13. Microscope for time lapse image acquisition: a DeltaVision microscope system (Applied Precision Instruments) operated by MetaMorph software (Molecular Devices) assembled around an Axiovert microscope (Zeiss) with a PXL CCD camera (Photometrics, Ltd.) and a Zeiss 63× plan Apochromat objective.

3. Methods

To monitor HSV replication and to observe replication compartments in infected cells, we constructed an HSV-1 recombinant virus **(Fig. 1)** that expressed the single-strand DNA binding protein, ICP8, fused to GFP (*see* **Note 4**). At 3–4 h postinfection, the ICP8-GFP fusion protein is diffusely distributed in the nucleus *(21)*. At 4 h postinfection, individual structures consisting of ICP8-GFP are observed that grow in size and number until the replication compartments nearly fill the entire volume of the nucleus **(Fig. 2)**. When viral DNA replication is inhibited, ICP8-GFP is distributed to prereplicative sites **(Fig. 2)**.

3.1. Preparation of Samples for Immunofluorescence Analysis

1. Sterilize 12 mm round coverslips by ethanol and flame and place into the wells of a 24-well cell culture tray. Handle individual coverslips using fine nose tweezers. Sterilize individual coverslips by dipping each coverslip into ethanol and then immediately passing through the flame of a Bunsen burner. Allow the coverslip to cool briefly and then lightly place into the well. Be cautious and remove all flammable

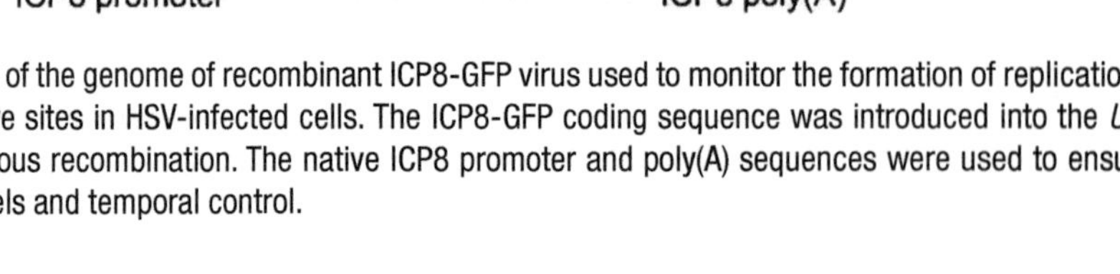

Fig. 1. Diagram of the genome of recombinant ICP8-GFP virus used to monitor the formation of replication compartments or prereplicative sites in HSV-infected cells. The ICP8-GFP coding sequence was introduced into the *U$_L$29* (*ICP8*) locus using homologous recombination. The native ICP8 promoter and poly(A) sequences were used to ensure physiological expression levels and temporal control.

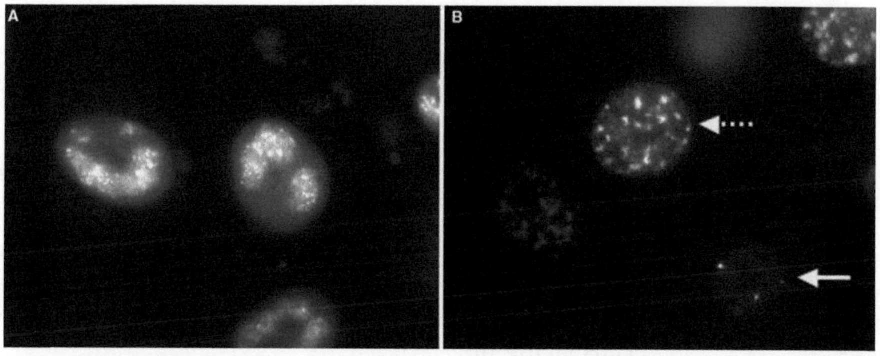

Fig. 2. Distribution of ICP8-GFP in infected cells. Cells were infected in the absence or presence of a viral replication inhibitor, PAA, with a recombinant HSV expressing an ICP8-GFP fusion protein. At 5.5 h postinfection, the cells were fixed to visualize GFP fluorescence. **A** In the absence of PAA, the ICP8-GFP molecule was distributed to viral replication compartments within the nuclei of infected cells. A low level of diffuse intranuclear ICP8-GFP was also observed. **B** In the presence of PAA, the ICP8-GFP was localized to prereplicative sites. The ICP8-GFP is distributed to two different sites: S-phase-dependent prereplicative sites (*hashed arrow*) and S-phase-independent prereplicative sites (*solid arrow*). The S-phase-dependent sites co-localize with regions of cellular DNA synthesis and are not believed to be involved in HSV replication. The S-phase-independent sites form adjacent to ND10 sites and are the precursors to replication compartments.

materials from the area as burning ethanol may drip from the coverslip.

2. Split cells into the 24-well trays the day before so that the cells are 90–95% confluent before use.
3. Infect cells at a multiplicity of infection (moi) of 20. Prepare recombinant GFP virus inoculum by rapid thaw of the viral stock vial in a 37 °C water bath and then dilute into a sufficient volume of ice-cold viral dilution media so that each well receives 200 μL of inoculum. Keep the inoculum on ice until added to the cells.
4. Remove the old media from the cells, and then wash the cells one time with PBS.
5. Add the diluted inoculum to the cells and then incubate at 37 °C for 1 h. Leave at least four wells as mock-infected controls containing only viral dilution media without virus (*see* **Note 5**).
6. Remove the inoculum, wash the cells briefly with PBS, and then overlay with 0.5 mL DMEM + 1% serum. Incubate at 37 °C. If you wish to observe prereplicative sites instead of replication compartments, add the viral DNA synthesis inhibitor, PAA, to wells at a final concentration of 400 μg/mL.
7. At 5.5 h postinfection, remove the coverslips using tweezers and place into a Chen coverslip rack in a glass dish containing enough ice-cold PBS to completely submerge the coverslips.

During the wash, fix, and permeabilization stages, handle the coverslips in the Chen rack.

8. Fix the cells for 10 min in 3.7% formaldehyde in PBS and then wash for 5 min in ice-cold PBS.
9. Permeabilize the cells in –20 °C acetone for 2 min and then wash with distilled water for 5 min.
10. Using tweezers, remove the coverslips from the Chen rack, lightly dab on a paper towel to remove excess moisture, and then place on the Parafilm in the humid chamber with the cells facing up.
11. Add 20 μL of the diluted primary antibodies 39S (1:200), 5H11D6 (1:50), and 58S (1:50) to the surface of the coverslip. Incubate in a humid chamber at 37 °C for 30 min.
12. Using tweezers, place the coverslips into the Chen rack in a glass dish containing enough ice-cold PBS to submerge the coverslips. Wash three times in PBS for 5 min each.
13. Repeat **steps 10** and **11,** but add 20 μL of diluted rhodamine-conjugated secondary antibody (1:200). Wash as described in step 12. Be cautious to limit light exposure once the fluorescent dye is added to minimize photobleaching.
14. Place a drop of prewarmed gelatin mounting media on a coverslip and then invert the coverslip onto the gelatin. Apply light pressure to the top surface of the coverslip to displace the air bubbles. The gelatin contains 1.3 mg/mL *p*-phenyldiamine as an antibleaching agent.
15. View GFP and the rhodamine fluorochrome using the appropriate filter sets. Overlay images of the GFP and rhodamine signal are used to determine co-localization (**Fig. 3**).
16. If unable to immediately view the samples, place them in a light tight container or slide folder at 4 °C.

3.2. Preparation of Samples for Short-Term Cultured Cell Analysis

1. Sterilize 22 × 22 mm glass coverslips as described in **Subheading 3.1.1**. Place coverslips into a 6-well tray.
2. Split cells into a 6-well tray the previous day so that the cells are 90–95% confluent before use.
3. Infect cells at an moi of 20. Prepare the viral inoculum with the recombinant GFP virus by rapidly thawing the viral stock vial in a 37 °C water bath and then dilute into viral dilution media so that each well receives 1 mL. Keep the inoculum on ice until use.
4. Remove the old media from the cells and wash once with ice-cold PBS.
5. Add the viral inoculum to the cells and incubate at 37 °C for 1 h. Leave at least one well uninfected as a mock control containing only the viral dilution media without virus.

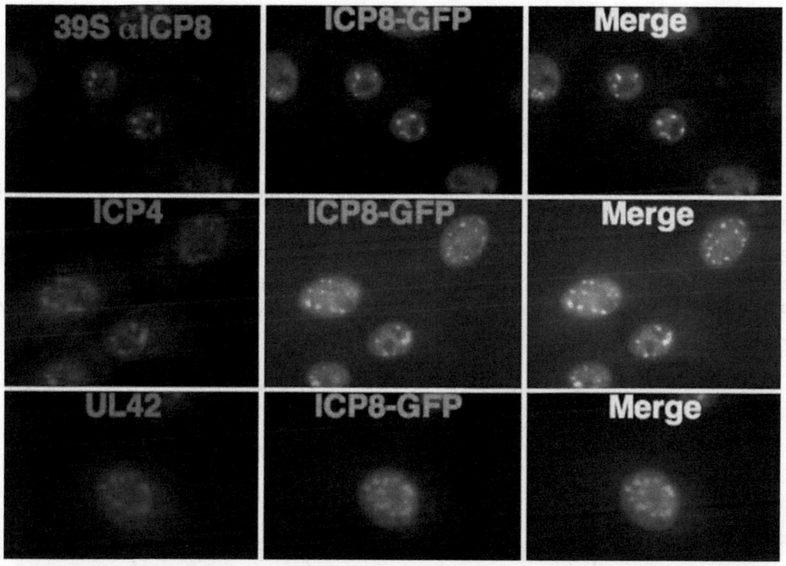

Fig. 3. Reactivity of the ICP8-GFP fusion protein with the 39S conformation-specific monoclonal antibody and co-localization with other HSV proteins typically found in replication compartments. Vero cells were infected with the recombinant HSV ICP8-GFP virus and processed for immunofluorescence as described. ICP8-GFP co-localized with both U_L42 and ICP4 indicating that ICP8-GFP was present in functional replication compartments and was not distributed to aberrant structures or protein aggregates. Adapted from *(21)*.

6. Remove the inoculum, wash the cells once with PBS, and then overlay with fresh 2 mL DMEM + 1% serum. Incubate at 37 °C. If you wish to observe prereplicative sites instead of replication compartments, add the viral DNA synthesis inhibitor, PAA, to the wells at a final concentration of 400 μg/mL.
7. At various times postinfection (4, 5.5, or 6.5 h postinfection), process cells for microscopy. Using tweezers, carefully remove the coverslip from the well, dip into PBS, and quickly dab on a paper towel to remove excess liquid. Lightly invert the coverslip onto a microscope slide. Remove excess liquid on the surface of the coverslip by laying a Kimwipe tissue on the coverslip, allowing it to absorb most of the moisture and then gently remove. Add a small drop of oil to the surface of the coverslip and then observe under a 63× objective. There is enough fluid trapped under the coverslip to allow 3–5 min of viewing and image acquisition for each coverslip **(Fig. 2)**.

3.3. Preparation of Samples for Time-Lapse Analysis

1. Sterilize 18 mm round glass coverslips as described in **Subheading 3.1.1**. Place coverslips into a 6-well tray.
2. Split cells into a 6-well tray the previous day so that the cells are 90–95% confluent before use.
3. Infect cells as described above in **steps 3.2.3–3.2.6.**

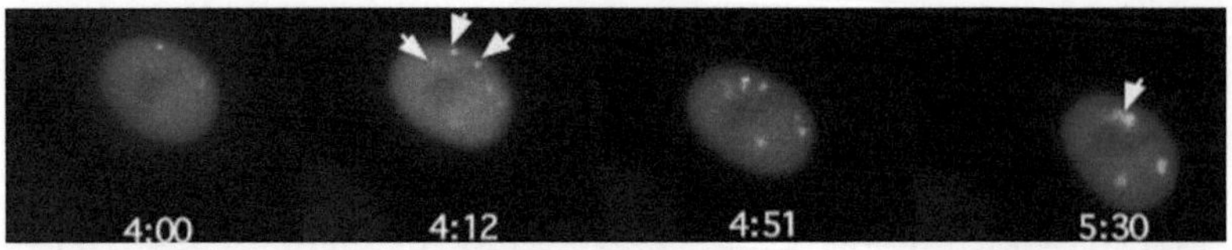

Fig. 4. Time-lapse analysis of replication compartment formation in infected cells. Cells infected with the ICP8-GFP recombinant virus were visualized starting at 4 h postinfection with images acquired at 3-min intervals. The composite images show selected images at the indicated times postinfection (hour:minute). The compartments initiate at distinct individuals sites (4:00) and enlarge as time progresses. The structures are dynamic with some merging to form larger compartments. For example, the three distinct sites seen at 4:12 (*arrows*) eventually merge to form the one compartment seen at 5:30 (*arrow*). Adapted from *(21)*.

4. At 5 h postinfection, remove the glass coverslip and assemble it into the microscope stage prewarmed to 37 °C. Use the CO_2-independent media during the image acquisition period. For best results, the microscope should be on an air table to isolate the microscope from room vibration. Acquire phase and fluorescent images every 3 min over a period of 60–90 min (**Fig. 4**).

Notes

1. Heat inactivation of the serum at 65 °C for 20 min destroys serum complement proteins that inactivate the virus. Once a 500-mL bottle of serum is inactivated, it can be aliquoted into 50 mL conical tubes and stored at −20 °C.
2. We used two different methods to provide wild-type protein function in *trans* to complement any possible defects: co-infection with wild-type virus and a complementing cell line that expressed ICP8 upon infection with HSV. We preferred the use of Vero-derived complementing cell lines for repeatability and consistency issues. The Vero-derived complementing cells lines are naturally inducible with ICP8 detected only after infection with HSV. Co-infection is recommended if using a new cell or tissue type or in vivo infections.
3. Cold acetone must be properly stored in an explosion-proof freezer. The acetone can be reused a number of times before being discarded appropriately as chemical waste.
4. The choice of the protein and the location of the GFP tag must be considered, as well as the addition of a flexible linker consisting of 10–15 amino acid residues. For example, we constructed both N- and C-terminal ICP8-GFP fusion proteins; however, the N-terminal GFP fusion formed aggregates

in transfection experiments and did not localize to replication compartments after superinfection with wild-type HSV-1. We used the native ICP8 promoter to construct the recombinant HSV expressing the ICP-8GFP fusion protein because it was more physiological in respect to temporal control and expression levels. For transfection experiments, a plasmid construct with a CMV promoter/enhancer element can be used; however, it is absolutely critical to titrate the amount of plasmid used to prevent artifact formation and cell death due to overexpression. We assessed the functionality of the fusion protein in several ways *(21, 22)*:

- distribution in transfected cells with or without HSV superinfection
- ability to complement the replication of an ICP8-deficient mutant virus
- ICP8-GFP recombinant virus growth kinetics in a single-cycle viral yield assay
- ability to bind UL9, which is known to interact with ICP8
- protein conformation using the anti-ICP8 conformation-specific 39S monoclonal antibody.

5. Titrate antibodies to determine the appropriate dilution required to maximize signal strength and to minimize background fluorescence. Also, include primary and secondary antibody only controls to determine background fluorescence levels.

References

1. Quinlan, M. P., Chen, L. B., and Knipe, D. M. (1984) The intranuclear location of a herpes simplex virus DNA-binding protein is determined by the status of viral DNA replication. *Cell* 36, 857–68.
2. Bush, M., Yager, D. R., Gao, M., Weisshart, K., Marcy, A. I., Coen, D. M., and Knipe, D. M. (1991) Correct intranuclear localization of herpes simplex virus DNA polymerase requires the viral ICP8 DNA-binding protein. *J Virol* 65, 1082–9.
3. Ruyechan, W. T. (1983) The major herpes simplex virus DNA-binding protein holds single-stranded DNA in an extended configuration. *J Virol* 46, 661–6.
4. Olivo, P. D., Nelson, N. J., and Challberg, M. D. (1989) Herpes simplex virus type 1 gene products required for DNA replication: identification and overexpression. *J Virol* 63, 196–204.
5. Liptak, L. M., Uprichard, S. L., and Knipe, D. M. (1996) Functional order of assembly of herpes simplex virus DNA replication proteins into prereplicative site structures. *J Virol* 70, 1759–67.
6. Lukonis, C. J., and Weller, S. K. (1996) Characterization of nuclear structures in cells infected with herpes simplex virus type 1 in the absence of viral DNA replication. *J Virol* 70, 1751–8.
7. Goodrich, L. D., Schaffer, P. A., Dorsky, D. I., Crumpacker, C. S., and Parris, D. S. (1990) Localization of the herpes simplex virus type 1 65-kilodalton DNA-binding protein and DNA polymerase in the presence and absence of viral DNA synthesis. *J Virol* 64, 5738–49.
8. Taylor, T. J., and Knipe, D. M. (2004) Proteomics of herpes simplex virus replication compartments: association of cellular DNA replication, repair, recombination, and chromatin remodeling proteins with ICP8. *J Virol* 78, 5856–66.
9. Wilcock, D., and Lane, D. P. (1991) Localization of p53, retinoblastoma and host replication

proteins at sites of viral replication in herpes-infected cells. *Nature* 349, 429–31.

10. de Bruyn Kops, A., and Knipe, D. M. (1988) Formation of DNA replication structures in herpes virus-infected cells requires a viral DNA binding protein. *Cell* 55, 857–68.
11. Uprichard, S. L., and Knipe, D. M. (1997) Assembly of herpes simplex virus replication proteins at two distinct intranuclear sites. *Virology* 229, 113–25.
12. Ishov, A. M., and Maul, G. G. (1996) The periphery of nuclear domain 10 (ND10) as site of DNA virus deposition. *J Cell Biol* 134, 815–26.
13. Maul, G. G., Ishov, A. M., and Everett, R. D. (1996) Nuclear domain 10 as preexisting potential replication start sites of herpes simplex virus type-1. *Virology* 217, 67–75.
14. Lukonis, C. J., Burkham, J., and Weller, S. K. (1997) Herpes simplex virus type 1 prereplicative sites are a heterogeneous population: only a subset are likely to be precursors to replication compartments. *J Virol* 71, 4771–81.
15. de Bruyn Kops, A., and Knipe, D. M. (1994) Preexisting nuclear architecture defines the intranuclear location of herpesvirus DNA replication structures. *J Virol* 68, 3512–26.
16. Dodson, A. W., Taylor, T. J., Knipe, D. M., and Coen, D. M. (2007) Inhibitors of the sodium potassium ATPase that impair herpes simplex virus replication identified via a chemical screening approach. In Press, *Virology* (in press).
17. Da Costa, X., Kramer, M. F., Zhu, J., Brockman, M. A., and Knipe, D. M. (2000) Construction, phenotypic analysis, and immunogenicity of a UL5/UL29 double deletion mutant of herpes simplex virus 2. *J Virol* 74, 7963–71.
18. Showalter, S. D., Zweig, M., and Hampar, B. (1981) Monoclonal antibodies to herpes simplex virus type 1 proteins, including the immediate-early protein ICP 4. *Infect Immun* 34, 684–92.
19. Uprichard, S. L., and Knipe, D. M. (2003) Conformational changes in the herpes simplex virus ICP8 DNA-binding protein coincident with assembly in viral replication structures. *J Virol* 77, 7467–76.
20. Gao, M., DiTusa, S. F., and Cordingley, M. G. (1993) The C-terminal third of UL42, a HSV-1 DNA replication protein, is dispensable for viral growth. *Virology* 194, 647–53.
21. Taylor, T. J., McNamee, E. E., Day, C., and Knipe, D. M. (2003) Herpes simplex virus replication compartments can form by coalescence of smaller compartments. *Virology* 309, 232–47.
22. Taylor, T. J., and Knipe, D. M. (2003) C-terminal region of herpes simplex virus ICP8 protein needed for intranuclear localization. *Virology* 309, 219–31.

Chapter 17

Modified Bovine Herpesvirus 1 for Protein Secretion

Günther M. Keil

Summary

The traditional way to utilize bovine herpesvirus 1 (BHV-1) and many other herpesviruses as vectors for synthesis of heterologous proteins like reporter proteins, antigens, or immunomodulatory active molecules was (and still is) the expression of the protein of interest from an entire gene consisting of promoter, 5′- and 3′-noncoding regions, the open reading frame (ORF), and a signal sequence for polyadenylation. This approach is doubtlessly appropriate especially in cases when expression of large proteins or of proteins that do not enter the secretory pathway is envisaged. My laboratory has developed an alternative expression strategy for secreted proteins and peptides that uses the essential BHV-1 glycoprotein B (gB) as transporter for a cargo protein that is embedded in gB as a furin-excisable polypeptide that is released from the gB precursor molecule in the *trans*-Golgi network by the ubiquitously present endoprotease furin. The general applicability of this novel expression strategy is demonstrated by using GFP as reporter protein to monitor secretion. We hypothesize that also other secreted or membrane-bound (glyco)proteins can be engineered to function as transporters for oligopeptides and also more complex larger proteins.

Key words: Bovine Herpesvirus 1, Glycoprotein B, Furin cleavage, gB-embedded protein, Furin-excisable GFP, Protein secretion.

1. Introduction

Bovine herpesvirus 1 (BHV-1), a member of the *Alphaherpesvirinae* subfamily of the family *Herpesviridae* has been frequently used to express heterologous (reporter) proteins to study, for example, the effect of specific viral gene deletion mutants *(1–3)*, the development of bivalent vaccines *(4–7)*, or the improvement of vaccine efficacy by expressing bovine cytokines *(8, 9)*. Initially, mainly the open reading frames (ORFs) encoding β-galactosidase or chloramphenicol acetyltransferase (CAT) were used to identify gene deletion mutants or to define BHV-1 gene loci

Barry W. Hicks (ed.), *Methods in Molecular Biology, Viral Applications of Green Fluorescent Protein, vol. 515*

DOI: 10.1007/978-1-59745-559-6_17

suitable for the integration of foreign ORFs *(2, 3, 10)*. These approaches required incubation of infected cell cultures under a BluoGal-containing agarose overlay which led to appearance of blue-staining cells 3–4 days after infection with recombinant virus or laborious determination of CAT activity in infected-cell extracts, respectively. Later, the time needed for identification, isolation, plaque purification, and infectious titer determination was significantly reduced by expressing the green fluorescent protein (GFP). The protein fluorescence is detectable a few hours after infection provided the human cytomegalovirus major immediate early enhancer/promoter (HCMVie) or the murine cytomegalovirus immediate early 1 enhancer/promoter (MCMVie1) was used to direct expression of the GFP ORF. GFP fluorescence was also used to show that fusion proteins containing the type II glycoprotein membrane anchor of the bovine respiratory syncytial virus (BRSV) attachment protein G can be incorporated into the envelope of extracellular BHV-1 virions which in addition demonstrated that GFP fluorescence is not impeded by protein modifications that might occur during intracellular transport from the endoplasmic reticulum into the *trans*-Golgi network *(11)*. Most of the above-mentioned transgene expression attempts required integration into the BHV-1 genome of an expression cassette consisting of a heterologous promoter, the ORF of interest, and a signal sequence for polyadenylation. In the absence of a positive selection marker, identification and isolation of the envisaged recombinants had to be done by dot blot screening of hundreds of single plaques of the virus progeny obtained after cotransfection of purified viral DNA and the respective recombination plasmids which was time consuming and laborious. We discovered that the fusion protein of BRSV, a class I glycoprotein from which a subsequently secreted intervening peptide is excised in the *trans*-Golgi network by furin *(12)* which cleaves after the consensus sequence RX(K/R)R *(13)* can be used as transporter for secreted peptides and proteins *(14)*. We also developed a similar transporter system using BHV-1 glycoprotein B (gB) that is only cleaved once by furin in its wild-type form *(15)*. BHV-1 gB is an abundant protein of the viral envelope. It is essential for membrane fusion processes during entry of BHV-1 into target cells and directs spreading from cell to cell. Mutant viruses that lack gB need gB-expressing complementing cells for infectious virus replication. Thus, infectious progeny of cotransfections with purified DNA of gB-negative BHV-1 with plasmids containing modified gB ORFs into normal cells can only be generated after recombination of the modified ORF into the gB^-BHV-1 genome *(15)*. To test whether introduction of a foreign protein sequence and a second furin cleavage site into gB is compatible with gB function for membrane fusion processes, we used GFP as cargo protein. Plaques with GFP fluorescence began to appear within 2 days of cotransfection of gB⁻BHV-1 DNA and the recombination plasmid pgB-Asc/2FuGFP which indicated that the integration of the second furin cleavage site and GFP did not impede infectious BHV-1 replication. Detailed analyses demonstrated that furin efficiently excises GFP out of gB and it is subsequently released into the culture medium. In further application of this approach, we successfully utilized bovine alpha interferon *(15)* and bovine lactoferricin as cargo proteins to study applicability for small peptides (Sadowski and Keil, unpublished observations). Although not yet demonstrated, we assume that this novel method for expression of secreted proteins and polypeptides can be transferred to other furin-cleaved (glyco) proteins.

2. Materials

2.1. PCR Reactions

1. FastStart Taq DNA polymerase including 10× reaction buffer (Roche, Mannheim, Germany) and 25 mM $MgCl_2$.
2. Dimethyl sulfoxide (DMSO) (Sigma, Munich, Germany), stored in aliquots at −20 °C (*see* **Note 1**).
3. dNTP set, PCR grade (Roche).
4. Primers (MWG, Ebersberg, Germany). The *Asc*I recognition sequence is given in italics, sequences derived from the GFP ORF are printed in bold.

   FuGFP+: 5′-TAA*GGCGCGCC*GC**ATGGTGAGCAAGGGCGAGGAGCTG**-3′ and

   FuGFP−: 5′-TAA*GGCGCGCC*CG**TGTACAGCTCGTCCATGCC**-3′.
5. Plasmid pEGFP-N1 (Clontech, Palo Alto, CA).
6. PCR grade water (Roche).
7. Thermocycler Eppendorf Mastercycler Gradient.
8. QIAquick PCR Fragment Purification kit (Qiagen, Hilden, Germany).

2.2. Agarose Gel Electrophoresis

1. Tris-acetate buffer (50×): 2 M Tris, 0.25 M sodium acetate, 0.05 M EDTA adjusted to pH 7.8 with acetic acid.
2. Agarose for gel electrophoresis (Invitrogen, Karlsruhe, Germany).
3. Ethidium bromide (Sigma), solution of 10 mg/mL in water (*see* **Notes 2** and **3**).
4. Electrophoresis buffer for agarose gel electrophoresis is 1× tris-acetate, 0.1 μg/mL ethidium bromide in water.

5. DNA loading buffer: 40% sucrose, 1 mM EDTA pH 7.0, 0.05% bromophenol blue, 0.1% SDS.
6. 1 kbp DNA ladder (Gibco BRL, Berlin, Germany).
7. Agarose gel electrophoresis equipment.
8. UV-transilluminator (254 and 302 nm).

2.3. Cloning Procedures

1. Restriction enzymes *Asc*I, *Bsr*GI, and *Nco*I including 10× reaction buffers (NEB, Frankfurt, Germany).
2. Plasmid pgB-Asc in which the codons for the furin recognition sequence were modified (**Fig. 1**) to create an *Asc*I cleavage site (available from the author).
3. Calf intestine alkaline phosphatase (CIP) including 10× reaction buffer (Roche).
4. Phage T4 DNA ligase including 10× reaction buffer (Roche).
5. QIAquick Gel Extraction kit (Qiagen, Hilden, Germany).
6. 60 mM ethylene glycol tetraacetic acid (EGTA), pH 7.0 (Serva, Heidelberg, Germany).
7. TE buffer: 10 mM Tris, 1 mM EDTA, pH 7.5.
8. TE-saturated phenol (Roth, Karlsruhe, Germany).
9. Chloroform-isoamyl alcohol 24:1 (v/v) (Chloro-Iso).
10. 3 M potassium acetate, adjusted to pH 5.5 with acetic acid.
11. 10× TA for ligation: 330 mM Tris, 660 mM potassium acetate, 100 mM magnesium acetate, adjusted to pH 7.9 with acetic acid, 1 mg/mL bovine serum albumin (BSA), 5 mM dithiothreitol.
12. Chemically competent bacteria, commercially available from different suppliers.
13. LB-medium: dissolve 10 g Bacto-Tryptone, 5 g Bacto Yeast-Extract, and 8 g NaCl/L. For petri dishes, add 15 g Bacto-Agar.

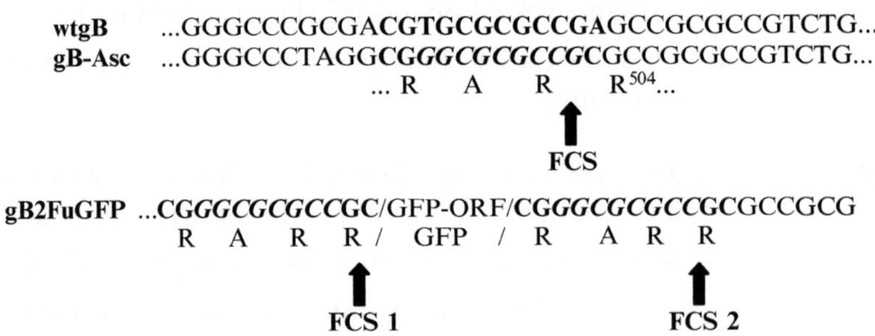

Fig. 1. Sequences encoding the furin cleavage site (*FCS*) in wild-type (*wt*) gB and after modification to contain an *Asc*I recognition sequence (gB-Asc) are typed in *bold*. The recognition sequence for AscI is in *italics* and the amino acid sequence is given in *one-letter code*. For gB2FuGFP, the sequences flanking the GFP ORF and GFP are indicated accordingly.

Autoclave. When required, after cooling to 56 °C, add ampicillin to a final concentration of 100 μg/mL.

14. Qiagen Plasmid Mini kit and Plasmid Midi kit (Qiagen, Hilden, Germany).

2.4. Generation of Recombinant BHV-1

1. Madin-Darby bovine kidney (MDBK) cells: Collection of Cell Cultures in Veterinary Medicine (FLI, Greifswald-Insel Riems, Germany). Incubate cell cultures at 37 °C in a humidified atmosphere containing 5% CO_2.
2. Dulbecco's modified Eagle's medium (DMEM) supplemented with 10% fetal calf serum (FCS) (Gibco-Invitrogen, Karlsruhe, Germany) and 2.4 mM L-glutamine, 100 U/mL penicillin, and 100 μg/mL streptomycin. 2× DMEM supplemented with 10% FCS, 4.8 mM L-glutamine, 200 U/mL penicillin, and 200 μg/mL streptomycin.
3. Cell culture flasks and multi-well plates (Corning Inc., Corning, NY).
4. Mammalian cell transfection kit (Stratagene, La Jolla, CA).
5. Purified DNA of gB-BHV-1 (available from the author).
6. For 500 mL glycerol shock solution, mix 125 g (100 mL) of glycerol (87%) with 250 mL 50 mM HEPES, 1.5 mM Na2HPO4, 280 mM NaCl, pH 7.13 with NaOH, and 150 mL water.
7. 1.2% Seakam Agarose (Marine Colloids Inc., Rockland, Maine) in water. Store this solution at room temperature after autoclaving.
8. Agarose overlay medium for plaque purification: place fluidized 1.2% Seakam Agarose in a 42 °C water bath. For one 6-well plate, pipette 9 mL of 2× DMEM with supplements into a 50-mL plastic tube. Incubate at 42 °C in a water bath for about 30 min. Add 9 mL of the agarose, mix and leave in the 42 °C water bath until use.

3. Methods

3.1. Synthesis of the GFP ORF Flanked by Sequences Encoding Furin Cleavage Sites (2FuGFP)

1. Set up a PCR reaction mixture consisting of 1 ng of plasmid pEGFP-N1, 5 μL of 10× reaction buffer, 5 μL DMSO, 1 μL 10 mM dNTPs, 1 μL 25 mM $MgCl_2$, 62.5 pM of primers FuGFP+ and FuGFP−, 1.25 Units FastStart Taq DNA polymerase and PCR grade water to a final volume of 50 μL. The primers incorporate the furin cleavage sites.
2. Start the reaction in the thermocycler for 2 min at 95 °C followed by 33 cycles of 95 °C for 30 s and 55 and 72 °C for 1 min and a final extension for 3 min at 72 °C.

3. Check for synthesis of the amplification product of ~750 bp by 0.6% agarose gel electrophoresis.
4. Purify the PCR product using the QIAquick PCR purification kit according to the instructions of the supplier.

3.2. Agarose Gel Electrophoresis

1. I use self-made agarose gel instruments with buffer recirculation. However, the instructions can easily be conferred to other formats. Assemble the gel chamber or casting tray as appropriate.
2. Prepare a 0.6% gel solution by boiling 3 g agarose in 490 mL water in a microwave oven until completely melted and cool to 56 °C in a water bath. Replenish evaporated water (*see* **Note 4**).
3. Add 10 mL 50× tris-acetate buffer and 5 μL of 10 mg/mL ethidium bromide solution and store in a 56 °C water bath until use.
4. Pour the gels according to the specifics of your system (*see* **Note 5**). For examination of the PCR reactions and to control restriction enzyme cleavage reactions, make small (6 cm × 4 cm (l × w)), ~5 mm thick gels. For purification of DNA fragments, make larger, 5 mm thick gels (25 cm × 15 cm (l × w)). Cover the gels with electrophoresis buffer after they have solidified and remove the comb. Pipette the samples into the wells and include one well with the 1 kbp DNA ladder.
5. Run the small gels at 8 V/cm distance between the electrodes for 25 min and the large gels at 4 V/cm distance between the electrodes for 3 h.
6. When the run is completed, place the gel on a UV transilluminator (254 nm for documentation only, 302 nm for excision of fragments) to visualize the DNA fragments. Make a photograph for documentation (*see* **Note 6**).

3.3. Insertion of the 2FuGFP Encoding Fragment into the Recombination Vector, pgB-Asc

1. Add to the purified 2FuGFP PCR product 5 μL of 10× *Asc*I reaction buffer, 1 μL *Asc*I enzyme, and water to a final volume of 50 μL. Incubate for 2 h at 37 °C.
2. Add 10 μL gel loading buffer and pipette 30 μL in each of two wells of a large 0.6% agarose gel. Electrophorese for 3 h with 4 V/cm distance between the electrodes. Place gel on a 302-nm transilluminator (*see* **Note 6**).
3. Cut out the 750 bp fragment with a scalpel or razor blade and isolate the DNA using the QIAquick Gel Extraction kit according to the manufacturer's instructions. Elute the DNA with 30 μL water.
4. For the vector preparation, cleave 5 μg of plasmid pgB-Asc using 5 μL of 10× *Asc*I reaction buffer, 1 μL *Asc*I enzyme, and water to a final volume of 50 μL for 2 h at 37 °C.

5. Add 25 µL 10× CIP buffer, 174 µL ultra-pure water, and 1 µL CIP. Incubate 30 min at 37 °C. Add 1 mL CIP and incubate 30 min at 56 °C. Add 50 µL of 60 mM EGTA and incubate 30 min at 65 °C to inactivate the phosphatase.
6. Add 300 µL of TE-saturated phenol to the DNA sample. Shake vigorously or vortex for 20–30 s. Centrifuge the sample for 2 min at room temperature to separate the phases (*see* **Note 7**).
7. Transfer the aqueous phase into a fresh microcentrifuge tube and add 300 µL of a 1:1 mixture TE-Phenol/Chloro-Iso. Mix thoroughly and centrifuge the sample for 2 min at room temperature to separate the phases.
8. Transfer the aqueous phase into a fresh microcentrifuge tube and add 1 mL Chlor-Iso (*see* **Note 8**). Mix thoroughly and centrifuge the sample for 2 min at room temperature to separate the phases.
9. Transfer the aqueous phase into a fresh microcentrifuge tube and determine the volume. Add 360 µL of TE to 40 µL 3 M sodium acetate, pH 5.5, and 1 mL 100% ethanol. Mix thoroughly and incubate at −70 °C for about 30 min.
10. Pellet the precipitated DNA by centrifugation for 15 min at room temperature and remove the ethanol.
11. Wash the pellet with 70% ethanol and centrifuge for 10 min. Remove the ethanol and dry the pellet for 5–10 min by incubating the open tube at 56 °C.
12. Resuspend the dried pellet in 50 µL of TE by incubation at 56 °C for 15 min.
13. Into a 1.5-mL microcentrifuge tube, pipette 5 µL of the purified vector, 25 µL of the purified PCR product, 5 µL BSA, 5 µL 10× TA, 5 µL 100 mM DTT, 5 mL 10 mM ATP, and 0.1 µL T4 ligase (*see* **Note 9**). As a control, prepare the same ligation mixture but use water instead of the purified PCR product. Incubate for 5 min at 37 °C, 1 h at 25 °C, and over night at 4 °C (*see* **Note 10**).
14. To transform chemically competent bacteria, incubate freshly thawed aliquots on ice for 5 min and add 1 and 10 µL of the ligation mixtures to the recommended amount of bacteria. Incubate for 20 min on ice, 2 min at 42 °C, and again on ice for 5 min.
15. Add 200 µL LB medium per tube, incubate for 1 h at 37 °C, and plate on LB-agar petri dishes containing ampicillin. Incubate overnight at 37 °C and count colonies the next day (*see* **Note 11**).
16. Pick 6–24 (ore even more) colonies (*see* **Note 12**) with sterile pipette tips, sterile toothpicks, or an inoculation loop into

5 mL LB-medium with ampicillin and incubate overnight on a gyratory shaker.

17. Prepare plasmid DNA using the Qiagen Plasmid Mini kit according to the manual.
18. Determine the DNA concentration by UV spectrophotometry at 260 nm (1 OD_{260} corresponds to a DNA concentration of 50 µg/mL).
19. Cleave 0.5 µg plasmid DNA with *Bsr*G1 and *Nco*I for 2 h in the recommended buffer for double cleavage.
20. Separate cleavage products on a large 0.6% agarose gel for 3 h at 4 V/cm. Correct insertion of the PCR product yields fragments of ~4.4, 1.45, 0.4, and 0.3 kbp. Plasmids in which the PCR product was inserted in the wrong orientation are cleaved into fragments of ~4.4, 1.05, 0.7, and 0.3 kbp.
21. Inoculate 50 mL LB-medium with ampicillin with 100 µL of the appropriate original culture and incubate overnight on a gyratory shaker.
22. Prepare plasmid DNA with the Qiagen Plasmid Midi kit according to the manual. Determine the DNA concentration by UV spectrophotometry and verify identity and purity of the preparation of pgB-Asc/2FuGFP by cleavage of 0.5 µg plasmid DNA with *Bsr*GI and *Nco*I for 2 h followed by 0.6% agarose gel electrophoresis.

3.4. Integration of the gB-2FuGFP Orf into the Genome of gB⁻ BHV-1

1. Split maintenance cultures of MDBK cells 1:4 on 35-mm cell culture dishes 1 day before transfection. Cultures should be nearly confluent.
2. In a sterile 1.5-mL microcentrifuge tube, pipette 1 µg of purified gB⁻BHV-1 genomic DNA and 5 µg of pgB-Asc/2FuGFP and sterile water to a final volume of 0.75 mL. Mix the solution by gently vortexing. Add 8.3 µL solution 1 (calcium) of the mammalian cell transfection kit, and again mix by vortexing. Finally, add 83.3 µL solution 2 (phosphate) and mix by gently vortexing the solution. Let the solution stand at room temperature for 10–20 min.
3. Remove the cell culture medium from the MDBK cells and add 2 mL fresh medium. Add the transfection mixture dropwise to the culture and gently swivel the dish to evenly distribute the calcium phosphate/DNA precipitates. Incubate the culture for 4 h at 37 °C.
4. Remove the medium and add 2 mL glycerol shock solution for 2 min (*see* **Note 13**). Remove the glycerol shock

solution and wash culture two times with 3 mL of fresh medium. Add 3 mL medium and incubate at 37 °C.

5. Monitor culture after 2–3 days for presence of cell foci or plaques that exhibit GFP fluorescence. Their appearance indicates successful generation of recombinant BHV-1 since only virus that had acquired the modified gB from the plasmid can produce infectious progeny (*see* **Note 14**).
6. Harvest culture supernatant together with the cells 4–5 days after transfection (*see* **Note 15**). Freeze at −70 °C, thaw, and clarify by low speed centrifugation at ~2,000 × *g*. For single plaque purification, make serial tenfold dilutions until 10^{-5}. Inoculate MDBK cells in 6-well dishes, split 1:4 the day before with 100 mL of the undiluted and diluted virus isolate in a total volume of 1 mL medium/well. Incubate cells with viral medium for 2–4 h at 37 °C, then remove the medium and overlay with 3 mL of 0.6% agarose medium (*see* **Note 16**).
7. Three to four days later, harvest 2–4 fluorescing plaques from wells inoculated with the highest dilution by aspirating the cells and the overlaying agarose plug with a 1-mL Gilson P1000 pipette adjusted to 50 μL. Transfer cell plaque and agarose to ~10^6 freshly trypsinized MDBK cells in T25 cell culture flasks.
8. When the CPE comes to completion, harvest medium and detached cells and store this primary stock of BHV-1/gB2FuGFP at −70 °C in aliquots.
9. Prepare working stock by inoculating MDBK cells 2 days after a 1:6 split with 100 mL of the primary stock and harvest detached cells and medium when the CPE is complete. Store the working stock in aliquots at −70 °C (*see* **Note 17**). Determine titer of the working stock.

3.5. Monitoring Secretion of GFP from Cells Infected with BHV-1/gB2FuGFP

1. Infect MDBK cells in 24-well plates with BHV-1/gB2FuGFP at a multiplicity of infection (M.O.I.) of 10 one day after a 1:4 split.
2. Wash cells two times with medium at 1 h after infection and add 250 μL medium per well.
3. Harvest culture supernatant at different times after infection (e.g., as in the example shown in **Fig. 2** at 6, 8, 10, 12, 14, and 16 h) and clarify by low speed centrifugation at 800 × μ.
4. Store clarified supernatant at −20 °C until all samples are harvested. Pipette 200 μL of each sample into 96-well cell culture plates and determine fluorescence with a fluorescence scanner.

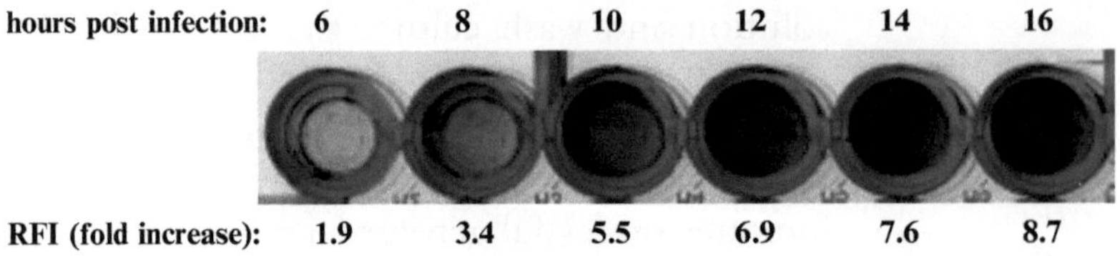

Fig. 2. Time course for secretion of green fluorescent protein (*GFP*) from cells infected with BHV-1/gB2FuGFP. MDBK cells were infected with the recombinant at a multiplicity of infection (*M.O.I.*) of 10. The inoculum was removed 1 h after infection, washed twice with medium, and culture supernatant was harvested at the times indicated. GFP fluorescence was quantitated with a Fuji FLA-3000 fluorescence scanner and relative fluorescence intensity (RFI) is given as fold induction over the background value of the culture supernatant from infected cells harvested at 1 h after infection. The result shows that secreted GFP is found already at 6 h after infection and that the amount continuously increases until 16 h after infection.

Notes

1. Melt the DMSO aliquot at 37 °C and put it on ice after use. If the DMSO does not solidify, discard it.
2. Solid ethidium bromide is highly toxic when inhaled. Ethidium bromide is mutagen. Thus, solutions should be regarded as hazardous. Wear gloves while handling.
3. When referred to water for solutions, buffers, and so on, then use ultra pure water with a conductivity of <0.06 μS/cm.
4. Do not use Erlenmeyer flasks to prepare agarose gels because superheating may cause elution of the agarose solution due to a chimney effect. Using a glass beaker is safer.
5. Air bubbles on the gel surface are an only aesthetical problem. They can be easily removed with the flame of a pocket lighter or a Bunsen burner.
6. Short-wave UV light is hazardous for your eyes. Wear safety goggles or a face shield when examining or cutting out DNA fragments on a transilluminator to prevent damage to the eyes.
7. For centrifugation use full speed of your microcentrifuge unless otherwise stated.
8. I use this relatively large amount of Chloro-Iso to more efficiently remove residual phenol from the aqueous phase. The erstwhile used ether for this purpose is not recommended.
9. Use of TA for ligation stems from the ancient times of cloning but gives a better cloning efficiency than the ligation buffer provided with the enzyme – at least in my hands.
10. Incubation at 4 °C over night originates also from the ancient times. Again, I have a better efficiency than by using 16 °C as frequently recommended.

11. The number of colonies from the ligations without insert should be less than 1/10 of the number of colonies from the ligations with insert. Otherwise it might be that the phosphatase treatment of the vector was insufficient.
12. The ratio between the number of colonies obtained from the ligations with and without insert determines the number of clones to tested. If the ratio is below 3, I test 24 colonies to have a good chance to obtain a plasmid with the correct orientation of the insert. At higher ratios, I reduce the number tested colonies down to 6 at ratios over 10.
13. Glycerol shock is essential for successful recovery of BHV-1 from transfected DNA.
14. Single fluorescing cells are not indicative for successful recombination because gB2FuGFP expression can be induced from the plasmid within the transfected cells.
15. If plaques appear only hesitant, it may be necessary to trypsinize the culture and reseed the cells into a 25-cm^2 tissue culture flask for preparing the primary stock.
16. For pipetting the agarose overlay medium, do not pass glass pipettes through a Bunsen Burner flame because the agarose medium may become too hot and kill the cells.
17. BHV-1 should not be stored at −20 °C because infectivity declines very rapidly. The virus is stable at −70 °C. BHV-1 is also relatively resistant to repeated freezing and thawing. No significant titer reduction was observed after 10 freeze/thaw cycles with wild-type virus.

Acknowledgments

I thank Katrin Giesow for expert technical assistance. This work was supported by the European Union Network of Excellence EPIZONE (Contract No FOOD-CT-2006-016236).

References

1. Liang, X., Babiuk, L. A., and Zamb, T. J. (1992) An in vivo study of a glycoprotein gIII-negative bovine herpesvirus 1 (BHV-1) mutant expressing beta-galactosidase: evaluation of the role of gIII in virus infectivity and its use as a vector for mucosal immunization. *Virology* 189, 629–639.
2. Fehler, F., Herrmann, J., Saalmüller, A., Mettenleiter, T. C., and Keil, G. M. (1992) Glycoprotein IV of bovine herpesvirus 1-expressing cell line complements and rescues a conditionally lethal viral mutant. *J. Virol.* 66, 831–839.
3. Chowdhury, S. I., Ross, C. S., Lee, B. J., Hall, V., and Chu, H. J. (1999) Construction and characterization of a glycoprotein E gene-deleted bovine herpesvirus type 1 recombinant. *Am. J. Vet. Res.* 60, 227–232.

4. Kit, M., Kit, S., Little, S. P., Di Machi, R. D., and Gale, C. (1991) Bovine herpesvirus-1 (infectious bovine rhinotracheitis virus)-based viral vector which expresses foot-and-mouth disease epitopes. *Vaccine* 9, 564–572.
5. Kit, S., Otsuka, H., and Kit, M. (1992) Expression of porcine pseudorabies virus genes by a bovine herpesvirus-1 (infectious bovine rhinotracheitis virus) vector. *Arch. Virol.* 124, 1–20.
6. Taylor, G., Rijsewijk, F. A., Thomas, L. H., Wyld, S. G., Gaddum, R. M., Cook, R. S., Morrison, W. I., Hensen, E., van Oirschot, J. T., and Keil, G. M. (1998) Resistance to bovine respiratory syncytial virus (BRSV) induced in calves by a recombinant bovine herpesvirus-1 expressing the attachment glycoprotein of BRSV. *J. Gen. Virol.* 79, 1759–1767.
7. Schmitt, J., Becher, P., Thiel, H.-J., and Keil, G. M. (1999) Expression of bovine viral diarrhoea virus glycoprotein E2 by bovine herpesvirus-1 from a synthetic ORF and incorporation of E2 into recombinant virions. *J. Gen. Virol.* 80, 2839–2848.
8. Raggo, C., Fitzpatrick, C. R., Babiuk, L. A., and Liang, X. (1992) Expression of bovine interleukin-1 beta in a bovine herpesvirus-1 vector: in vitro analysis. *Virology* 221, 78–86.
9. König, P., Beer, M., Makoschey, B., Teifke, J. P., Polster, U., Giesow, K., and Keil, G. M. (2003) Recombinant virus-expressed bovine cytokines do not improve efficacy of a bovine herpesvirus 1 marker vaccine strain. *Vaccine* 22, 202–212.
10. Bello, L. J., Whitbeck, J. C., and Lawrence, W. C. (1992) Bovine herpesvirus 1 as a live virus vector for expression of foreign genes. *Virology* 199, 666–673.
11. Keil, G. M. (2000) Fusion of the green fluorescent protein to amino acids 1 to 71 of bovine respiratory syncytial virus glycoprotein G directs the hybrid polypeptide as a class II membrane protein into the envelope of recombinant bovine herpesvirus-1. *J. Gen. Virol.* 81, 1051–1055.
12. Zimmer, G., Rohn, M., McGregor, G. P., Schemann, M., Conzelmann, K. K., and Herrler, G. (2003) Virokinin, a bioactive peptide of the tachykinin family, is released from the fusion protein of bovine respiratory syncytial virus. *J. Biol. Chem.* 278, 46854–46861.
13. Klenk, H. D., and Garten, W. (1994) Host cell proteases controlling virus pathogenicity. *Trends Microbiol.* 2, 39–43.
14. König, P., Giesow, K., Schuldt, K., Buchholz, U. J., and Keil, G. M. (2004) A novel protein expression strategy using recombinant bovine respiratory syncytial virus (BRSV): modifications of the peptide sequence between the two furin cleavage sites of the BRSV fusion protein yield secreted proteins, but affect processing and function of the BRSV fusion protein. *J. Gen. Virol.* 85, 1815–1824.
15. Keil, G. M., Höhle, C., Giesow, K., and König, P. (2005) Engineering glycoprotein B of bovine herpesvirus 1 to function as transporter for secreted proteins: a new protein expression approach. *J. Virol.* 79, 791–799.

Chapter 18

Identification of Viral Peptide Fragments for Vaccine Development

Zhanglin Lin, Shuang Li, and Yong Chen

Summary

We report a simple method for identifying foldable viral surface protein fragments in a random but systematic manner. The method involves digestion and reassembly of a target gene to generate a pool of smaller DNA fragments with random ends but controllable lengths, followed by screening for foldable fragments using green fluorescent protein (GFP) as a folding reporter. The surface glycoproteins of SARS-CoV and HIV-1 were used as model proteins. Two foldable fragments for SARS-CoV spike protein were identified, which coincide with various anti-SARS-CoV peptides. A similar treatment of the HIV-1 gp120 yielded a number of fragments that are associated with the critical CD4 binding site, or the partially buried CCR5 binding site of the protein. The random dissection approach described here should be applicable to other viral proteins for isolating soluble viral surface protein fragments, and may provide alternatives to the full-length proteins (subunits) or linear short peptides in search for antigen or vaccine candidates.

Key words: Vaccine development, Peptide fragment, Surface glycoprotein, SARS, CoV, HIV-1, gp120.

1. Introduction

The entry of viruses into target cells depends on sequential interactions of the viral surface proteins with the cellular receptors, and thus these surface proteins are important for developing antiviral treatments *(1)*. However, it is often difficult to express viral surface proteins in facile recombinant hosts such as *Escherichia coli* (*E. coli*), and this hinders antibody or vaccine development. We set out to find a way to identify smaller but soluble viral surface protein fragments for use in screening for possible antigen or vaccine candidates. Compared with linear peptides derived from

Barry W. Hicks (ed.), *Methods in Molecular Biology, Viral Applications of Green Fluorescent Protein, vol. 515*

DOI: 10.1007/978-1-59745-559-6_18

the surface proteins, soluble fragments may be advantageous as they have the potential to provide discontinuous epitopes. Additionally, it has been suggested that the binding of viral surface proteins to cell receptors may expose sites in the viral proteins otherwise inaccessible *(2, 3)*. In particular, the binding of HIV gp120 to CD4 seems to expose a site that is critical for subsequent binding to the CCR receptor, but which is partially buried in the resting state of the HIV surface protein. This has been implicated as a possible mechanism adopted by the virulent virus to evade the immune surveillance *(2, 3)*. We reasoned that by producing discrete foldable fragments which are taken out of the context of the full-length protein, some of these hidden critical sites might become exposed in the fragments thus obtained. Consequently, these discrete foldable polypeptide fragments might facilitate screening for more effective vaccines or antigen candidates.

Searching for soluble fragments of a polypeptide chain may be achieved by genetic cleavage at predetermined sites based on detailed analysis of the target structure, or by the more traditional protease digestion method *(4–7)*. However, the former approach is limited by the generally poor understanding of the structure–function relationship of proteins, while for the latter only a small number of exposed surface sites are accessible to proteases. A random method that in theory can thoroughly and individually analyze all possible folding fragments would be more desirable. To this end, we formulated a dissection scheme that is capable of searching for soluble fragments of a given target in a random but systematic manner, and which is independent of protein function. The solubility of the dissected protein fragments was accessed through expression analysis of these fragments as N-terminal fusions to green fluorescent protein (GFP), which has been shown to be a good indicator for the "foldability" of the upstream polypeptide partner *(8, 9)*. This method using GFP as folding reporter is generally applicable because it does not require structural or functional information on the target proteins. Results from dissection of the severe acute respiratory syndrome coronavirus (SARS-CoV) spike protein and HIV-1 gp120 are presented.

2. Materials

2.1. Reagents, Buffers, and Solutions

1. PEG 8000 (Ameresco, Farmingham, MA).
2. dNTPs (TaKaRa, Dalian, China).
3. LB medium.
4. Kanamycin (BBI, Ont., Canada).
5. IPTG (isopropylthio-β-D-galactoside) (Ameresco).
6. Lysozyme (Sigma).

7. Coomassie brilliant blue R250.
8. Lysis buffer: 50 mM Tris-HCl, pH 7.2, 50 mM NaCl, and 5% glycerol.
9. DNase I random digestion buffer: 50 mM Tris-HCl, pH 7.5, 10 mM $MnCl_2$.
10. Fragment reassembly buffer: 10 mM Tris-HCl, pH 8.3, 2 mM $MgCl_2$, 50 mM KCl, 0.2 mM each dNTP.
11. Loading buffer: 50 mM Tris-HCl, pH 6.8, 0.1% bromophenol blue, 2% SDS, 1% 2-mercaptoethanol, and 10% glycerol.
12. Prestained protein markers (New England Biolabs, Beverly, MA).

2.2. Strain and Plasmid

1. *E. coli* strains BL21(DE3) (Novagen, Madison, WI).
2. pET30a(+) expression system (Novagen).
3. Oligonucleotide primers (TaKaRa).

2.3. Enzymes

1. Restriction enzymes (New England Biolabs).
2. Taq polymerase (TaKaRa).
3. Deep Vent® polymerase (New England Biolabs).
4. T4 DNA ligase (New England Biolabs).
5. DNase I (Worthington, Lakewood, NJ).
6. T4 DNA polymerase (New England Biolabs).
7. T4 polynucleotide kinase (New England Biolabs).
8. Shrimp alkaline phosphatase (Promega, Madison, WI).

2.4. Kits and Apparatus

1. QIAquick Gel Extraction Kit, QIAquick PCR Purification Kit, QIAprep Spin Miniprep Kit (Qiagen, Valencia, CA).
2. Thermocycler (MJ research PTC-225, Miami, FL).
3. Electroporator (Eppendorf, Hamburg, Germany).
4. Agarose gel equipment.
5. Sodium dodecyl sulfate-polyacrymide gel electrophoresis (SDS-PAGE) equipment.
6. Microcon® YM-30 (Millipore, Billerica, MA).

3. Methods

The methods described below outline (1) the construction of the expression plasmid, (2) viral protein fragment library construction, (3) screening of fragments, and (4) expression analysis of viral protein fragments.

3.1. Expression Plasmid

The construction of the expression plasmid pET30-linker-GFP for viral gene fragments is described in the **Subheadings 3.1.1–3.1.2**. This includes (a) the description of the expression vector of pET30a(+), (b) the description of the GFP gene, and (c) the cloning strategy.

3.1.1. pET30a(+) Expression Vector and cDNA of GFP

1. Use the T7 promoter-driven system originally developed by Studier and colleagues *(10–12)*, pET30a(+) (*see* **Fig. 1a**). It is a powerful vector for the cloning and expression of recombinant proteins in *E. coli.* Target genes can be inserted into the multiple cloning region and placed under the control of strong bacteriophage T7 transcription and translation signals, and expression is induced by isopropylthio-β-D-galactoside (IPTG). The pET30a(+) expression vector carries an N-terminal His•Tag configuration plus an optional C-terminal His•Tag sequence which can be used to assay expression levels and purify proteins. The pET30a(+) expression vector also contains a kanamycin resistance gene.

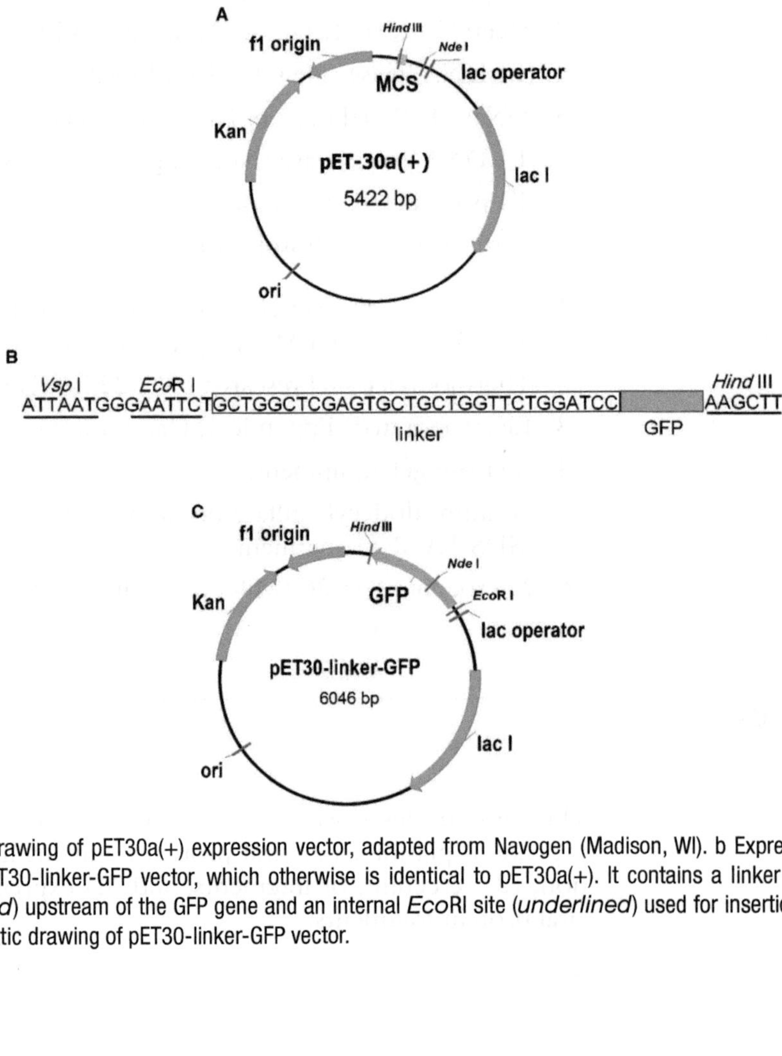

Fig. 1a. Schematic drawing of pET30a(+) expression vector, adapted from Navogen (Madison, WI). b Expression feature contained in pET30-linker-GFP vector, which otherwise is identical to pET30a(+). It contains a linker sequence AGSSAAGSGS (*boxed*) upstream of the GFP gene and an internal *EcoRI* site (*underlined*) used for insertion of gene fragments. c Schematic drawing of pET30-linker-GFP vector.

2. Subclone the GFP gene from a commercial source. We used an in-house GFP-containing vector pET30a-*hydA*, which was constructed from pQB-2 *(13)*. This GFP variant carries 11 extra amino acid residues at the C-terminus which help emit a stronger fluorescent signal than the wild-type GFP *(13)*.
3. Amplify the GFP gene using *LA* Taq polymerase (TaKaRa) and the following primers (forward and reverse, respectively):

5′-CATCGTT**ATTAAT**GGG**GAATTC**T<u>GCTGGCTC-GAGTGCTGCTGGT-</u>3′, and 5′-TAG**AAGCTT**AGCTAAT-TCA-GCTTGGCTGC-3′.

The restriction endonuclease sites in these primers are *Vsp* I, *Eco*RI, and *Hin*dIII (shown in bold), respectively, and the partial linker sequence is underlined) (**Fig. 1b**).

3.1.2. pET30-Linker-GFP Construction

1. Doubly digest the amplified GFP gene with *Vsp*I and *Hin*dIII.
2. Ligate the insert into the pET30a(+) plasmid which has been doubly digested with *Nde*I and *Hin*dIII to yield pET30-linker-GFP (*see* **Note 1**).
3. Transform the expression vector into *E. coli* BL21(DE3) cells by standard methods *(14)*.
4. Plate the transformed *E. coli* BL21(DE3) cells on LB plates containing kanamycin (50 μg/mL) and incubate overnight at 37 °C.
5. Select single colonies and grown overnight in LB medium with kanamycin. This can be done in 96-well plates.
6. Isolate the plasmid DNA using a QIAprep Spin Miniprep Kit (QIAGEN), and verify sequence of the region flanking GFP (**Fig. 1c**).

3.2. Fragment Library Construction

Described below are the steps for the construction of a viral surface protein fragment library (**Fig. 2**).

3.2.1. cDNA

SARS Spike Protein

1. Amplify the gene of interest for vaccine development by PCR using Taq polymerase. The SARS-CoV spike gene (GenBank No. AY278488) we used was amplified by PCR reaction from a cDNA sample kindly donated by Huada Beijing Genomics Institute containing the full length cDNA for SARS-CoV spike protein.
2. The primers used were (forward and reverse, respectively): 5’-CGGAATTCCATATGTTTATTTTCTTATTATTTC-3′ and 5′-CCGGATCCTTAGTGGTGGTGGTGGTGGTGT-GTGTAATGTAATTTGACACC-3′.

HIV-1 gp120

If a vector containing the gene is not available but its sequence is known, it can be synthesized according to DNAworks *(15, 16)* and confirmed by sequencing (*see* **Note 2**). We used this strategy to obtain the HIV-1 gp120 gene.

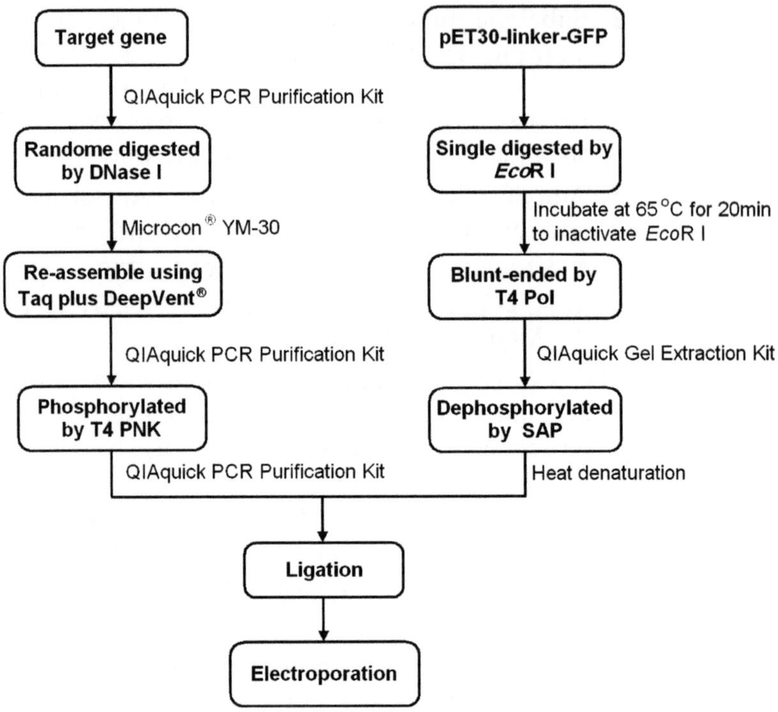

Fig. 2. Schematic drawing of fragment library construction.

3.2.2. Random Dissection of Target Genes

Fragmentation of both of the target genes (i.e., SARS-CoV spike and HIV-1 gp120) is performed by digestion with a nonselective DNase I (*see* **Note 3**) *(17)*.

1. Purify 2–4 μg of the target gene using a QIAquick PCR Purification Kit.
2. Incubate the purified DNA in 50 μL DNase I random digestion buffer for 10 min at 15 °C.
3. Add 0.075 U of DNase I and incubate at 15 °C for 4 min.
4. Add 10 μL of 0.5 M EDTA to terminate the DNA digestion.
5. A pool of short DNA segments ranging from ~50 to 100 bp should be prepared. Verify this by running a 1–4% agarose gel with a DNA ladder containing a 100-bp marker (**Fig. 3a**).

3.2.3. Reassembly of Gene Fragments

1. Purify the randomly digested gene fragments by passing the reaction mixture through a Microcon® YM-30 column (*see* **Note 4**).
2. Resuspend the product in the fragment reassembly buffer at a concentration of 1 ng/μL.
3. Add 2.5 U/100 μL of both Taq polymerase and Deep Vent® polymerase.

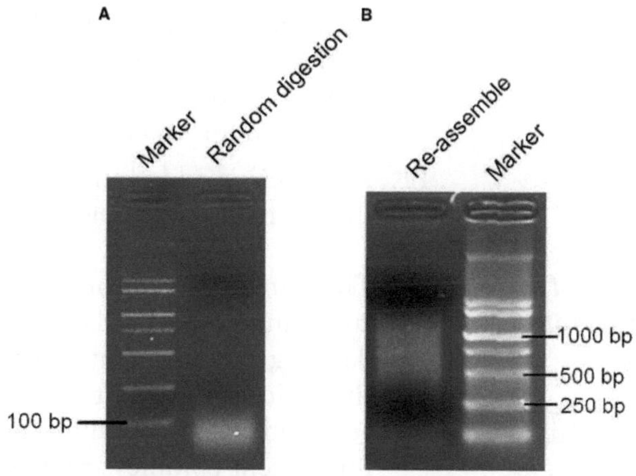

Fig. 3. Fragment library construction for SARS-CoV spike. **a** Fragmentation of target gene digested by DNase I. **b** Reassembled gene fragments.

4. Using an MJ Research PTC-225 thermocycler or comparable instrument, run a PCR program (*18*) consisting of 10–20 cycles of 94 °C for 1 min, 55 °C for 1 min, and 72 °C for 1 min + 5 s/cycle (*see* **Note 5**).
5. This should generate gene fragments largely in the range of 200–1,200 bp (**Fig. 3b**). Verify this by agarose gel electrophoresis with a ladder spanning 100–2,000 bp.

3.2.4. Cloning

1. Purify the reassembled DNA sample using a QIAquick PCR Purification Kit.
2. Phosphorylate the DNA with T4 polynucleotide kinase (T4 PNK) at a ratio of 5 U/μg DNA by incubating at 37 °C for 30 min.
3. Digest the backbone vector (pET30-linker-GFP) with *Eco*RI.
4. Make blunt-ended vector DNA using T4 DNA polymerase in the presence of 0.1 mM of each dNTP (*see* **Note 6**).
5. Purify the DNA with a QIAquick Gel Extraction Kit to remove residual enzyme activity.
6. Dephosphorylate the linearized and blunt-ended vector twice with shrimp alkaline phosphatase (SAP) at a ratio of 10 U/μg DNA by incubating at 37 °C for 45 min in the supplied SAP buffer.
7. Incubate at 70 °C for 20 min to inactivate the alkaline phosphatase enzyme (*see* **Note 7**).
8. Ligate the gene fragments to the backbone vector at 12 °C overnight in the presence of 5% PEG 8000.
9. Transform *E. coli* BL21(DE3) competent cells by electroporation.

10. Plate transformed *E. coli* BL21(DE3) cells on LB agar supplemented with 50 μg/mL kanamycin and grow overnight at 37 °C. Continue growing cells at ambient temperature on the bench for about 20 h (*see* **Note 8**).
11. No IPTG is added in these experiments, as it would inhibit the formation of fluorescent colonies (*see* **Note 9**).

3.3. Screening of Soluble Viral Protein Fragments

3.3.1. Screening Scheme

1. The transformed fluorescent colonies are picked under a UV lamp and tested with standard colony PCR using primers flanking the fragment inserts, and sequenced. Alternatively, the colonies are left on bench at room temperature for additional 24 h to allow for color development, and then fluorescent colonies are picked directly.
2. The primers we used in the colony PCR were (forward and reverse, respectively): 5′-TAAGAAGGAGATATACATAATG-3′ and 5′-AGAACCAGCAGCACTCGAGCCA-3′.

These primers allowed determination of the size of the inserted fragment.

3.3.2. Primary Analysis of Screened Fragments

SARS-CoV Spike Protein

1. Screen as many clones as necessary/possible. For example, from ~4,300 clones screened, only 230 clones were found to be fluorescent (*see* **Fig. 4a**).
2. Most of the fluorescent clones will contain empty vectors or vectors with fragments smaller than 100 base pairs (bp) as judged by colony PCR and sequencing.

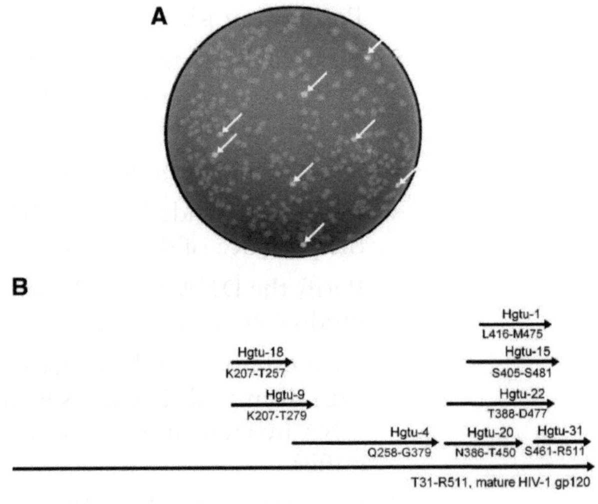

Fig. 4a. Colonies obtained from inserting and expression of SARS-CoV spike gene fragments in pET30-linker-GFP using *E. coli* BL21(DE3) as the host. The fluorescent clones were indicated by *arrows*. b Fragments of HIV-1 gp120 protein as deduced from the sequences of the inserts contained in fluorescent GFP-fusion clones. The numbering of the residues is that used in the X-ray structure of a truncated HIV-1 gp120 *(2)*. Only fragments larger than 50 deduced amino acid residues are shown. The start and stop positions of the deduced amino acid sequence for each fragment are indicated beneath the respective *arrow*.

3. In addition, ~20 of the 230 fluorescent colonies were found to contain vectors with inserts in the reverse orientation or not in frame, so sequencing all of the colonies with appropriately sized fragments is absolutely necessary.
4. Screen the peptides by SDS-PAGE.
5. Judged by SDS-PAGE results, many of the peptides encoded by these gene inserts will be degraded in the corresponding fusion proteins (data are not shown).
6. We were able to locate two inserts larger than 150 nucleotides (50 amino acid residues). These were labeled ssPtu-15 (residues 1,118–1,175 of the original protein) and ssPtu-16 (residues 1,129–1,186).

HIV-1 gp120

Out of 2,800 clones that we screened, 115 were found to be fluorescent, from which eight fragments containing greater than 50 deduced amino acid residues were isolated (**Fig. 4b**).

3.4. Expression Analysis of Viral Protein Fragments

3.4.1. Preliminary Structural Examination

SARS-CoV Spike

Several studies *(19–21)* have reported that SARS-CoV S-mediated fusion can be inhibited by heptad repeat region 2 (HR2) but not HR1-derived peptides, most likely by interfering with the six-helix bundle formation, a process essential to drive the membrane fusion reaction and to initiate infection *(1)*.

1. By alignment with the HR2 region using Clustal W *(22)*, we later found that the SARS fragment ssPtu-15 overlaps with the HR2 (residues 1,147–1,185) of SARS-CoV spike protein *(23)*, while fragment ssPtu-16 contains the whole SARS HR2 (**Fig. 5**).
2. Given the high similarity of ssPtu-15 and ssPtu-16 with these peptides derived from the HR2 region *(19–21)*, ssPtu-15 and ssPtu-16 may share potential as therapeutic agents for the direct inhibition of SARS-CoV cell entry, as anti-SARS vaccines, and high throughput assay reagents for screening for small molecule inhibitors of SARS envelope-mediated cell fusion.

```
                                               **** *****:************
       CP-1 -----------------------------------GINASVVNIQKEIDRLNEVAKNLNESLIDLQELGKYE----   37
        HR2 -------------------------------GDISGINASVVNIQKEIDRLNEVAKNLNESLIDLQELG-------   38
 GST-HR2-38 -------------------------------GDISGINASVVNIQKEIDRLNEVAKNLNESLIDLQELG-------   38
 GST-HR2-44 -------------------------------GDISGINASVVNIQKEIDRLNEVAKNLNESLIDLQELGKYEQYI-   44
     SHR2-2 ------------PKEELDKYFKNHTSPDVDLGDISGINASVVNIQKEIDRLNEVAKNLNESLIDLQELGKYE----   60
     SHR2-1 --------ELDSPKEELDKYFKNHTSPDVDLGDISGINASVVNIQKEIDRLNEVAKNLNESLIDLQELGKYE----   64
     SHR2-8 --------ELDSFKEELDKYFKNHTSPDVDLGDISGINASVVNIQKEIDRLNEVAKNLNESLIDLQELGKYEQYIK   68
     SHR2-9 --------ELDSFKEELDKYFKNHTSPDVDLGDISGINASVVNIQKEIDRLNEVAKNLNESLIDLQEL--------   60
   ssPtu-16 -----------SFKEELDKYFKNHTSPDVDLGDTSGINASVVNIQEEIDRLNEVAKNLNESLIDLQELG-------   58
   ssPtu-15 TVYDPLQPELDSFKEELDKYFKNHTSPDVDLGDISGINAYVVNIQEEIDRLNEVAKNL------------------   58
      ruler 1.......10........20........30........40........50........60........70......
```

Fig. 5. Clustal W *(22)* alignment of ssPtu-15 and ssPtu-16 with HR2-derived peptides which interfere SARS-CoV S-mediated fusion to host cells: peptide CP-1 *(20)*, peptides HR2, GST-HR2-38, GST-HR2-44 *(21)*, peptides SHR2-1, SHR2-2, SHR2-8, SHR2-9 *(19)*.

Table 1
Deduced HIV-1 gp120 fragments and their locations in the protein structure

Fragment number	Number of residues	Location in the protein structure[a]
Hgtu-1	60	β19 – loop(β24~α5)
Hgtu-4	122	loopB – β16
Hgtu-9	72	loop (β3~β4) – β9
Hgtu-15	77	loopV4 – α5
Hgtu-18	51	loop (β3~β4) – loopB
Hgtu-20	65	β17 – β23
Hgtu-22	90	α4 – α5
Hgtu-31	51	loopV5 – C-terminal

[a]According to the assignment of structural elements used in the X-ray structure of HIV-1 gp120 *(2)*

HIV-1 gp120

The eight fragments were mapped onto the HIV-1 gp120 structure as shown in **Table 1**. A preliminary structural examination suggested that most of these fragments are associated with the proposed binding sites for CD4 and/or the chemokine receptor CCR5 *(2)*.

3.1.2. Fragment Expression Analysis

Induction and Protein Extraction

1. Dilute saturated overnight cultures 100-fold into fresh LB media containing 50 μg/mL kanamycin and grow at 37 °C for about 2 h to reach an optical density at 600 nm (OD_{600}) of 0.5–0.6.
2. Initiate protein expression by adding 0.2 mM IPTG and culture for 24 h at 23 °C.
3. Collect about 20 OD_{600} of the cells by centrifugation (3,500 rpm, 10 min) and resuspend in 1 mL of lysis buffer.
4. Sonicate for 66 pulses of 3 s each with a 3-s interval in an ice-water bath.
5. Supplement the lysate with 0.2 mg/mL of lysozyme and shake gently for 1 h at room temperature.
6. Centrifuge at 16,100 rpm for 5 min to separate soluble protein from cell debris and inclusion bodies.

Fragment-GFP Fusions Characterized by SDS-PAGE

1. Collect the supernatants, and resuspend the pellets in 1 mL of 2× SDS polyacrylamide gel electrophoresis (SDS-PAGE) loading buffer. Likewise prepare supernatant samples for SDS-PAGE.

2. Resolve protein samples using a 12% acrylamide gel, and stain with coomassie brilliant blue dye. Typically, 10 μL aliquots were loaded.
3. Analysis by SDS-PAGE of fragment-GFP fusions found from both of SARS spike protein fragments (ssPtu-15, ssPtu-16) and three of the eight HIV-1 gp120 fragments (Hgtu-1, Hgtu-4, and Hgtu-15) to be partially soluble (*see* **Fig. 6**). Under the conditions we have tested so far, no significant soluble expression was seen for the other five HIV-1 gp120 fragments as determined by coomassie brilliant blue staining, while inclusion bodies were observed. Thus, Hgtu-1, Hgtu-4, and Hgtu-15 might be more useful as possible antigen and vaccine candidates.

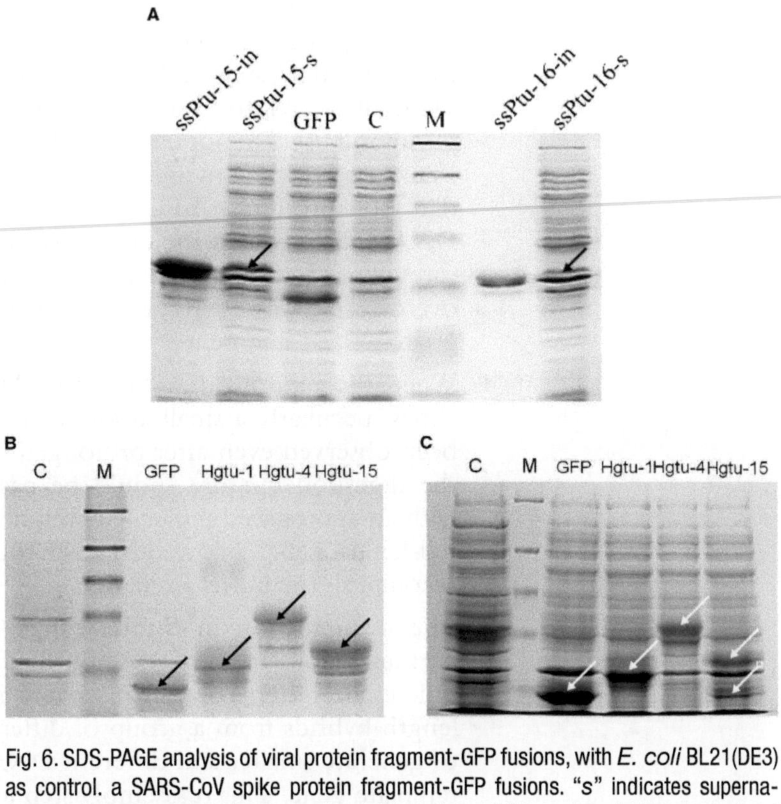

Fig. 6. SDS-PAGE analysis of viral protein fragment-GFP fusions, with *E. coli* BL21(DE3) as control. a SARS-CoV spike protein fragment-GFP fusions. "s" indicates supernatants of lysates, and "*in*" denotes insoluble pellets of the lysates. b Supernatants of the lysates for selected HIV-1 gp120 fragment-GFP fusions. c Pellets of the lysates for selected HIV-1 gp120 fragment-GFP fusions. Calculated molecular weights for GFP, ssPtu-15, ssPtu-16, Hgtu-1, Hgtu-4, and Hgtu-15 were 29.9, 36.3, 36.3, 36.5, 43.5, and 38.6 kDa, respectively. Corresponding band positions are indicated by *arrows*. *C*: control; *M*: protein marker, broad range (NEB), whose bands were 175, 83, 62, 48, 33, 25, and 17 kDa, respectively.

Notes

1. The 5′ cohesive end of the PCR product containing the GFP gene and digested with *Vsp*I is compatible with the 3′ end of pET30a(+) plasmid digested with *Nde*I. The reason for the use of *Vsp*I is that there is an *Nde*I restriction site in the GFP gene.
2. By inputting the gene sequence of HIV-1 HBc2 gp120 (1,446 bp) into the free software DNAworks (http://mcl1.ncifcrf.gov/dnaworks/dnaworks2.html), 66 oligo sequences were returned, and synthesized accordingly *(15)*. These oligos were assembled to gp120 by the standard overlapping procedure *(24)*. Then two flanking primers (5′-ATGACCGAAAAACTGTGGGTGA-3′ and 5′-AGCGCTTCTCACGTTGAACAACACG-3′) were used to amplify the full-length of gp120 by touch-down PCR.
3. In the random digestion step, the use of DNase I in the presence of $MnCl_2$ is critical as this protocol will generate DNA fragments of relatively uniform sizes, which facilitates the reassembly step (**17**, also *see* **Note 5**). Other digestion methods often lead to smears of fragments that are difficult to purify and use. Do not use the buffer supplied with the DNase I enzyme, which contains $MgCl_2$ and would lead to smears of digestion products. Additionally, the amounts of the target DNA, DNase I, and the digestion time should be controlled carefully as indicated above. Otherwise, the DNA would be overdigested or underdigested and could not produce appropriately sized fragments.
4. When the random digestion by DNase I is used for some genes, peculiarly a small amount of the full-length gene has been observed even after prolonged incubation. In this case, the digestion mixture should be passed through a column with an appropriate molecular weight cut-off to filter out the full-length gene, followed by a Microcon® YM-30 column to accomplish the buffer exchange.
5. The manipulation in **Subheadings 3.2.2** and **3.2.3** is in part analogous to the DNA shuffling protocol *(24, 25)*, but unlike the latter, the purpose here is not to produce full-length hybrids from a group of different parental genes, but to generate smaller and different DNA fragments for a single template gene. This reassembly step following DNase I treatment is necessary in order to prepare a large amount of DNA sequences with controlled lengths, which is achieved by tailoring the PCR cycle number used in the reassembly.
6. In our work, several enzymes, that is, T4 DNA polymerase, S1 Nuclease, Mung Bean Nuclease, and Deep Vent® polymerase,

were tested for blunt-ending gene fragments. T4 DNA polymerase yielded the best results.

7. In order to improve the final yield of backbone vector, the manipulation steps should be as few as possible. The heat-sensitive shrimp alkaline phosphatase was used to reduce one purification procedure after dephosphorylation.
8. In order to improve the percentage of positive clones containing larger inserts (inserts of more than 50 amino acids), three principal improvements were made. First, gene fragments resulting from reassembly were phosphorylated using T4 polynucleotide kinase. Second, T4 polymerase was denatured with a QIAquick Gel Extraction Kit upon the completion of the blunt-ending reaction to prevent residual T4 DNA polymerase activity from degrading DNA in later steps. Third, a second step of the dephosphorylation by heat-liable SAP was incorporated to improve dephosphorylation efficiency.
9. We found that the basal expression level often observed for the pET vectors was sufficient for the expression of gene fragment-GFP fusion. Low IPTG concentrations (lower than 0.1 mM) did not enhance green fluorescent colonies, while higher concentrations only inhibited fluorescent colonies or generated abnormal hallow colonies.

Acknowledgments

The authors thank Dr. K. Kohno and Dr. X. Xing for the GFP gene, the Huada Beijing Genomics Institute, and Dr. Y. Ruo for the SARS-CoV spike gene, and Prof. Rolf D. Schmid and Prof. S. Yang for useful comments. The work was supported by grants from the National Key Basic Research Program of China (2003CB716002) and Tsinghua University SARS special fund, and by a Bairen Program faculty award of Tsinghua University (to Z.L.).

References

1. Eckert, D. M., and Kim, P. S. (2001) Mechanisms of viral membrane fusion and its inhibition. *Annu. Rev. Biochem.* 70, 777–810.
2. Kwong, P. D., Wyatt, R., Robinson, J., Sweet, R. W., Sodroski, J., and Hendrickson, W. A. (1998) Structure of an HIV gp120 envelope glycoprotein in complex with the CD4 receptor and a neutralizing human antibody. *Nature* 393, 648–659.
3. Rizzuto, C. D., Wyatt, R., Hernandez-Ramos, N., Sun, Y., Kwong, P. D., Hendrickson, W. A., and Sodroski, J. (1998) A conserved HIV gp120 glycoprotein structure involved in chemokine receptor binding. *Science* 280, 1949–1953.
4. Collinet, B., Herve, M., Pecorari, F., Minard, P., Eder, O., and Desmadril, M. (2000) Functionally accepted insertions of proteins

within protein domains. *J. Biol. Chem.* 275, 17428–17433.

5. Hocker, B., Beismann-Driemeyer, S., Hettwer, S., Lustig, A., and Sterner, R. (2001) Dissection of a (βα)8-barrel enzyme into two folded halves. *Nat. Struct. Biol.* 8, 32–36.
6. Smith, V. F., and Matthews, C. R. (2001) Testing the role of chain connectivity on the stability and structure of dihydrofolate reductase from *E. coli*: fragment complementation and circular permutation reveal stable, alternatively folded forms. *Protein Sci.* 10, 116–128.
7. Soberon, X., Fuentes-Gallego, P., and Saab-Rincon, G. (2004) *In vivo* fragment complementation of a (β/α)8 barrel protein: generation of variability by recombination. *FEBS Lett.* 560, 167–172.
8. Waldo, G. S., Standish, B. M., Berendzen, J., and Terwilliger, T. C. (1999) Rapid protein-folding assay using green fluorescent protein. *Nat. Biotechnol.* 17, 691–695.
9. Waldo, G. S. (2003) Genetic screens and directed evolution for protein solubility. *Curr. Opin. Chem. Biol.* 7, 33–38.
10. Studier, F. W., and Moffatt, B. A. (1986) Use of bacteriophage T7 RNA polymerase to direct selective high-level expression of cloned genes. *J. Mol. Biol.* 189, 113–130.
11. Rosenberg, A. H., Lade, B. N., Dao-shan, C., Lin, S.-W., Dunn, J. J., and Studier, F.W. (1987) Vectors for selective expression of cloned DNAs by T7 RNA polymerase. *Gene* 56, 125–135.
12. Studier, F. W., Rosenberg, A. H., Dunn, J. J., Dubendorff, J. W., and David,V. G.(1990) Use of T7 RNA polymerase to direct expression of cloned genes, in Meth. Enzymol., San Diego, CA: Academic Press. 185, 60–89.
13. Kimata,Y., Iwaki, M., Lim, C. R., and Kohno, K. (1997) A novel mutation which enhances the fluorescence of green fluorescent protein at high temperatures. *Biochem. Biophys. Res. Commun.* 232, 69–73.
14. Sambrook, J., and Russell, D. W. (2001) *Molecular Cloning: A Laboratory Manual*, 3rd ed. Cold Spring Harbor, New York: Cold Spring Harbor Laboratory Press.
15. Hoover, D. M., and Lubkowski, J. (2002) DNAWorks: an automated method for designing oligonucleotides for PCR-based gene synthesis. *Nucleic Acids Res.* **30**.
16. Stemmer, W. P. C., Crameri, A., Ha, K. D., Brennan, T. M., and Heyneker, H. L. (1995) Single-step assembly of a gene and entire plasmid from large numbers of oligodeoxyribonucleotides. *Gene* 164, 49–53.
17. Lorimer, I. A. J., and Pastan, I. (1995) Random recombination of antibody single-chain Fv sequences after fragmentation with DNase I in the presence of Mn^{2+}. *Nucleic Acids Res.* 23, 3067–3068.
18. Prodromou, C., and Pearl, L. (1992) Recursive PCR: a novel technique for total gene synthesis. *Protein Eng.* 5, 827–829.
19. Bosch, B. J., Martina, B. E. E., van der Zee, R., Lepault, J., Haijema, B. J., Versluis, C., Heck, A. J. R., de Groot, R., Osterhaus, A., and Rottier, P. J. M. (2004) Severe acute respiratory syndrome coroavirus (SARS-CoV) infection inhibition using spike protein heptad repeat-derived peptides. *Proc. Natl. Acad. Sci. USA* 101, 8455–8460.
20. Liu, S. W., Xiao, G. F., Chen, Y. B., He, Y. X., Niu, J. K., Escalante, C. R., Xiong, H. B., Farmar, J., Debnath, A. K., Tien, P., and Jiang, S. B. (2004) Interaction between heptad repeat 1 and 2 regions in spike protein of SARS-associated coronavirus: implications for virus fusogenic mechanism and identification of fusion inhibitors. *Lancet* 363, 938–947.
21. Zhu, J. Q., Xiao, G. F., Xu, Y. H., Yuan, F., Zheng, C. Y., Liu, Y. Y., Yan, H. M., Cole, D. K., Bell, J. I., Rao, Z. H., Tien, P., and Gao, G. F. (2004) Following the rule: formation of the 6-helix bundle of the fusion core from severe acute respiratory syndrome coronavirus spike protein and identification of potent peptide inhibitors. *Biochem. Biophys. Res. Commun.* 319, 283–288.
22. Thompson, J. D., Higgins, D. G., and Gibson, T. J. (1994) CLUSTAL W: improving the sensitivity of progressive multiple sequence alignment through sequence weighting, position-specific gap penalties and weight matrix choice. *Nucleic Acids Res.* 22, 4673–4680.
23. Tripet, B., Howard, M. W., Jobling, M., Holmes, R. K., Holmes, K. V., and Hodges, R. S. (2004) Structural characterization of the SARS-coronavirus spike S fusion protein core. *J. Biol. Chem.* 279, 20836–20849.
24. Joo, H., Lin, Z. L., and Arnold, F. H. (1999) Laboratory evolution of peroxide-mediated cytochrome P450 hydroxylation. *Nature* 399, 670–673.
25. Stemmer, W. P. C. (1994) Rapid evolution of a protein *in vitro* by DNA shuffling. *Nature* 370, 389–391.

Chapter 19

Use of GFP to Investigate Expression of Plant-Derived Vaccines

Hong-Ye Li and Mee-Len Chye

Summary

Plants are low-cost bioreactors for the production of various biopharmaceuticals including oral vaccines. Plant-derived oral vaccines are potentially useful in combating viral infections involving mucosal immunity. Transgenic plants have been generated to successfully produce mucosal vaccines against cholera, hepatitis B, foot-and-mouth disease, and Norwalk virus. As a first step toward the generation of oral vaccines against the severe acute respiratory syndrome coronavirus (SARS-CoV), we have expressed a recombinant S1 protein of the SARS-CoV in transformed tobacco. Since plant transformation and regeneration of stable transformants require considerable time, we initially used a green fluorescent protein (GFP) to tag the antigen in transient expression. GFP was fused to the carboxy-terminus of S1 for expression of S1-GFP to show expression of recombinant S1 by agroinfiltration of tobacco leaves. The GFP tag enables a relatively quick confirmation of antigen expression in plant cells by fluorescent microscopy. Such analysis using GFP that precedes stable plant transformation will enable the rapid screening of multiple constructs to attain optimal recombinant protein expression. Furthermore, this approach determines the subcellular localization of the recombinant protein in plant cells, providing information on optimal subcellular targeting for production in plant bioreactors.

Keywords: Autofluorescent protein tag, Plant bioreactors, Plant transformation, Recombinant subunit vaccines, Severe acute respiratory syndrome coronavirus, Subcellular localization.

1. Introduction

Transformed plants can be easily converted to become effective bioreactors for the production of desired recombinant biopharmaceuticals. To this end, the foreign DNA encoding the relevant recombinant protein can be expressed from the nuclear genome or the chloroplast genome of the genetically engineered plant.

Barry W. Hicks (ed.), *Methods in Molecular Biology, Viral Applications of Green Fluorescent Protein, vol. 515*

DOI: 10.1007/978-1-59745-559-6_19

Transgenic plants have already been generated by nuclear transformation to express mucosal vaccines against cholera, Norwalk virus, hepatitis B, foot-and-mouth disease, and severe acute respiratory syndrome, SARS *(1, 2)*. These viral antigens generated in transgenic plants are effective in inducing mucosal and serum immune responses in animals.

In the case of SARS, nuclear-transformed tobacco and tomato expressing the SARS-CoV viral antigens induced immunogenic responses in mice; SARS-CoV-specific IgA increased in mice following oral ingestion of S1-expressing tomato fruits *(2)*. Such edible vaccines deliverable as fruits (e.g., tomato), leaves (e.g., lettuce), tubers (e.g., potato), seeds (e.g., rice or corn), stems, or roots, would obliterate purification procedures required for microbial-derived recombinant vaccines. Further added advantages include the easy storage, transport, and administration by direct ingestion of these oral vaccines. Represented in the form of plant organs (fruits, leaves, tubers, seeds, stems, or roots), these vaccines can be easily grown and distributed, omitting the need for refrigeration in delivery or the requirement for injection by trained health personnel, thereby reducing the costs of immunization programs in developing countries.

Green fluorescent protein (GFP) has been routinely used in laboratories worldwide for tagging proteins in both plant and animal cells. It has been proven easy and convenient to use in transient expression and stable transformation of plant cells. Further, it is pertinent to note that in its application for tagging of viral antigens in plants, GFP can be processed by the ubiquitin-proteasome pathway *(3)* since ubiquitinated antigens are a prerequisite to processing into antigenic peptides for presentation to the major histocompatibility complex *(4)*.

We describe here several techniques that utilize *Agrobacterium tumefaciens* for the introduction of foreign DNA into plant cells. In nature, *Agrobacterium tumefaciens* is a Gram-negative soil bacterium that transfers and incorporates its transfer-DNA (T-DNA) into the plant nuclear genome. This results in an expression of genes on the T-DNA that will cause crown gall in the infected plant *(5)*. It has been demonstrated in *Agrobacterium*-mediated transformation that following co-cultivation of tobacco BY-2 cells with an *Agrobacterium* harboring a GFP fusion construct, the rate of fusion protein synthesis markedly increased between 24 and 36 h, after which rapid decrease occurred *(6)*. Agroinfiltration is a more recent technique that can be applied to investigate transient expression in plant cells by which an *Agrobacterium* liquid culture is infiltrated into intact plant leaves *(7)*. The feasibility of rapid analyses of foreign protein expression in transiently expressed leaves following agroinfiltration presents a major advantage in its use. The transiently expressed foreign protein is probably derived from transiently synthesized T-DNA containing the construct that had been mobilized from *Agrobacterium* cells to the tobacco nuclei but has yet to be integrated into the tobacco nuclear genome *(8)*.

Such analysis using *Agrobacterium*-mediated transient expression with GFP-tagged constructs will enable the rapid screening of multiple constructs so that the construct that gives optimal recombinant protein expression can be quickly identified. Upon the identification of such a high-expressing construct, it can then be selected for use in stable plant transformation to generate transgenic plants, a more time-consuming process. This approach by using GFP as an autoflourescent tag can further determine the subcellular localization of recombinant-tagged protein expressed in plant cells, thus providing information for optimal subcellular targeting of such recombinant proteins produced in plant bioreactors.

2. Materials

2.1. Constructs for the Expression of the SARS-Cov S1-GFP Fusion

1. Plasmid pCRII-S1, a pCRII (Invitrogen) derivative containing the *SARS-CoV S1* cDNA (encoding amino acids 1–658 of the S protein) amplified by the Polymerase Chain Reaction (gift of Dr. L.L.M. Poon and Prof. J.S.M. Peiris, Department of Microbiology, University of Hong Kong).
2. QuikChange Multi Site-directed Mutagenesis Kit with *Pfu-Turbo* DNA polymerase (Stratagene) to obtain an “optimized codon-usage” S1 for cloning into plant transformation vectors.
3. Polymerase chain reaction (PCR) for mutagenesis (50 μL): 5 μL of 10× reaction buffer, 5–50 ng of double-stranded (ds) DNA template, 125 ng of oligonucleotide primer #1 for introduction of nucleotide change, 125 ng of oligonucleotide primer #2 for introduction of nucleotide change, 1 μL of dNTP mix (10 mM each), double-distilled water (ddH_2O) to a final volume of 50 μL, Then add 1 μL of *Pfu-Turbo* DNA polymerase (2.5 U/μL).
4. *DpnI* and buffers supplied with enzyme.
5. Supercompetent *Escherichia coli* DH5α cells.
6. Agarose gel electrophoresis reagents and apparatus.
7. GFP gene from plasmid pGFP2 (gift of Prof. Nam-Hai Chua, Rockefeller University, New York).
8. Plant nuclear transformation (binary) vector pGD *(9)* (gift of Dr. M.M. Goodin, University of California, Berkeley) derived from a commonly used binary vector pCAMBIA-1301.

2.2. Agroinfiltration of Tobacco for Expression of the S1:GFP Antigen

1. Tobacco (*Nicotiana tabacum*) potted plants grown in soil.
2. *Agrobacterium tumefaciens* strain LBA4404 *(10)*.
3. *Agrobacterium* culture medium: yeast extract peptone (YEP) medium (pH 7.0) containing 10 g/L yeast extract, 10 g/L Bacto-peptone, and 5 g/L NaCl supplemented with 10 mM $MgSO_4$, 100 μg/mL of streptomycin and, in the case of a derivative harboring pGD, plus 50 μg/mL of kanamycin.
4. Infiltration medium: 10 mM 2-[*N*-morpholino]ethanesulfonic acid (MES) pH 5.5 plus 10 mM $MgSO_4$ supplemented with 100 μM acetosyringone (AS, Aldrich).

2.3. Analyses of Transiently Expressed Proteins

2.3.1. Laser-Scanning Confocal Microscopy

1. Fluorescence microscopy or confocal laser-scanning microscope equipped with helium/neon lasers.
2. Image processing software LSM 510 (Zeiss) and Photoshop 6.0 (Adobe).

2.3.2. Protein Extraction from Plant Tissues

1. Liquid nitrogen.
2. Mortar and pestle.
3. Protein homogenizing buffer: 50 mM Tris-HCl, pH 7.5, 2 mM EDTA, and freshly add 28 mM β-mercaptoethanol and 2 mM phenylmethanesulfonylfluoride (PMSF, Sigma).

2.3.3. Western Blot Analysis

1. Sodium dodecyl sulfate (SDS)-polyacrylamide gel electrophoresis (PAGE).
2. Separation buffer (3×): 1.125 M Tris-HCl, 0.3% SDS, pH 8.8.
3. Stacking buffer (5×): 0.625 M Tris-HCl, 0.5% SDS, pH 6.8.
4. Stacking gel (5 mL) 5%: acryl-bis solution (40% acrylamide) 0.625 mL, stacking buffer (5×) 1 mL, ddH_2O 3.375 mL, ammonium persulfate (APS) (30%, prepared fresh) 20 μL, *N,N,N,N′*-tetramethyl-ethylenediamine (TEMED, Bio-Rad) 5 μL.
5. Separation gel (10 mL) 10%: Acryl-bis solution (40% acrylamide) 2.5 mL, separation buffer (3×) 3.33 mL, ddH_2O 4.17mL, APS (30%) 20 μL, TEMED 5 μL.
6. Electrode buffer (10×): 250 mM Tris-HCl, 129 mM glycine, 1% SDS.
7. Sample loading buffer (5×): 250 mM Tris-HCl, 10% SDS, 0.5% bromophenol blue, 50% glycerol, pH 6.8.
8. Mini-PROTEAN II slab gel unit (Bio-Rad) or comparable apparatus.
9. Rainbow™ colored protein molecular weight marker (Amersham).
10. Hybond-C (Amersham) supported nitrocellulose membranes.

11. Trans-Blot cell (Bio-Rad) or comparable apparatus.
12. Transfer buffer: 25 mM Tris-base, 192 mM glycine, 20% methanol.
13. Primary antibodies against GFP (Clontech) to detect the presence of the GFP fusion recombinant proteins.
14. Secondary antibodies: alkaline phosphatase conjugated anti-mouse IgG (whole molecule) (Sigma).
15. Blocking solution: 5% nonfat milk in Tris-buffered saline (TBS; 20 mM Tris-HCl, pH 7.5, 150 mM NaCl).
16. TBS with Tween (TTBS): 20 mM Tris-HCl, pH 7.5, 150 mM NaCl, 0.05% (v/v) Tween-20.

3. Methods

The GFP tag enables relatively quick confirmation of antigen expression in plant cells by fluorescent microscopy. By simple infiltration of *Agrobacterium* cells carrying appropriate gene constructs into tobacco plants leaves, transient expression assays can be performed within 3 days without using expensive instruments or complicated procedures. Two days after agroinfiltration, expression and subcellular localization of the GFP fusion proteins in tobacco leaves can be determined by simple observation under fluorescence or confocal laser scanning microscopy. In addition, RNA and protein can be extracted from the agroinfiltrated leaves for reverse transcription-polymerase chain reaction (RT-PCR), northern blot analysis, and western blot analysis.

3.1. Construction of Vector for Expression of the SARS-CoV S1-GFP Fusion

The *SARS-CoV S1* cDNA (encoding amino acids 1–658 of the S protein) was amplified by PCR using template derived from total RNA of Vero cells infected with the *SARS-CoV (11)*. Plasmid pCRII-S1 containing the *SARS-CoV S1* cDNA was used as a mutagenesis template for generating derivatives with optimized codon usage for heterogeneous gene expression in plants. Mutagenesis was carried out by PCR with *Pfu-Turbo* DNA polymerase using the QuikChange Multi Site-directed Mutagenesis Kit (Stratagene), according to the manufacturer's specifications.

1. Perform PCR using the following conditions for amplification of derivatives with optimized codon usage: 95 °C for 30 s followed by 12–18 cycles of 95 °C 30 s, 55 °C for 1 min, 68 °C for 1 min/kb of plasmid length.
2. After PCR, add 1 μL of *Dpn*I restriction enzyme (10 U/μL) directly to the amplification reaction and incubate at 37 °C for 1 h to digest the nonmutated supercoiled dsDNA.

3. Transform supercompetent *Escherichia coli* DH5α bacteria using 1 μL of the *Dpn*I-treated DNA from each reaction.
4. Select clones for extraction of plasmid DNA to verify the change in DNA sequence.
5. The oligonucleotides we used in site-directed mutagenesis are shown in **Table 1**.
6. A total of 13 nucleotide changes were incorporated for an "optimized" S1. Use the optimized S1 for cloning into nuclear transformation vector.
7. Generate plasmid pCV12 (**Fig.1**) or similar nuclear transformation vector for expression of a fusion protein consisting of the SARS-CoV S1 and GFP. This was generated by cloning a 0.8-kb *Xho*I-*Spe*I GFP fragment from plasmid pGFP2 in the *Sal*I-*Xba*I sites of pCRII-S1 to create an S1-GFP fusion *(12)*. Subsequently,

Table 1
Oligonucleotides used for site-directed mutagenesis of S1 for codon optimization (reproduced from ref. *(12)* with permission from the Society for Experimental Biology and Medicine)

Affected residue	Sequence of primer
R18	5′-GTAGTGACCTTG*ACA*GATGCACCACTTTTGAT-3′
T75	5′-GGGTTTCATACTATTAATCAT*ACT*TTTGGCAACCCTGTCATAC-3′
S113	5′-CCATGAACAACAAGTCACAGTC*T*GTGATTATTATTAACAATTCTACT-3′
S169	5′-AGTACATATCTGATGCCTTTTC*T*CTTGATGTTTCAGAAAAGTC-3′
L209	5′-CCTATAGATGTAGTTCGTGATCT*T*CCTTCTGGTTTTAACACTTTG-3′
T247	5′-CAAGACATTTGGGGCAC*T*TCAGCTGCAGCCTAT-3′
A398	5′-GATGATGTAAGACAAATAGC*T*CCAGGACAAACTGG-3′
P507	5′-TCTTTTGAACTTTTAAATGCACC*T*GCCACGGTTTGTGGACC-3′
T509	5′-CTTTTAAATGCACCTGCCAC*T*GTTTGTGGACCAAAATTATC-3′
L597	5′-CTTCATCTGAAGTTGCTGTTCT*T*TATCAAGATGTTAACTGCAC-3′
R620	5′-CAACTCACACCAGCTTGG*AGA*ATATATTCTACTGGAAACAATG-3′

Nucleotides in italics are mutated; the altered codons are underlined

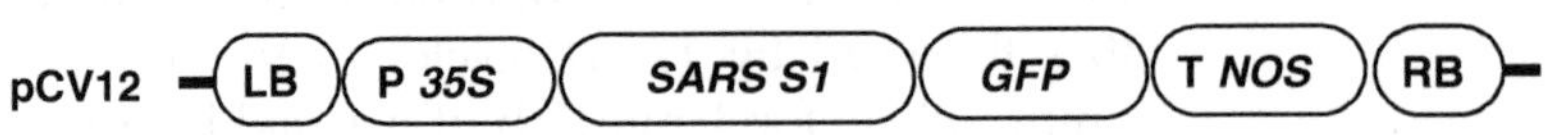

Fig. 1. Plasmid pCV12 that was used for expression of S1 tagged to GFP in plant cells. The plasmid pCV12 is a plant nuclear transformation vector used for the expression of a protein fusion consisting of the SARS-CoV S1 protein fused translationally with GFP. RB and LB represent right and left borders of T (transfer)-DNA for random insertion into the plant nuclear genome; P*35S*, Cauliflower Mosaic Virus *35S* promoter; T*NOS*, *NOS* terminator.

a 2.8-kb *Bam*HI-*Apa*I fragment containing the S1 cDNA (encoding amino acids 1–643) and *GFP* cDNA was cloned into the *Bgl*II-*Apa*I sites of a plant binary vector, pGD *(12)*.

3.2. Agroinfiltration of Tobacco for Expression of the S1:GFP Antigen

Production of tobacco leaves transiently expressing a protein fusion consisting of the SARS-CoV S1 protein fused with the GFP was carried out using *Agrobacterium*-mediated transformation with plasmid pCV12.

1. Grow *Agrobacterium* LBA4404/pCV12 for infiltration overnight with shaking (200 rpm) at 28 °C in YEP medium supplemented with 10 mM $MgSO_4$, 100 μg/mL of streptomycin and, in the case of pGD, plus 50 μg/mL of kanamycin.
2. Subsequently, initiate a new culture by inoculating fresh medium with the overnight culture at 1:10 dilution.
3. After growing for 5–7 h (OD_{600} should be ~0.8–1.0), harvest the bacterial cells by centrifugation for 10 min at 1,000 × *g* at 4 °C.
4. Wash the cells twice with 10 mM $MgSO_4$, and then resuspend in 10 mM MES (pH 5.5), 10 mM $MgSO_4$ supplemented with 100 μM AS to form a dilute bacterial suspension with an absorbance of ~0.6 at 600 nm. Keep the suspension for 2–4 h at room temperature before infiltration.
5. Make some tiny wounds using a needle in the region where infiltration will apply. Using a syringe lacking a needle, press the syringe tip on the wounds against the lower side of a tobacco leaf, meanwhile applying counterpressure to the upper leaf, so that the bacterial suspension can be efficiently introduced through the stomata to the leaf lamina. It generally takes a few minutes to do one infiltration. There is no need to wash off the bacterial solution outside the leaves following the procedure. Freshly expanded young leaves are preferentially used for better expression (*see* **Notes 1–4**).

3.3. Analyses on Transient-Expressed Proteins

3.3.1. Laser-Scanning Confocal Microscopy

1. Two days after agroinfiltration, the expression and subcellular localization of transiently-expressed GFP fusion proteins can be determined. Representative tobacco leaf epidermal cells are selected and observed by fluorescence microscopy or confocal microscopy.
2. Compare tobacco leaf cells subjected to agroinfiltration using *Agrobacterium* LBA4404 harboring plasmid pCV12 (expressing the S1:GFP-fusion protein) and those agroinfiltrated with LBA4404 harboring pGDG (control expressing GFP alone).
3. Detect fluorescence and collect digital images using a Zeiss LSM 510 inverted confocal laser-scanning microscope equipped with helium/neon lasers or comparable microscope. GFP fluorescence excites at 488 nm, is filtered through a primary dichroic (UV/488/543), a secondary dichroic of 545 nm, and subsequently through BP505-530 nm emission filters to the photomultiplier tube detector.

4. The images can be processed using the LSM 510 software (Zeiss) and Photoshop 6.0 (Adobe) or comparable software. Images of transient expression of S1:GFP antigen in the cytosol of agroinfiltrated tobacco leaves are shown in **Fig. 2.** Expression of the construct with GFP alone was not confined to the cytosol. Nuclear expression is seen because of the accumulation of passive diffusion of GFP into the nucleus *(13)*. The S1:GFP fusion was localized to the cytosol, and also to the periphery of the nucleus as well, implicating that S1:GFP was secreted through the endoplasmic reticulum network. Depending on total experimental conditions, the expression of the target gene at the infiltrated area can vary. Under optimal conditions, almost all cells at the infiltrated area show expression, like in **Fig. 2.** Generally, it should not be difficult to get expression in more than 50% of plant cells at the infiltrated area.

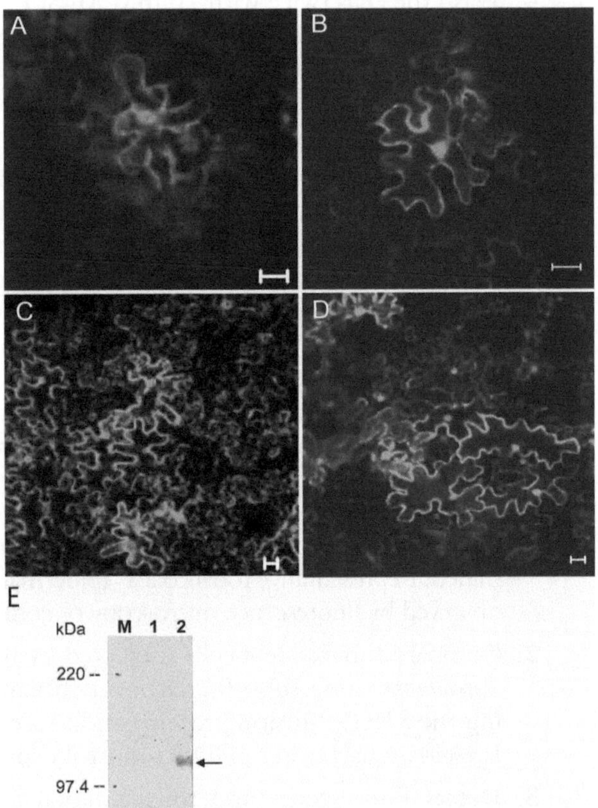

Fig. 2. Transient expression of S1:GFP in agroinfiltrated tobacco leaves. Representative tobacco leaf epidermal cells are shown by confocal microscopy 2 days following agroinfiltration of *Agrobacterium tumefaciens* LBA4404 harboring plasmid pCV12 expressing S1:GFP fusion protein (**A, C**) or LBA4404 harboring pGDG expressing GFP alone (**B**, **D**). Bar represents 20 μm. Western blot analysis using antibodies against GFP shows transient expression of S1:GFP in tobacco leaves following agroinfiltration. GFP alone (*lane 1*) or plasmid pCV12 expressing S1:GFP fusion (*lane 2*). Arrow indicates cross-reacting S1:GFP band (calculated size 99.1 kDa). **M**, molecular mass markers. This figure is reproduced from **ref.** *(12)* with permission from the Society for Experimental Biology and Medicine.

3.3.2. Western Blot Analysis

To confirm the transient expression of S1:GFP in tobacco leaves following agroinfiltration, western blot analysis was carried out using antibodies against GFP. Total proteins were extracted from tobacco leaves infiltrated with plasmid pGDG expressing GFP alone or plasmid pCV12 expressing the S1:GFP fusion.

1. Harvest plant tissues and immediately freeze in liquid nitrogen. Grind using a mortar and pestle in protein homogenizing buffer. Use about 10 mL of ice-cold homogenizing buffer per 1 g of tissue.
2. Boil ground tissue at 100 °C for 10 min, centrifuge at 14,000 rpm for 5 min, and transfer the supernatant into a new Eppendorf and store at –20 °C. Determine the protein content.
3. Denature the protein samples in 1× SDS sample loading buffer by heating at 95 °C for 5 min before loading onto the gel. Separate protein samples (20 μg) by SDS-PAGE using a 10% gel using a Mini-PROTEAN II slab gel unit or comparable apparatus.
4. Use Rainbow™ colored protein molecular weight markers in gels destined for transfer to Hybond-C membrane.
5. Electrophoretically transfer proteins to Hybond-C membrane from SDS-PAGE gel using a Trans-Blot cell at 50 V for 1.5 h in cold transfer buffer. A piece of Hybond-C membrane, cut to the size of the gel and wetted with distilled water, was placed on top of the gel without inclusion of any air bubbles. The gel and the membrane were then sandwiched between two pieces of Whatman No. 3 paper that were also cut to the size of the gel and were presoaked in transfer buffer (*see* **Note 5**).
6. For detection using anti-GFP antibodies in western blot analysis, incubate the blot for 1 h at room temperature (RT) in blocking solution with gently shaking.
7. After washing the blots twice for 10 min in TTBS, then incubated for 2 h at room temperature in monoclonal anti-GFP antibodies diluted 1:5,000 in TTBS with gentle agitation.
8. Wash the blot twice for 10 min with TTBS, followed by incubation in anti-mouse IgG alkaline phosphatase diluted 1:30,000 in TTBS for 1 h at room temperature with gentle shaking.
9. Wash in TTBS four times for 10 min, and stain the blot in freshly prepared color development solution at room temperature until the reacting protein bands are visible.
10. Stop the reaction by washing the membrane twice for 5 min in water. Results of western blot analysis using antibodies against GFP of total plant proteins extracted from tobacco leaves infiltrated with plasmid pCV12 (expressing the S1:GFP fusion) are shown in **Fig. 2e** (lane 2 shows an expected band of calculated size 99.1 kDa, demonstrating that the S1:GFP fusion protein was successfully expressed).

Notes

1. The technique is suitable for leaves or petals. *Agrobacterium* has to be introduced into the leaves through the stomata or via wounding. In transient gene expression, the levels in expression may be age dependent and may differ in leaf tissues at varying stages of development. Generally, the highest level of transient expression is observed in tissues that contain recently expanded leaf cells.
2. Agroinfiltration can also be used in conjunction with vacuum infiltration on either detached or attached plant leaves. With the use of detached leaves, the leaves should be best incubated on wet filter paper rather than on solidified MS medium following agroinfiltration since incubation on MS often gave lower expression levels *(8)*.
3. The pH of the *Agrobacterium* resuspension can be within a range of pH 5.0–5.8. We usually use pH 5.5. The concentration of *Agrobacterium* cells is important for infiltration. The optimal concentration of the *Agrobacterium* suspension is OD_{600} = 0.4–0.6. The *Agrobacterium* culture is grown to log phase and then resuspended to OD_{600} = 0.4–0.6 for better expression. If the bacterial cells are out of log phase, the efficiency could be much decreased.
4. Acetosyringone is usually added in the agroinfiltration medium to increase the efficiency. It has been demonstrated in Chinese Cabbage that agroinfiltration can be improved by adding hydrogen peroxide *(14)*. Since the plasmid vector for agroinfiltration does not necessarily carry an antibiotic-resistance selectable marker gene, a smaller vector can be considered for use to enhance T-DNA transfer in plant cells by agroinfiltration.
5. Make sure there are no air bubbles between nitrocellulose membrane and gel or else certain regions will not be transferred to the membrane. Alternatively, the blot can be assembled under transfer buffer to avoid trapping bubbles.

Acknowledgments

We thank Prof. J.S.M. Peiris and Dr. L.L.M. Poon (Department of Microbiology, University of Hong Kong) for gift of plasmid pCRII-S1, Prof. N.-H. Chua (Rockefeller University, New York) for plasmid pGFP-2, and Dr. M.M. Goodin (University of California, Berkeley) for plasmids pGD and pGDG. We are grateful to the Society for Experimental Biology and Medicine for permission to use previously published figures that had appeared in **ref.** *(12)*.

References

1. Walmsley, A.M. and Arntzen, C.J. (2003) Plant cell factories and mucosal vaccines. *Curr. Opin. Biotechnol.* 14, 145–150.
2. Porgrebnyak, N., Golovkin, M., Andrianov, V., Spitsin, S., Smirnov, Y., Egolf, R., and Koprowski, H. (2005) Severe acute respiratory syndrome (SARS) S protein production in plants: development of recombinant vaccine. *Proc. Natl. Acad. Sci. USA* 102, 9062–9067.
3. Dantuma, N.P., Lindsten, K., Glas, R., Jellne, M., and Masucci, M.G. (2000) Short-lived green fluorescent proteins for quantifying ubiquitin "proteasome-dependent proteolysis in living cells". *Nat. Biotechnol.* 18, 538–543.
4. Kloetzel, P.M. (2004) Generation of major histocompatibility complex class I antigens: functional interplay between proteasomes and TPPII. *Nat. Immunol.* 5(7), 661–669.
5. Gelvin, S.B. (2003) *Agrobacterium*-mediated plant transformation: the biology behind the "gene-jockeying" tool. *Microbiol. Mol. Biol. Rev.* 67, 16–37.
6. Yoshioka, Y., Takahashi, Y., Matsuoka, K., Nakamura, K., Koizumi, J., Kojima, M., and Machida, Y. (1996) Transient gene expression in plant cells mediated by *Agrobacterium tumefaciens*: application for the analysis of virulence loci. *Plant Cell Physiol.* 37, 782–789.
7. Schob, H., Kunz, C.,andMeins, F., Jr. (1997) Silencing of transgenes introduced into leaves by agroinfiltration: a simple, rapid method for investigation of sequence requirements for gene silencing. *Mol. Gen. Genet.* 256, 581–585.
8. Kapila, J., De Rycke, R., Van Montagu, M., and Angenon, G. (1997) An *Agrobacterium-mediated* transient gene expression for intact leaves. *Plant Sci.* 122, 101–108.
9. Goodin, M.M., Dietzgen, R.G., Schichnes, D., Ruzin, S., and Jackson, A.O. (2002) pGD vectors: versatile tools for the expression of green and red fluorescent protein fusions in agroinfiltrated plant leaves. *Plant J.* 31, 375–383.
10. Hoekema, A., Hirsch, P.R., Hooykas, P.J.J., and Schilperoort, R.A. (1983) A binary plant vector strategy based on separation of vir and T-region of the A. tumefaciens Ti plasmid. *Nature* 303, 179–180.
11. Poon, L.L.M., Wong, O.K., Luk, W., Yuen, K.Y., Peiris, J.S.M., and Guan, Y. (2003) Rapid diagnosis of a coronas associated with severe acute respiratory syndrome (SARS). *Clin. Chem.* 49, 953–955.
12. Li, H.Y., Ramalingam, S., and Chye, M.L. (2006) Accumulation of recombinant SARS-CoV spike protein in plant cytosol and chloroplasts indicate potential for development of plant-derived oral vaccines. *Exp. Biol. Med.* 231, 1346–1352.
13. Gorlich, D. and Mattaj, I.W. (1996) Nucleocytoplasmic transport. *Science* 271, 1513–1518.
14. Shin, D.-I. and Park, H.-S. (2005) Transient expression in Chinese Cabbage by hydrogen peroxide-aided agroinfiltration. *Agric. Chem. Biotechnol.* 4, 229–230.

References

Chapter 20

Studies of the Role and Function of Barley Stripe Mosaic Virus Encoded Proteins in Replication and Movement Using GFP Fusions

Sophie Haupt, Angelika Ziegler, Graham Cowan, and Lesley Torrance

Summary

This chapter describes techniques to investigate the localisation and function of virus-encoded proteins in plants using green fluorescent protein (GFP) transiently expressed from plasmids or infectious cDNA reporter clones of barley stripe mosaic virus. Virus movement and the localisation of GFP-tagged proteins in living cells were monitored by confocal laser scanning microscopy (CLSM). In addition, GFP expression was imaged in transgenic plants where specific organelles or subcellular structures such as endoplasmic reticulum were labelled with another fluorophore (e.g., monomeric red fluorescent protein). Using these approaches we discovered evidence for additional roles played by virus encoded movement protein TGB2 and γb protein in virus replication. Methods are described for clone construction and mutagenesis, and for transient expression (biolistic bombardment or agrobacterium infiltration) in the epidermal cells of *Nicotiana benthamiana* or barley. In addition, techniques for chloroplast isolation and imaging of the different fluorescent proteins, and the avoidance of interference from autofluorescence, are described.

Key words: Barley stripe mosaic virus, Confocal laser scanning microscopy, Virus-based vectors, Virus protein function, Virus replication, Chloroplasts, Microprojectile bombardment, Binary vectors, Run-off transcripts, In vivo imaging.

1. Introduction

Most plant virus genomes are relatively small (~6–10 kb) comprising positive-sense single-stranded RNA that encode a few structural and non-structural proteins *(1)*. Non-structural proteins include those for genome replication as well as movement, both cell-to-cell and long distance in the vascular system, for the

Barry W. Hicks (ed.), *Methods in Molecular Biology, Viral Applications of Green Fluorescent Protein, vol. 515*

DOI: 10.1007/978-1-59745-559-6_20

systemic infection of host plants. In addition, virus-encoded proteins can play a role in the regulation of genome translation or replication and the avoidance of host defence responses, for example, by suppression of RNA silencing *(2, 3)*. Different approaches have been used to identify the role and function of these proteins and this work has been greatly facilitated by the availability of infectious cDNA clones. Viral clones can be engineered to express GFP or other fluorophores, either as free soluble markers or fused to specific virus proteins. The GFP-tagged proteins can be expressed transiently after biolistic bombardment in epidermal cells or from the virus vector. Examination of epidermal cells by confocal laser scanning microscopy (CLSM) enables monitoring of intracellular and systemic movement of the virus infection process (free GFP) or identification of sites of localisation of individual GFP-tagged proteins *(4, 5)*. In addition, mutant clones can be prepared where a protein has been deleted or modified to examine the effect of a particular protein or amino acid motif on infection or localisation. Transient expression of such clones or fusion proteins in transgenic plants where specific organelles have been labelled with GFP (or red fluorescent protein; RFP) has revealed information on sub-cellular organelle targeting and association with plant proteins *(4)*.

These approaches have been used to examine the role and function of virus-encoded proteins of barley stripe mosaic virus (BSMV). The BSMV has a tripartite genome that encodes seven or eight proteins (**Fig. 1**). Viral movement requires the coordinated action of three proteins encoded by a conserved genetic module called the triple gene block (TGB). The BSMV also encodes capsid and replicase proteins and a protein called γb that enhances pathogenicity and suppresses host defence responses *(3, 6)*. We studied the localisations of GFP-TGB2 and GFP-γb when expressed either alone, or from a modified virus vector.

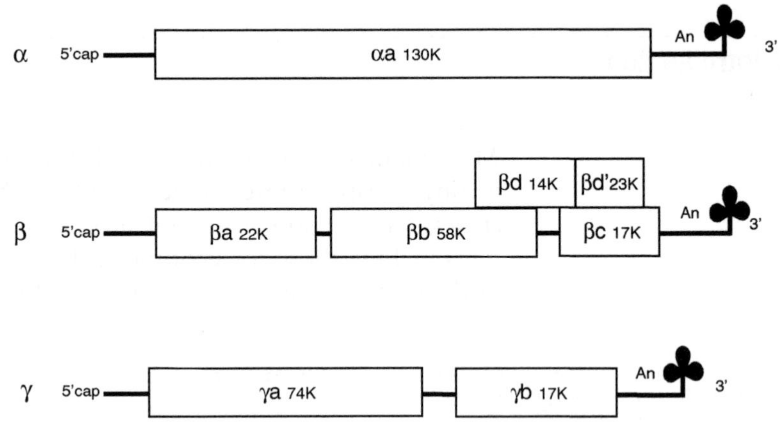

Fig. 1. Diagram of BSMV genome.

Green fluorescence from GFP-γb was cytosolic when expressed transiently under the control of the cauliflower mosaic virus 35S promoter but localised to the chloroplast envelope and inclusions within the chloroplast when expressed together with virus RNA *(5)*. Furthermore, green fluorescence from GFP-TGB2 was seen in membranes of the endoplasmic reticulum, motile spots, and vesicles when expressed alone but was also observed at the chloroplast envelope when coexpressed with virus RNA *(5)*. To verify association with chloroplasts we used standard methods to isolate leaf plastids and detect the presence of virus genome by RT-PCR and coat protein by western blots *(7)*. In this work we showed that in addition to the known functions in virus movement and enhancement of pathogenicity, these two proteins may play a role at the site of virus replication, possibly by helping to establish virus replication complexes in chloroplasts.

2. Materials

2.1. Chloroplast Isolation

1. Grinding buffer: 0.35 M Sorbitol, 0.05 M HEPES adjusted to pH 7.5 with KOH, 2 mM EDTA, 0.5 mM $MgCl_2$, 1 mM DTT, 10 mg/mL BSA.
2. Sorbitol buffer: 0.35 M Sorbitol, 35 mM HEPES adjusted to pH 8.3 with KOH, 10 mM K_2HPO_4, 0.5 mM $MgCl_2$, 1 mM DTT.
3. Percoll buffer: 40 mM HEPES adjusted to pH 7.5 with KOH, 0.05 mM $MgCl_2$, 0.35 M Sorbitol, 1 mM DTT.

2.2. RNA Extraction, cDNA Synthesis, and RT-PCR

1. Qiagen RNeasy plant mini kit.
2. Transcriptor reverse transcriptase for cDNA synthesis (Roche) (*see* **Note 1**).
3. Standard PCR reagents.
4. Standard apparatus and reagents for agarose gel electrophoresis.

2.3. Production of GFP-Expressing BSMV

2.3.1. Virus-Based Reporter Clones

1. BSMV clones are based on the ND18 strain of BSMV *(8)*.
2. The construct encoding RNA γ must be modified so that GFP is fused to the C-terminus of γb as in **ref.** *(6)*. For transcription, these cDNAs are linearised for use as templates (*see* **Subheading 2.3.4**).
3. Also, a second set of BSMV clones must be constructed under the control of the cauliflower mosaic virus 35S promoter and the gene encoding γb was fused in frame to the sequence of the foot-and-mouth virus 2a peptide *(9)* and GFP. When expressed, a proportion of the γb remains fused to GFP at the N-terminus (Pogue, personal communication).

2.3.2. Fusion Constructs

Splicing by overlap extension involves three separate PCR reactions *(10)*. Using this method, both 5′ and 3′ fusions to the gene of interest can be produced. The two fragments produced in the first-stage reactions will have overlapping sequence and they form the template for the second stage. Four primers are required for each construct, two flanking primers and two hybrid primers. When the two first-stage products are mixed, they can partially anneal and contribute to the amplification of the hybrid gene.

1. Primers: Provided lyophilised by supplier. Make stock solutions in distilled sterile water and keep frozen.
2. dNTPs: Keep concentrated frozen stocks in the freezer. Keep working stock solutions of 10 mM dNTP in small aliquots to avoid repeated cycles of thawing and freezing.
3. PCR reaction buffer: Use the buffers supplied with the polymerase enzyme.
4. $MgCl_2$: The amount of $MgCl_2$ required must be optimised for each experiment so it is advisable to use a supplier that provides the PCR reaction buffer with separate $MgCl_2$ solutions.
5. Polymerase: There are many good quality heat-stable DNA polymerases available from different suppliers, including proofreading enzymes (e.g., *Pfu* (Stratagene) or Phusion High Fidelity (New England Biolabs)).
6. TBE buffer.
7. TBE agarose gel.
8. Vectors. We routinely use the plasmid vector pRTL2 *(11)* for transient expression from a CaMV 35S promoter. In addition, virus-based vectors such as TMV30B *(12)* and binary vectors such as pGREEN *(13)* can be used.
9. LB medium for bacterial growth.
10. Antibiotics (e.g., ampicillin). Keep stocks at –20 °C.
11. Competent *E. coli* cells.
12. Ligase and ligase buffer.

2.3.3. Mutagenesis

The introduction of mutations in reporter clones can help with elucidating the role and function of a protein of interest. The Stratagene QuikChange Site-Directed Mutagenesis Kit (Cat #200523) is based on synthetic oligonucleotide primers and *Dpn*I digestion of methylated plasmid DNA. This results in a very low background of wild-type plasmid. The system allows point mutations and deletions or insertions of single or multiple adjacent amino acids. Because of a high fidelity DNA polymerase (*Pfu* Turbo), unwanted second-site errors are virtually eliminated.

2.3.4. Transcription

1. RiboMAX™ Large Scale RNA Production system-T7 (Promega). Transcripts are synthesised from linearised full-length viral cDNA using the manufacturer's protocol.

2.4. Transformation and Cultures of Agrobacterium; Leaf Infiltration

1. *Agrobacterium* strain LBA4404 (Clontech Laboratories, Inc., Palo Alto, CA).
2. SOC medium.
3. Sterile distilled water.
4. 10% glycerol.
5. Electroporator.
6. Selective plates.
7. LB containing antibiotics, 10 mM MES, and 20 μM acetosyringone.
8. 10 mM $MgCl_2$, 10 mM MES pH 5.6, and 150 μM acetosyringone.
9. 2 mL syringe.

2.5. Biolistic Inoculation

1. Plasmid DNA or transcript RNA from viral constructs.
2. 100% ethanol.
3. Micro-gold carrier (Bio-Rad Laboratories, Hercules, CA): for preparation of a stock solution wash about 50 mg of the microcarrier in 1 mL of 100% ethanol for 1 h, then wash four times in sterile water, pellet the clean microcarrier, and resuspend in 1 mL sterile water.
4. Discharge assembly (13-mm Plastic Swinney Filter Holder, PALL Gelman Laboratory, Ann Arbor, MI).
5. 6- to 8-week-old plants.
6. Handgun *(14)* (**Fig. 2a**).

Fig. 2. Biolistic bombardment and chloroplast isolation. **A** Purpose-built device for biolistic bombardments. **B** Fully mature source leaf from *Nicotiana benthamiana* after biolistic inoculation. **C** Fully mature source leaf from *Nicotiana benthamiana* showing area after infiltration. **D** Close-up images of vesicles, showing fluorescence as a ring surrounding discrete patches of fluorescence; the excitation wavelength is 584 nm and emission wavelength is: **i** 600–620 nm, **ii** 660–700 nm, and **iii** overlay. **E** Emission wavelength scan (495–755 nm) of isolated vesicles (**Dii**) showing emission peak at 680 nm (typical of chlorophyll *a*). **F** Chloroplast isolation after Percoll gradient. The *arrow* indicates the band of intact chloroplasts; the *arrowhead* indicates a disrupted membrane fraction.

2.6. Imaging and Image Processing

1. Double-sided adhesive tape (Sellotape GB, Dunstable, UK).
2. Microscope glass slides.
3. Confocal laser scanning microscope Leica SP2 with Leica water immersion lenses for live cell work (CLSM, Leica Microsystems, Heidelberg, Germany).
4. Leica SP2 confocal software (Leica Microsystems, Heidelberg, Germany) or comparable.
5. Adobe Photoshop 8.0 (Mountain View, CA) or comparable imaging software.

3. Methods

3.1. Chloroplast Isolation (15)

1. Grind 1 g of leaf material in 7 vol. of ice-cold grinding medium and filter through four layers of muslin (*see* **Notes 2–4**).
2. Centrifuge the extract at 1,000 × *g* for 3 min at 4 °C with the brake on. The chloroplasts are in the pellet which is very loose, so, carefully decant and discard the supernatant before gently, but fully, resuspending in 1 mL Sorbitol medium (*see* **Note 5**).
3. Layer the sample onto a Percoll step gradient comprising 1 mL of 40% and 1 mL 85% Percoll (prepared in Percoll buffer) and centrifuge at 13,000 × *g* for 7 min at 4 °C with the brake off. Recover the intact chloroplasts from the interphase between the 40 and 85% Percoll using a Pasteur pipette and dilute with 5 vol. of Sorbitol medium, then centrifuge at 4,000 × *g* for 5 min at 4 °C with the brake on (**Fig. 2f**).
4. Resuspend the intact chloroplasts in the same volume of Sorbitol medium and centrifuge as before, completing the washing step, then resuspend in 0.5 mL Sorbitol medium.

3.2. RNA Extraction, cDNA Synthesis, and RT-PCR

1. Extract RNA using a Qiagen RNeasy plant mini kit, following the protocol supplied by the manufacturer. Prior to cDNA synthesis, RNA was treated with RNase-free DNaseI (DNA-free™ kit, Ambion Ltd., Ambion Inc., Austin, TX), following the manufacturer's recommendations.
2. Synthesise first-strand cDNA from positive-strand viral RNA by using an antisense primer, 1 μg DNase I-treated total RNA Transcriptor Reverse Transcriptase (Roche) according to the manufacturer's instructions.
3. Add 20 μL TE buffer to the cDNA, and use 1 μL of the diluted cDNA in a PCR reaction with specific primers.
4. Analyse PCR product on a 1.2% TBE agarose gel.

3.3. Production of GFP-Expressing Constructs

3.3.1. PCR Overlap Extension and Cloning of PCR Products for Transient Expression

1. Perform two separate 50 μL PCR reactions to amplify the GFP sequence (primers 1 and 2) and the target gene (primers 3 and 4). Each reaction should contain 5 μL of 10× PCR reaction buffer (without $MgCl_2$), 1 μL of template, 1 μL of 50 mM $MgCl_2$, 2 μL of 10 mM dNTP mix, 2 μL of the relevant primers (10 mM), and sterile distilled water to a final volume of 50 μL in a 0.2-mL PCR tube. Lastly, add 0.5 μL Taq polymerase (Roche).
2. Use a thermocycler (e.g., Eppendorf Mastercycler personal) to perform 25 cycles of amplification (94 °C for 30 s, 60 °C for 1 min, 72 °C for 2 min), followed by an extension at 72 °C for 10 min.
3. Load a 5-μL aliquot onto a 1.2% TBE agarose gel. Stain gel in ethidium bromide solution. The GFP amplification should give a 700-bp band. The no-template control reaction should not produce a PCR product.
4. Set up a PCR reaction containing 2 μL of each primary PCR product, 10× buffer, primers 1 and 4, dNTP mix, $MgCl_2$, water, and polymerase.
5. Set up a negative control reaction without primary PCR products.
6. Use a thermocycler and perform 25 cycles of amplification (as **step 2**).
7. Load a 5-μL aliquot onto a 1.2% TBE agarose gel. The expected PCR product should be the combined size of the target gene and the GFP sequence (*see* **Note 6**). For the negative control reaction, no product should be visible.
8. Digest PCR product with the relevant restriction enzymes. Load reaction mix onto 1.2% TBE-agarose gel. After visualisation with ethidium bromide, gel-purify the DNA. The Qiagen Min Elute Gel Extraction Kit and the Zymogen Zymoclean Gel DNA Recovery Kit both work well (*see* **Note 7**).
9. Ligate digested PCR product and vector DNA of choice (pRTL-2 for transient expression, or a virus-based vector such as TMV 30B, or a binary vector such as pGREEN).
10. Transform competent cells (e.g., XL10 Gold, Stratagene) and plate on selective medium. Incubate plates overnight at 30 °C (*see* **Note 8**).
11. Check colonies for presence of the insert and expand in 100 mL of LB medium. Make a stock plasmid DNA preparation using a Qiagen Plasmid Hispeed Midiprep Kit. Use plasmid DNA for transformation of *Agrobacterium* or for biolistic inoculation.

3.3.2. Synthesis of RNA Transcripts from Full-Length Viral cDNA

1. Linearise viral cDNA by restriction enzyme digestion. Analyse an aliquot on a 1.2% TBE Agarose gel.
2. Use the linearised cDNA as template in a RiboMAXTM Large Scale RNA Production System-T7 (Promega) according to the manufacturer's instructions.
3. Analyse an aliquot of the transcription reaction on a 1.2% TBE Agarose gel.
4. Use transcript RNA for biolistic inoculation (see **Subheading 3.5**).

3.4. Transformation of Agrobacterium by Electroporation, Agrobacterium Cultures and Leaf Infiltration

1. Grow an overnight culture of *Agrobacterium* at 28 °C in LB.
2. Dilute the overnight culture ten times in fresh LB medium and grow until OD_{600nm} of 0.5.
3. Centrifuge at 4,000 rpm for 10 min at 4 °C.
4. Wash cells in cold sterile distilled water and centrifuge as above.
5. Wash cells in 10% glycerol (cold) and centrifuge as above.
6. Resuspend cells in 10% glycerol (1/100 original volume). Aliquots of competent cells can be stored frozen at –80 °C for several months.
7. Mix 1 μL of DNA with a 100-μL aliquot of competent cells.
8. Electroporate at 2.5 kV, 200 W, and 25 μF.
9. Add SOC medium and let the cells recover at 28 °C for 1 h with shaking.
10. Plate on selective medium.
11. Grow *Agrobacterium* culture overnight at 28 °C in 5 mL LB containing the relevant antibiotic.
12. Use this culture to inoculate 50 mL of LB containing antibiotics, 10 mM MES, and 20 μM acetosyringone (*see* **Note 9**).
13. Pellet cells by centrifugation and resuspend pellet in a solution containing 10 mM $MgCl_2$, 10 mM MES, and 150 μM acetosyringone. The concentration of *Agrobacterium* should be 0.5 OD_{600}.
14. Leave the solution at room temperature for 2–3 h, then load it into a 2-mL syringe.
15. Choose a fully mature "source" leaf on a 6- to 8-week-old plant.
16. Pierce a small hole into the leaf at the tip avoiding major veins (intercostal fields, **Fig. 2B**), using a small pipette tip.
17. Infiltrate by pressing a syringe without a needle firmly against the leaf surface while supporting the leaf with your hand

during the process to avoid tearing. The infiltrated area will appear shiny.

18. Leave plant under light source and well watered for 1–2 days.
19. Detach leaf and examine under the confocal microscope.

3.5. Biolistic Inoculation

1. Mix 0.5–1.0 µL of DNA or RNA transcript (40–50 ng/µL) with 5–10 µL of 100% ethanol (add 1 µL at a time).
2. Vortex 11 µL of tungsten or gold particles into the nucleic acid mixture.
3. Load between 2 and 4 µL onto the nozzle and let the ethanol evaporate.
4. Put nozzle into the holder and screw it tight into the gun.
5. Raise plant on stage until target leaf is 2–3 cm under the nozzle (*see* **Note 10**).
6. Use up to 4 shots per loading inoculating areas at the tip of the leaf while avoiding the major veins (**Fig. 2A, B**).
7. Support the leaf with your hand when discharging the gun to avoid tearing.

3.6. Imaging and Image Processing

1. Cover microscope slide with double-sided adhesive tape.
2. Cut out leaf area of interest and press gently but firmly onto microscope slide until it is completely flattened using your thumb or palm of your hand.
3. View under a confocal microscope (Leica SP2) using the following excitation and emission wavelengths in **Table 1** (*see* **Notes 11–17**).
4. Images are processed for size and resolution using Adobe Photoshop.

Table 1
Excitation and emission wavelengths for confocal fluorescence microscopy

	PMT position	Excitation (nm)	Emission (nm)
GFP	1	488	500–530
mRFP	2	561	570–635
Chlorophyll	3	488	670–700

Notes

1. This transcriptase works well; its themostability helps to avoid problems due to RNA secondary structure and GC content. The incubation temperature can be increased up to 65 °C.
2. For optimal yield of chloroplasts, use actively growing plants.
3. Plants are best kept in the dark for 18 h before chloroplast isolation to avoid increased levels of starch granules which can rupture the chloroplast envelope during centrifugation.
4. The ratio of grinding buffer to amount of leaf material is dependent on the source of plant material used.
5. Perform all steps on ice or at 4 °C.
6. The position of the GFP (e.g., N- or C-terminal) can have an effect on the function of the gene studied.
7. The Qiagen Min Elute Gel Extraction Kit and the Zymogen Zymoclean Gel DNA Recovery Kit both work well. Both have the advantage that the purified DNA can be recovered in a very small amount of buffer (8 μL), which results in a higher concentration of DNA. Ethidium bromide is toxic and a potent mutagen; avoid any direct contact.
8. In our hands, incubation at 30 °C as opposed to 37 °C increases the plasmid stability.
9. Acetosyringone actively supports the transfer of T-DNA from *Agrobacterium* to the plant by inducing the *vir* functions.
10. Keep plant 3 days prior to bombardment in a growth chamber with high humidity to increase infectivity.
11. When using combinations of GFP- and RFP-labelled fusion genes, sequential scanning must be used to avoid cross-talking because of overlapping emission spectra.
12. Lambda scanning can be used to identify the emission spectrum of a specific fluorochrome (e.g., autofluorescence (**Fig. 2e**) or GFP) and help to optimise the detection. This helps as well in discriminating autofluorescence from fluorescence originated by fluorophores like GFP or mRFP.
13. When imaging moving objects like ER or Golgi, it is not advisable to use the "Average" function to improve the signal-to-noise ratio. However, decreasing the laser speed from 400 to 200 Hz will considerably improve image quality.
14. Using the 488 nm excitation wavelength for GFP results in autofluorescence from chloroplasts. By imaging the chloroplasts with a separate photo multiplier tube (PMT), the autofluorescence can be subtracted from the images.
15. Keeping the gain as low as possible will increase image quality. This will also keep the background noise low. For optimisation of gain and threshold, use "Qlut" function.

16. For sequential scanning, separate PMT have to be used for emission spectra.
17. Good tutorials websites for the Leica SP2:

    http://www.hi.helsinki.fi/amu/AMU%20Cf_tut/cf_tut_part2–6c.htm

    http://www.confocal-microscopy.com/WebSite/SC_LLT.nsf

Acknowledgement

We thank Andy Jackson, Greg Pogue and Merete Albrechtsen for providing BSMV clones.

References

1. Harrison, B. D., and Robinson, D. J. (2005) Another quarter century of great progress in understanding the biological properties of plant viruses. *Ann Appl Biol.* 146, 15–37.
2. Donald, R. G. K., and Jackson, A. O. (1996) RNA-binding activities of barley stripe mosaic virus &ggr;b fusion proteins. *J Gen Virol.* 77, 879–888.
3. Bragg, J. N., and Jackson, A. O. (2004) The C-terminal region of the Barley stripe mosaic virus &ggr;b protein participates in homologous interactions and is required for suppression of RNA silencing. *Mol Plant Pathol.* 5, 465–481.
4. Haupt, S., Cowan, G. H., Ziegler, A., Roberts, A. G., Oparka, K. J., and Torrance, L. (2005) Two plant–viral movement proteins traffic in the endocytic recycling pathway. *Plant Cell.* 17, 164–181.
5. Torrance, L., Cowan, G. H., Gillespie, T., Ziegler, A., and Lacomme, C. (2006) Barley stripe mosaic virus-encoded proteins triple-gene block 2 and γb localize to chloroplasts in virus-infected monocot and dicot plants, revealing hitherto-unknown roles in virus replication. *J Gen Virol.* 87, 2403–2411.
6. Lawrence, D. M., and Jackson, A. O. (2001) Interactions of the TGB1 protein during cell-to-cell movement of Barley stripe mosaic virus. *J Virol.* 75, 8712–8723.
8. Petty, I. T. D., Hunter, B. G., and Jackson, A. O. (1988) A novel strategy for one-step cloning of full-length cDNA and its application to the genome of barley stripe mosaic virus. *Gene.* 74, 423–432.
9. Ryan, M. D., King, A. M. O., and Thomas, G. P. (1991) Cleavage of foot-and-mouth-disease virus polyprotein is mediated by residues located within a 19 amino acid sequence. *J Gen Virol.* 72, 2727–2732.
10. Horton, R. M., Hunt, H. D., Ho, S. N., Pullen, J. K., and Pease, L. R. (1989) Engineering hybrid genes without the use of restriction enzymes: gene splicing by overlap extension. *Gene.* 77, 61–68.
11. Restrepo, M. A., Freed, D. D., and Carrington, J. C. (1990) Nuclear transport of plant potyviral proteins. *Plant Cell.* 2, 987–998.
12. Shivprasad, S., Pogue, G. P., and Lewandowski, D. J. (1999) Heterologous sequences greatly affect foreign gene expression in tobacco mosaic virus-based vectors. *Virology.* 255, 312–323.
13. Hellens R. P., Edward, A., Leyland, N. R., Bean, S., and Mullineaux, P. M. (2000) pGREEN: a versatile and flexible binary Ti vector for Agrobacterium-mediated plant transformation. *PMB.* 42, 819–832.
14. Gal-On, A., Meiri, E., Elman, C., Gray, D. J., and Gaba, V. (1997) Simple hand-held devices for the efficient infection of plants with viral-encoding constructs by particle bombardment. *J Virol Methods.* 64, 103–110.
15. Nivison, H. T., Fish, L. E., and Jagendorf A.T. (1986) Translation by isolated pea chloroplasts. *Methods Enzymol.* 118, 282–295.

Chapter 21

Sendai Virus for Cancer Immunotherapy

Yasuji Ueda, Mamoru Hasegawa, and Yoshikazu Yonemitsu

Summary

Dendritic cells (DCs) have a crucial role to play in fighting nonself organisms and cells, including tumors. Clinically, numerous DC vaccinations have been attempted for cancer immunotherapy since the first trial, published in 1995, but with limited success. We found that Sendai virus (SeV) vector infection induces maturation of DCs and produces more powerful antitumor immunity against DCs in mouse models. We used a SeV vector as an immune booster for tumors and believe that this novel therapy, designated as "immunostimulatory virotherapy," will offer potent treatment for tumors.

Key words: Sendai virus vector, Dendritic cells, Immunostimulatory virotherapy.

1. Introduction

Mammals have a multilayered defense system against invading microbes, including fungi, bacteria, and viruses. Immune cells "seek out," "uncover," and "take on" the invading enemy, and then "notify" other cells in the body about the invasion by nonself elements, signaling these cells to "attack" those elements. Many types of cells participate in this immune mechanism and dendritic cells (DCs), which have been described as the conductors of total immunity, play a crucial role in defense. DCs not only stimulate T-cell-mediated adaptive immunity *(1)* but also innate immunity by activating natural killer (NK) cells *(2)* or NKT cells *(3)*.

Immunologically, tumor cells are occasionally recognized as "nonself" and numerous specific tumor-associated antigens have been identified in various tumors *(4, 5)*. Prior to identification of these antigens, it was well known that stimulation of the immune

Barry W. Hicks (ed.), *Methods in Molecular Biology, Viral Applications of Green Fluorescent Protein, vol. 515*

DOI: 10.1007/978-1-59745-559-6_21

system enhances antitumor efficacy in both clinical and experimental settings *(6, 7)*. Clinical treatment of DCs began with high expectations about 10 years ago. Although numerous DC treatments against malignant tumors have been performed, the clinical response has been limited *(8)*. One reason for unsatisfactory outcomes is probably that most DCs used for cancer immunotherapy were in a low state of activity *(9)*. We recently found that Sendai virus (SeV) vector infection strongly activated DCs, and this activation stimulated DC-enhanced antitumor immunity *(10, 11)*.

Nontransmissible SeV vectors have been developed by deletion of a viral envelope gene from the wild-type SeV genome *(12–14)*. The SeV vector has some unique properties. Unlike other vectors, such as retrovirus or adenovirus, genetic information remains in the cytoplasm in the host cell and does not affect the host genome sequence after infection. This deficient vector can transduce a wide range of cells such as neural cells, muscular cells, and blood cells, including DCs (**Fig. 1**), and the infected cells do not produce any transmissible virions. The SeV vector is thus a novel cytoplasmic vector that can stimulate DCs to the same extent as the wild-type virus. Because of safety concerns, we used this new nontransmissible vector as a stimulant for DC cancer immunotherapy and found that DCs that are highly stimulated with Sendai virus produce strong regression of various types of subcutaneous tumors, including melanomas and squamous cell carcinomas, in a mouse model. SeV infection of DCs enhances expression of costimulatory molecules such as CD80, CD86, and CD40 and production of several cytokines such as interferon-β and IL-12. Recent studies have successively revealed key molecules to recognize single-stranded RNA viruses, including SeV, and immunostimulatory mechanisms against them. At present, the most important molecules in antiviral immune induction are

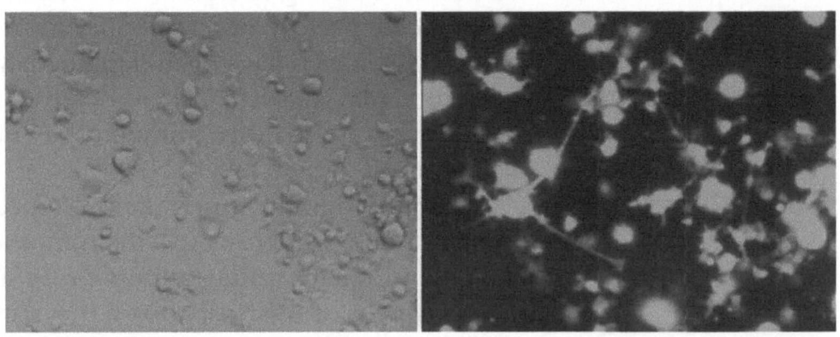

Fig. 1. Transduction of mouse dendritic cells (DCs) with *GFP* gene by F-deficient SeV vector. DCs differentiated from bone marrow progenitor cells after being in culture for 7 days in the presence of GM-CSF and IL-4. These cells were then transduced with GFP-expressing F-deficient SeV vector at MOI 40. Most cells (*left:* bright field) expressed GFP 2 days after transfection (*right:* fluorescence micrograph).

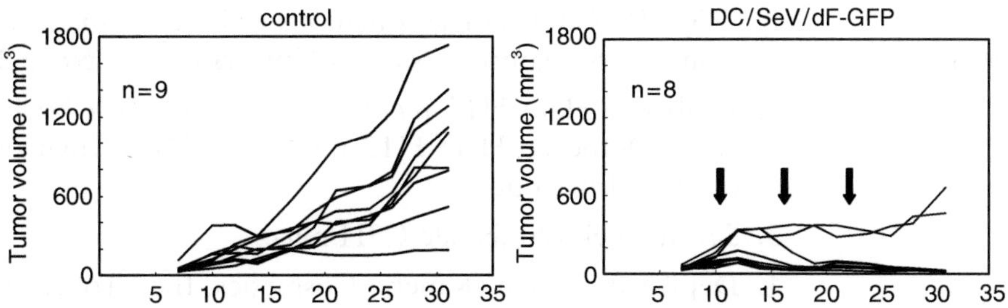

Fig. 2. Antitumoral effects of intratumoral administration of tumor lysate-pulsed dendritic cell (*DC*)-activated SeV/dF-GFP. This figure shows the time course of change in tumor volumes (*left:* nontreated, *right:* SeV/DC treated). A million cells of SCCVII were subcutaneously inoculated at a site on the abdomen (day 0), and 10^6 DCs were pulsed with tumor cell lysates and infected with GFP-expressing F-deficient SeV vector administrated intratumorally three times (days 10, 17, and 23; *arrows*).

toll-like receptor 3 (TLR-3) and retinoic acid-inducible gene I (RIG-I), a member of the RNA helicases *(15, 16)*. We consider RIG-I to be a more important molecule in stimulation of the immune system against tumors, because DC maturation by SeV infection occurs in a TLR-independent manner. In our animal study, SeV-infected DCs caused regression of several tumors (**Fig. 2**) more effectively than immature DCs or DCs that had been stimulated with lipopolysaccharide. Therefore, we advocate this novel antitumor immunotherapy as an "immunostimulatory virotherapy" and consider it to be potent cancer treatment.

2. Materials

2.1. DC Preparation

1. RPMI containing 2% fetal bovine serum (FBS): RPMI 1640 medium (Gibco/Invitrogen, Bethesda, MD), FBS (Cambrex Charles City, IA).
2. RPMI containing 10% FBS.
3. SpinSep™ mouse hematopoietic progenitor enrichment kit (StemCell Technologies, Canada).
4. Mouse interleukin-4 (mIL-4; Peprotech, Rocky Hill, NJ) and granulocyte-macrophage colony-stimulating factor (mGM-CSF; Peprotech) are separately dissolved at 2 µg/mL in PBS (Gibco/Invitrogen) containing 0.1% BSA (Wako Chemicals, Osaka, Japan) and stored in aliquots at –80 °C.
5. Seven-week-old female C3H syngenic mice of a tumor cell line to be used (*see* **Note 1**).
6. Six-well plastic plate (Nunc, Roskilde, Denmark, #140675).

2.2. Tumor Establishment

1. SCCVII (C3H mouse squamous cell carcinoma cell line, gift from Dr. Shibamoto, Nagoya University) (*see* **Note 2**).
2. Dulbecco-Ham H12 medium containing 10% bovine serum (BS): Dulbecco-Ham H12 medium (Gibco/Invitrogen), BS (Gibco/Invitrogen).
3. Seven-week-old female C3H mice.
4. Depilatory cream (Kanebo Cosmetics, Inc., Tokyo, Japan).
5. Methoxyflurane.
6. Insulin syringe with a 29-gauge needle (Termo, Tokyo, Japan).

2.3. Pulsing with Tumor Lysates (Optional)

1. SCCVII.
2. Liquid nitrogen.

2.4. SeV Vector Infection

1. Green fluorescent protein (GFP)-expressing F-defected SeV vector (*see* **Note 3**).
2. F-defected SeV null vector.

2.5. Flow Cytometry Analysis

1. Phycoerythrin (PE)-conjugated antimouse CD11c (BD Pharmingen, San Diego, CA).
2. PE-conjugated isotype control IgG (BD Pharmingen).
3. Low cell-binding, round-bottom 96-well plate (Nunc).

2.6. Measurement of Tumor Volume

1. Digital slide caliper (Mitsutoyo Corp., Kawasaki, Japan).

3. Methods

DCs used in cancer immunotherapy are differentiated from bone marrow hematopoietic progenitor cells and are negatively separated using the SpinSep™ mouse hematopoietic progenitor enrichment kit, according to the manufacturer's instructions (*see* **Note 1**). When cultured with GM-CSF and IL-4 progenitor cells, differentiate into DCs within 7 days.

Mature DCs usually lose the ability to capture antigens, unlike SeV-infected DCs *(11)*. In this section, we describe the method to use tumor lysates to pulse immature DCs before infection using the SeV vector. However, this step can be skipped as SeV-infected DCs will nevertheless capture tumor antigen in the tumor.

The purity of DCs obtained, efficiency of SeV vector infection, and enhancement of expression of costimulatory molecules can be assessed by flow cytometry. The efficiency of SeV vector infection of DCs can be easily tested using a GFP-expressing SeV vector. The maturation state of DCs is assessed by expression of cell surface markers such as CD40, CD80, CD86, and class II antigen.

Efficacy of DC treatment against tumor immunity depends on the number and the state of DCs, frequency and timing of DC injections, and the size and location of the tumor. It is important to check the expression of class I antigen in the tumor cells, because its expression level affects recognition of target cells by cytotoxic lymphocytes. For the tumor cells that express low levels of this antigen, an interferon β-expressing SeV vector is much more effective *(10)*.

3.1. DC Preparation

1. Mix SpinSep density medium and allow it equilibrate to room temperature for a minimum of 30 min before use.
2. Fill a 2.5-mL syringe with RPMI containing 2% FBS and attach 22-gauge needle.
3. Sacrifice a C3H mouse by cervical dislocation and sterilize the hind legs with 70% ethanol. To obtain viable cells from bone marrow, harvest bone marrow from the sterilized legs in a sequential manner.
4. Dissect the femur and tibia and cut both ends of the bones with scissors.
5. Hold the bone in a 15-mL conical tube with forceps and insert the tip of the 22-gauge needle into one end of the bone, and flush the bone marrow with 1–1.5 mL of medium.
6. After flushing all the bone marrow, disperse the cells by pipetting using a 5-mL pipette, pass the cells through a cell strainer (BD Bioscience, Franklin Lakes, NJ), and centrifuge at 1,500 rpm at 4 °C for 5 min.
7. Count the number of cells and dilute them to 2×10^7–5×10^7 mL^{-1} in RPMI containing 2% FBS in a 5-mL FACS tube. Up to 2 mL of suspended cells should be transferred into one tube.
8. Add the antibody cocktail to the cell suspension (10 μL of cocktail for every 1 mL of suspension). Mix well and incubate on ice for 30 min.
9. Wash the cells with RPMI containing 2% FBS, and centrifuge the cells at 1,500 rpm at 4 °C for 5 min.
10. Decant the supernatant and resuspend the pellet of washed cells in 2 mL of RPMI containing 2% FBS.
11. Gently vortex the tube of SpinSep dense particles for about 30 s or until there are no clumps.
12. Add 500 μL of dense particle suspension to 10^8 cells. Mix well by tapping the tube bottom, and incubate on ice for 20 min.

13. After the cells have been incubated with dense particles, dilute them with 6 mL of medium and transfer the cell suspension into a 15-mL conical tube.
14. Insert a sterile Pasteur pipette into the tube, and pour 4 mL of density medium into the pipette using a syringe with an 18-gauge needle to produce a layer of density medium beneath the cell suspension. Gently remove the pipette after all the density medium has moved underneath the cell suspension layer (*see* **Note 4**).
15. Centrifuge for 10 min at 2,300 rpm (1,200× g) at room temperature.
16. In a separate 15-mL tube, collect the enriched progenitor cells from the interface layer that has formed between the RPMI medium and the density medium, and wash the cells with 5 mL of RPMI containing 2% FBS.
17. Culture the progenitor cells in RPMI containing 10% FBS and supplemented with mGM-CSF (20 ng/mL) and mIL-4 (20 ng/mL) in a 6-well plate at a cell density of 10^5 cells/mL (*see* **Note 5**).
18. Three days after the start of culture, change the medium and replenish with fresh medium containing GM-CSF and IL-4.

3.2. Tumor Establishment

1. Harvest exponentially growing SCCVII cells (*see* **Note 6**). Wash the cells three times with PBS to remove bovine serum contained in the culture medium.
2. Count the number of cells, and suspend the cells at 2×10^7 cells/mL in PBS.
3. Mix well and aliquot 1×10^6 cells into 1 mL insulin syringes outfitted with 29-gauge needles (*see* **Note 7**).
4. Apply depilatory cream 30 min before injection on the abdomen of mice. After depilation is complete, anesthetize the animal using methoxyflurane or another inhalant anesthetic.
5. Inoculate 1×10^6 cells by subcutaneous injection of 50 μL of cell suspension into the abdomen of the mice (*see* **Note 8**). Mix the suspension well by inverting each syringe before injection.

3.3. Pulsing with Tumor Lysates (Optional)

1. Harvest SCCVII with 0.1% trypsin and 1 mM EDTA solution, and wash the cells with PBS three times. Place 3×10^7 cells into a 15-mL conical tube and resuspend in 1.5 mL of PBS.
2. Freeze the cell suspension in liquid nitrogen and quickly thaw in a water bath at 37 °C.
3. Repeat the freezing and thawing procedure two more times.

4. Place cell lysates totaling three times as many tumor cells as DCs into the DC culture (i.e., lysates from 3× 10^6 cells added to 1× 10^6 DCs).
5. Incubate the DC culture with tumor lysates in a CO_2 incubator for 12 h.

3.4. SeV Infection and DC Injection

1. After 7 days of cell culture, harvest and count the DCs.
2. Replate the DCs in the 6-well plate, and load tumor lysates into the DC culture.
3. Thaw the frozen stock (−80 °C) of SeV vector at 37 °C. SeV vector titers are generally 10^9–10^{10} CIU (cell infectious units)/mL. Dilute the SeV vector with serum-free RPMI (1:100).
4. Plate the cells in the 6-well dish at 2× 10^5–10× 10^5 cells/well. Add diluted SeV vector to aliquots of the culture medium to produce infection at MOI 40. For 5× 10^5 DCs, resuspend the cells in 2.5 mL of culture medium, place 2× 10^7 CIU of SeV vector into 2.5 mL of aliquot of the medium, and mix well. Then, add this vector solution to the DC suspension and mix well (*see* **Note 9**).
5. Immediately refreeze the rest of the SeV vector at −80 °C; it can be used for several months with only a slight decrease of titer.
6. Place the plate in a CO_2 incubator for 8 h.
7. After incubation for 8 h, harvest the cells, wash with PBS three times, and suspend in PBS at 1× 10^7 cells/mL.
8. Inject the SeV vector-infected DCs into the established tumor (*see* **Note 10**).

3.5. Flow Cytometry Analysis

1. Place an aliquot of about 1× 10^5 cells infected with GFP-expressing SeV vector or immature DCs into a well of a 2-methacryloylocyethyl phosphorylcholine (MPC)-treated, round-bottom 96-well plate with up to 200 μL of cell suspension (*see* **Note 11**).
2. Centrifuge the DCs on a plate rotor and discard supernatant by quickly inverting the plate at the sink.
3. Suspend cell precipitates with FITC-conjugated antimouse CD11c (0.5–1.0 μg/mL) in PBS and a PE-conjugated anti-costimulatory molecule such as CD80 or CD86.
4. Protect from light and incubate the plate at 4 °C for 30 min, then centrifuge it at 1,500 rpm for 5 min.
5. Discard the supernatant and wash with PBS. Repeat the wash. Suspend the cells in 500 μL of PBS and analyze them by flow cytometry.
6. If it is necessary to measure transduction efficiency, SeV vector-expressing GFP is convenient. In such cases, PE-conjugated CD11c staining is needed.

3.6. Measurement of Tumor Volume

1. Measure the tumor volumes every second or third day. Three axes of the diameter, that is, long and short axis and height, should be measured with a slide caliper and calculate the tumor volume (mm^3) using the following formula: 0.5236× (long axis)× (short axis)× (height) *(17*, *see* **Note 12**).

Notes

1. Usually, 3×10^5–4×10^5 progenitor cells can be obtained from one 7-week-old C3H mouse using the kit and following manufacturer's instructions.
2. Contamination of mycoplasma in the cell lines you are culturing should be checked periodically. Mycoplasma will sometimes cause rodent pneumonia or other respiratory disorders. The contamination is easily tested by PCR analysis of culture supernatants or by commercially available detection kits.
3. SeV vectors were developed and produced by DNAVEC Corporation (Tsukuba, Japan; http://www.dnavec-corp.com) and now can be purchased from Medical and Biological Laboratories Co., Ltd. (Nagoya, Japan; http://www.mbl.co.jp).
4. Alternatively, the cell suspension can be layered on top of the density medium as described in the instruction manual; however, for separation of numerous cells, the use of several tubes is required. In these cases, the method using Pasteur pipettes might be more convenient.
5. Several cocktails of cytokines have been used to obtain differentiated DCs from bone marrow-derived progenitor cells, GM-CSF alone, GM-CSF and Flt3-L, GM-CSF and tumor necrosis factor α (TNF-α), and so on. One method that is widely used is the combination of GM-CSF and IL-4.
6. It is very important to use exponentially growing cells for establishing the tumor. To obtain actively growing cells, you should keep the cells 10–80% confluent during culture.
7. To control the size of the tumor, the number of cells and injection site should be uniform. Injection of a small volume using a fine needle prevents the subcutaneous spread of tumor cells.
8. Alternatively, tumor cells can be subcutaneously injected into the back of a mouse. However, inoculation of the tumor in the abdomen usually prevents the site from being bitten by other mice.
9. Unlike conventional retroviral vectors, the SeV vector can effectively infect nonadhesive cells such as hematopoietic

stem/progenitor cells without any pretreatment or centrifugation. In the mouse experiment, MOI 40 might be sufficient for 70–80% of transduction efficiency within some hours (**Fig. 1**). Sendai virus stocks are usually more concentrated than stocks of other vectors. It is difficult to dilute a concentrated vector sample in the culture. We always dilute the vector solution first with culture medium, and then mix the diluted vector and cell culture.

10. Many solid tumors are so rigid that it is difficult to inject them. In such cases, we usually inject the DCs around the tumor. We injected DCs into the tumor from a distant site to prevent the DCs leaking from the needle hole after injection.
11. Treatment of cells with antibodies in a 96-well round-bottom plate has the following advantages: (i) it permits treatment of numerous samples at the same time with quick washing and (ii) it allows reduction of the reaction volume, thus conserving antibodies. After centrifugation, remove the plate lid and flip the plate above sink quickly to discard the supernatants. DCs are quite adhesive due to which they can easily attach to plastic matrix such as that found in a 96-well, nontreated, round-bottom plate or a microcentrifuge tube. However, most cell types will not attach to a low cell-binding culture plate coated with poly(MPC), and most cells can be taken up from the wells after centrifugation.
12. When tumor height cannot be precisely measured, the tumor volume can be calculated using the following formula: 0.5236× (long axis)× (short axis)^2. The measuring method should be consistent during the series of experiments.

References

1. Steinman, R. M. (1991) The dendritic cell system and its role in immunogenicity. *Annu. Rev. Immunol.* 9, 271–296.
2. Fernandez, N. C., Lozier, A., Flament, C., Ricciardi-Castagnoli, P., Bellet, D., Suter, M., et al. (1999) Dendritic cells directly trigger NK cell functions: cross-talk relevant in innate anti-tumor immune responses in vivo. *Nat. Med.* 5, 405–411.
3. Kadowaki, N., Antonenko, S., Ho, S., Rissoan, M. C., Soumelis, V., Porcelli, S. A., et al. (2001) Distinct cytokine profiles of neonatal natural killer T cells after expansion with subsets of dendritic cells. *J. Exp. Med.* 193, 1221–1226.
4. Knuth, A., Wolfel, T., Klehmann, E., Boon, T., and Meyer zum Buschenfelde, K. H. (1989) Cytolytic T-cell clones against an autologous human melanoma: specificity study and definition of three antigens by immunoselection. *Proc. Natl. Acad. Sci. USA* 86, 2804–2808.
5. Fisk, B., Blevins, T. L., Wharton, J. T., and Ioannides, C. G. (1995) Identification of an immunodominant peptide of HER-2/neu protooncogene recognized by ovarian tumor-specific cytotoxic T lymphocyte lines. *J. Exp. Med.* 181, 2109–2117.
6. Mathe, G., Amiel, J. L., Schwarzenberg, L., Schneider, M., Cattan, A., Schlumberger, J. R., et al. (1969) Active immunotherapy for acute lymphoblastic leukaemia. *Lancet* 1, 697–699.
7. Takeda, K., Kikuchi, Y., Yamawaki, S., Ueda, T., and Yoshiki, T. (1968) Treatment of artificial metastases of methylcholanthrene-induced rat sarcomas by autoimmunization of the autochthonous hosts. *Cancer Res.* 28, 2149–2154.

8. Rosenberg, S. A., Yang, J. C., and Restifo, N. P. (2004) Cancer immunotherapy: moving beyond current vaccines. *Nat. Med.* 10, 909–915.
9. Lutz, M. B. and Schuler, G. (2002) Immature, semi-mature dendritic cells: which signals induce tolerance or immunity? *Trends Immunol.* 23, 445–449.
10. Shibata, S., Okano, S., Yonemitsu, Y., Onimaru, M., Sata, S., Nagata-Takeshita, H., et al. (2006) Induction of efficient antitumor immunity using dendritic cells activated by Sendai virus and its modulation of exogenous interferon-b gene. *J. Immunol.* 177, 3364–3576.
11. Yoneyama, Y., Ueda, Y., Akutsu, Y., Matsunaga, A., Shimada, H., Kato, T., et al. (2007) Development of immunostimulatory virotherapy using non-transmissible Sendai virus-activated dendritic cells. *Biochem. Biophys. Res. Commun.* 355, 129–135.
12. Li, H. O., Zhu, Y. F., Asakawa, M., Kuma, H., Hirata, T., Ueda, Y., et al. (2000) A cytoplasmic RNA vector derived from nontransmissible Sendai virus with efficient gene transfer and expression. *J. Virol.* 74, 6564–6569.
13. Inoue, M., Tokusumi, Y., Ban, H., Shirakura, M., Kanaya, T., Yoshizaki, M., et al. (2004) Recombinant Sendai virus vectors deleted in both the matrix and the fusion genes: efficient gene transfer with preferable properties. *J. Gene Med.* 6, 1069–1081.
14. Yoshizaki, M., Hironaka, T., Iwasaki, H., Ban, H., Tokusumi, Y., Iida, A., et al. (2006) Naked Sendai virus vector lacking all of the envelope-related genes: reduced cytopathogenicity and immunogenicity. *J. Gene Med.* 8, 1151–1159.
15. Akira, S., and Takeda, K. (2004) Toll-like receptor signalling. *Nat. Rev. Immunol.* 4, 499–511.
16. Yoneyama, M., Kikuchi, M., Natsukawa, T., Shinobu, N., Imaizumi, T., Miyagishi, M., Taira, K., Akira, S., and Fujita, T. (2004) The RNA helicase RIG-I has an essential function in double-stranded RNA-induced innate antiviral responses. *Nat. Immunol.* 5, 730–737.
17. Spang-Thomsen, M., Nielsen, A., and Visfeldt, J. (1980) Growth curves of three human malignant tumors transplanted to nude mice. *Exp. Cell Biol.* 48, 138–154.

Chapter 22

Poxvirus Tropism for Primary Human Leukocytes and Hematopoietic Cells

Qigui Yu, Ningjie Hu, and Mario Ostrowski

Summary

Poxviruses including canarypox (ALVAC) and vaccinia viruses have, in recent years, received considerable attention as live vectors for the development of vaccines against infectious diseases such as AIDS, malaria, and tuberculosis. However, the cellular targets for viral infection within the human immune system and the consequences of infection for cells involved in the generation of immune responses have not been clearly delineated. Using recombinant enhanced green fluorescence protein (EGFP)-expressing ALVAC and vaccinia viruses, we have focused here on a side-by-side comparison of ALVAC and vaccinia virus tropism for cells from human peripheral blood and bone marrow. Both ALVAC and vaccinia viruses showed a strong bias toward monocyte infection. ALVAC minimally infected CD19^{+} B cells and was unable to infect ex vivo NK cells and T lymphocytes, whereas vaccinia virus could infect B lymphocytes and NK cell populations. Vaccinia virus was also able to infect T lymphocytes at low but detectable levels which could be enhanced upon their activation. Both ALVAC and vaccinia viruses could infect immature monocyte-derived dendritic cells (MDDCs), but only ALVAC infection induced their subsequent maturation. Infection in human bone marrow cells showed that ALVAC infection was restricted to a myelomonocytoid cell-specific CD33^{+} cell population, while vaccinia virus showed a strong, but not exclusive, preference for these cells.

Key words: Virus, ALVAC, Vaccinia, Peripheral blood mononuclear cell, Monocyte, Dendritic cell, Myeloid cell, Enhanced green fluorescence protein.

1. Introduction

Members of the poxvirus family (*poxviridae*) including canarypox (ALVAC) and vaccinia viruses have, in recent years, received considerable attention for the development of vaccine vectors that can induce humoral and cellular immunity against virus infections as well as immunotherapy for cancer *(1–3)*. Several

Barry W. Hicks (ed.), *Methods in Molecular Biology, Viral Applications of Green Fluorescent Protein, vol. 515*

DOI: 10.1007/978-1-59745-559-6_22

advantages of these vectors for vaccine development include strict cytoplasmic replication that is primarily abortive in human cells, and their good long-term safety profile *(4, 5)*. The most notorious member of *poxviridae* family, variola virus, caused smallpox and has claimed a greater number of human lives over the span of recorded history than all other infectious diseases combined *(6)*. Smallpox was finally eradicated in 1977, nearly two centuries after the introduction of prophylactic inoculations with vaccinia virus and cowpox *(7)*. The highly attenuated canarypox virus strain, ALVAC, with a genome of ~300 kbp, has been extensively developed to express genes from rabies virus *(5)*, canine distemper virus *(8)*, feline leukemia virus *(9)*, human immunodeficiency virus (HIV) *(10–16)*, simian immunodeficiency virus (SIV) *(17, 18)*, human cytomegalovirus *(19)*, melanoma *(20)*, and Japanese encephalitis virus *(21)*. Safe delivery of these antigens is insured by the inability of the avian host-restricted ALVAC virus to replicate in mammals. While human vaccines of most poxvirus vectors remain in experimental or clinical trial phases, ALVAC veterinary vaccines are now commercially available.

Understanding immunogenicity and reactogenicity of these vectors may also be obtained from studies of their effects on dendritic cells. For example, it has been determined that ALVAC infects immature myeloid-derived dendritic cells (DCs) more easily than their mature counterpart, and this infection induces DC maturation as measured by marked upregulation of the DC maturation markers CD80, CD83, CD86, CD25, and DC-LAMP *(22)*. Maturation is driven by tumor necrosis factor alpha (TNF-α) secretion in response to ALVAC infection, as well as by the ingestion of cellular debris from apoptotic ALVAC-infected immature DCs *(22)*. A direct comparison with vaccinia virus has not been reported, although such information may be useful to understanding the comparable biological effects of these two vaccines.

Surprisingly, despite extensive ongoing work in human clinical trials, there has, until recently *(23)*, been little information available regarding the range of infective tropism these virus vectors have in primary human cells. A recent paper surveyed the susceptibility of different leukocyte cell types in peripheral blood to vaccinia virus infection. Strong bias was shown toward the infection of peripheral monocytes, followed by B lymphocytes and NK cells. Ex vivo T lymphocytes were infected at low, but detectable levels *(24)*. Considering the substantial differences in host tropism, replicative capacity, and genome sequences between the vaccinia and canarypox viruses, it cannot be assumed that this specificity is paralleled by ALVAC. In addition to this consideration, the infection tropism for vertical cell lineages, that is, progenitors of leukocytes in bone marrow and monocyte-derived dendritic cells (MDDCs), has not yet been addressed for either vaccinia or canarypox virus. This knowledge is requisite for the rational development of an ALVAC vector and other poxvirus vectors in order to have enhanced immunogenicity and uncompromised safety. We have focused here on a side-by-side comparison of vaccinia and ALVAC capacities to infect leukocyte subsets in human bone marrow, peripheral blood cells, and MDDCs. We demonstrate that ALVAC and vaccinia viruses exhibit cellular infection tropism for human myeloid lineage cell types, with a more restricted pattern of tropism for ALVAC. In addition, the tropism and effects of these two viruses on myeloid-derived dendritic cells are compared and contrasted.

2. Materials

2.1. Viruses, Cell Culture, and Cytokines

1. Parental ALVAC virus, a highly attenuated strain of canarypox virus, and recombinant ALVAC virus vCP1540 expressing enhanced green fluorescent protein (EGFP) were obtained from Sanofi-Pastuer (Toronto, Canada).
2. A recombinant vaccinia virus expressing EGFP (EGFP-Vaccinia) was a gift from Dr. Jonathan Yewdell and Dr. Bernard Moss (NIH, Bethesda, MD). EGFP-Vaccinia was derived from a WR strain of vaccinia virus and has a fusion protein inserted consisting of the nucleoprotein of influenza A/Puerto Rico/8/34 fused to chicken ovalbumin 257–264 ($OVA_{257-264}$) epitope, followed by a COOH-terminal fusion to EGFP from *Aequorea victoria (25)*.
3. Eagle's minimal essential medium (EMEM) (Sigma, St. Louis, MO) supplemented with 10% fetal bovine serum (FBS) (Gibco/BRL, Bethesda, MD) and 2 mM L-glutamine (Gibco/BRL).
4. RPMI 1640 medium (Gibco/BRL) supplemented with 10% FBS plus 2 mM glutamine, 25 mM HEPES (Gibco/BRL), 100 U/mL penicillin, and 100 U/mL streptomycin sulfate (Sigma).
5. Iscove's modified Dulbecco's medium (IMDM; Gibco Laboratories, Grand Island, NY).
6. Primary chicken embryo fibroblasts (Charles River Laboratories, WI) in EMEM with 5% FBS are used for growth and titration of parental ALVAC, vCP1540, and EGFP-Vaccinia viruses (*see* **Note 1**).
7. Recombinant human granulocyte-macrophage colony-stimulating factor (rhGM-CSF) and interleukin-4 (rhIL-4) (PeproTech, Rocky Hill, NJ) are reconstituted in tissue-culture water at 50 µg/mL and stored in single use aliquots at −20 °C (*see* **Note 2**).

8. Ficoll-Hypaque solution (Amersham Pharmacia Biotech AB, Uppsala, Sweden).
9. Percoll gradient (Sigma-Aldrich).
10. T-cell activators: anti-human antibodies of CD3, CD28, CD49d (BD PharMingen, San Diego, CA).
11. Monocyte Negative Isolation Kit (Dynal, Oslo, Norway).
12. GolgiStop Protein Transport Inhibitor (containing monensin) (BD PharMingen).
13. CD40L trimer (CD40LT) protein (a gift from Immunex, Seattle, WA).

2.2. Labeled or Unlabeled Antibodies and Flow Cytometric Analysis

1. The following anti-human monoclonal antibodies (mAbs) or polyclonal Abs conjugated with fluorochrome were purchased from BD PharMingen: anti-CD1a^{FITC}, anti-CD3^{APC}, anti-CD4^{PerCP}, anti-CD8^{PE}, anti-CD10^{PE}, anti-CD14^{APC}, anti-CD19^{PE}, anti-CD33^{PerCP}, anti-CD56^{PE}, anti-CD80^{PE}, anti-CD83^{PE}, anti-CD86^{APC}, anti-TNF-α^{APC}, and matched-isotype control antibodies conjugated with FITC, PE, PerCP, or APC.
2. Purified monoclonal antibodies of anti-human CD3, anti-human CD28, and anti-human CD49d were also purchased from BD PharMingen.
3. Flow cytometry staining buffer: PBS + 0.02 (w/v) NaN_3 + 2% FBS (w/v). Store at 4 °C.
4. Cytofix/Cytoperm™ solution (BD PharMingen).
5. Perm/Wash™ solution (BD PharMingen).
6. Flow cytometry fixation solution (2% paraformaldehyde): 2 g paraformaldehyde (Sigma) + 100 mL PBS. To dissolve paraformaldehyde, heat solution to 70 °C in a fume hood for about 1 h, and then cool to room temperature before adjusting pH to 7.2 with concentrated NaOH. Store protected from light ≤1 month at 4 °C.

2.3. Reagents for DNA Extraction and PCR

1. QIAamp DNA Blood Mini Kit (QIAGEN, Valencia, CA) is used for DNA extraction. Purified DNA is quantified spectrophotocally using SmartSpec 3000 Spectrophotometer (Bio-Rad Laboratories, Inc. Hercules, CA) or comparable equipment.
2. PLATINUM *Taq* system (Invitrogen, Carlsbad, CA) including platinum DNA *Taq* polymerase, 10× PCR buffer (200 mM Tris-HCl pH 8.0, 500 mM KCl, and 50 mM $MgCl_2$).
3. 100 mM dNTP (2′-deoxynucleoside 5′-triphosphate) set consisting of all four deoxynucleotides (dATP, dCTP, dGTP, dTTP), each at a concentration of 100 mM, was purchased from the Invitrogen.

4. Following oligonucleotides were purchased from Invitrogen and used as PCR primers to detect ALVAC and vaccinia virus sequences, respectively.

   CNPV136F: (5′-ATTGCGCGATGTAGATAAATGTTACA-AAC-3′) and

   CNPV136R: (5′-GCATCAAAGAGTATAGCTTCATACCC-TG-3′) or

   VACVF4L.F389-409: (5′-CGTTGGAAAACGTGAGTCC-GG-3′) and

   VACVF4L.R774–754: (5′-ATTGGCGTTTTGCAGCC-AG-3′).
5. Oligonucleotides corresponding to human housekeeping gene glyceraldehyde-3-phosphate dehydrogenase (GAPDH) were also purchased from Invitrogen and used for PCR controls. The two oligonucleotides, GAPDH.For and GAPDH.Rev, are derived from GAPDH exons 7 and 8 *(26)*:

   GAPDH.For: 5′-CCTGGCCAAGGTCATCCATG-3′ and

   GAPDH.Rev: 5′-CTGCTTCACCACCTTCTTGA-3′.
6. Electrophoresis equipment and reagents and agarose gels.
7. GeneRuler 100 bp DNA ladder plus (MBI Fermentas).

2.4. Dot Blot Hybridization

1. 20× SSC (Sigma): 3 M NaCl plus 0.3 M sodium citrate, pH 7.0.
2. 10% Sodium dodecyl sulfate (SDS) (Sigma).
3. 0.7N NaOH solution.
4. Low Stringency Wash Buffer (2× SSC, 0.1% SDS): to 500 mL molecular biology grate water, add 100 mL of 20× SSC stock solution and 10 mL of 10% SDS stock solution, and then adjust volume to 1 L with water.
5. Ultra-High Stringency Wash Buffer (0.1× SSC, 0.1% SDS): to 500 mL molecular biology grate water, add 5 mL of 20× SSC stock solution and 10 mL of 10% SDS stock solution, and then adjust volume to 1 L with water.
6. PerfectHyb™ Plus hybridization buffer (Sigma).
7. Prehybridization solution is the hybridization solution without the labeled probe.
8. Blocking buffer: Low Stringency Wash Buffer containing 5% dry milk.
9. Nylon membrane (Boehringer Mannheim) and Whatman 3MM filter paper to cover the wells of the apparatus.
10. HYBRI-SLOT filtration manifold (Life Technologies, Gaithersburg, MD), or comparable equipment.
11. 2 M ammonium acetate.

12. UV transilluminator (Stratalinker, Stratagene, La Jolla, CA).
13. DIG-labeled nucleic acids with Anti-Dig-alkaline phosphatase.
14. Alkaline phosphatase bioluminescent substrate, CSPD ready-to-use (Boehringer Mannheim).
15. Kodak XAR-5 film.

3. Methods

3.1. Cell Culture (See Note 3)

3.1.1. Preparation of Peripheral Blood Mononuclear Cells and Monocytes

1. Obtain heparinized human peripheral blood from healthy blood donors.
2. Isolate peripheral blood mononuclear cells (PBMCs) by centrifugation of blood on Ficoll-Hypaque gradient.
3. For activating T cells, incubation of PBMCs with plate-bound anti-CD3/anti-CD28/anti-CD49d antibodies, or with medium alone as a control, at 37 °C for 24 h in a 5% CO_2 incubator (*see* **Note 4**).
4. Separate monocytes from PBMCs using multistep Percoll (Sigma) gradient centrifugation *(27, 28)*. Further purification of monocytes by depletion of contaminating B cells, T cells, NK cells, and granulocytes using antibody-conjugated magnetic beads in the Monocyte Negative Isolation Kit according to the manufacturer's guidelines.
5. The resulting monocyte preparation contains more than 95% monocytes, and <0.1% T and B lymphocytes, as assessed by CD14, CD4, CD8, and CD19 staining and flow cytometric analysis (*see* **Notes 5** and **6**).

3.1.2. Preparation of MDDCs

Generation of MDDCs from purified monocytes using a modification of a method previously described *(27–29)*.

1. Culture purified peripheral monocytes at 1×10^6 cells/mL in fresh complete RPMI 1640 medium (RPMI 1640 medium supplemented with 10% FBS plus 2 mM glutamine, 25 mM HEPES, 100 U/mL penicillin, and 100 U/mL streptomycin sulfate).
2. Addition of 50 ng/mL rhGM-CSF and 100 ng/mL rhIL-4 to the culture medium on day 0.
3. On days 3 and 5, rhGM-CSF and rhIL-4 were added again with the fresh complete RPMI 1640 medium.
4. After 7 days of culture, more than 50% of the cells will be $CD1a^{high}$, MHC class II^+, $CD80^{low}$, and $CD14^-$, which represents an immature DC (iDC) phenotype (*see* **Note 7**).

5. Infection of the immature MDDCs directly with vCP1540 or EGFP-Vaccinia virus, or grow the cells for another 2–3 days with stimulation of soluble recombinant human CD40LT at a concentration of 2 μg/mL to generate mature MDDCs *(27, 28)*.

3.1.3. Preparation of Bone Marrow Cells

1. Collect human bone marrow (BM) aspirations from the posterior iliac crest of three healthy donors. *Donors do not have a* history of previous hematologic disorders, previous chemotherapy, or radiation therapy.
2. Separate BM mononuclear cells (BMMNCs) by Ficoll-Hypaque density gradient (1.077 g/mL).
3. Wash the cells twice with Iscove's modified Dulbecco's medium.
4. Resuspension of isolated BMMNCs with complete RPMI 1640 medium. Infection of isolated BMMNCs directly with vCP1540 or EGFP-Vaccina virus.

3.2. Virus Infection In Vitro

1. Culture PBMCs, monocytes, MDDCs, or BMMNCs in 96-well round-bottom plates at 0.2 × 10^6 cells/200 μL medium/well in complete RPMI 1640 medium.
2. Add virus at a multiplicity of infection (MOI) of 10 (10 viral particles per cell) with parental ALVAC, recombinant ALVAC virus vCP1540, or EGFP-Vaccinia virus.
3. After 1 h adsorption at 37 °C, remove free virus particles by washing three times by centrifugation with ice-chilled 2% FBS/PBS (*see* **Note 8**).
4. After the final wash, resuspend the cell pellets in complete RPMI 1640 medium and culture at 37 °C in a 5% CO_2 incubator for various time intervals.

3.3. Flow Cytometric Analysis

Performance of flow cytometric analysis using a FACSCalibur flow cytometer (BD Biosciences, San Diego, CA).

3.3.1. Multicolor Surface Staining of a Single Cell Suspension for Flow Cytometry

1. Harvest the infected cells by centrifugation in a 5-mL polypropylene conical tube.
2. Gently resuspend the cell pellet and wash once with 1× PBS buffer + 2% FBS.
3. Gently resuspend the washed pellet in 1× PBS + 2% FBS, then aliquot to 5-mL polystyrene round-bottom tubes (BD Falcon) at 50 μL per sample with at least 100,000 cells per sample.
4. Add a pretitrated optimal concentration (≤0.5 μg) of a fluorochrome-conjugated mAb specific for a cell surface antigen, such as CD3, CD4, CD8, CD10, CD14, CD19, CD34, or CD56 to provide control for setting proper compensation of the brightest fluorescent signals. Multicolor staining of different cell surface antigens can be carried out in a single tube.

5. Incubate on ice for 25–30 min in the dark.
6. Wash two times with ice-cold PBS + 2% FBS at 4 °C to remove unbound antibody.
7. Gently resuspend pellet and add cold 2% paraformaldehyde in 1× PBS (200 μL per sample) to fix the stained cells.
8. Store in the dark at 4 °C.
9. Acquire at least 50,000 events per sample using BD FACS-Calibur flow cytometer.
10. Collect data for analysis using FlowJo software (Tree Star, San Carlos, CA) (**Fig. 1–3**).

3.3.2. Intracellular Staining of a Single Cell Suspension for Flow Cytometry (See Note 9)

1. Wash the cells two times with Staining Buffer, pellet cells by centrifugation (250 × *g*), and remove supernatant.
2. Thoroughly resuspend cells in 100 μL of Cytofix/Cytoperm™ solution, and incubate for 20 min at 4 °C to permeabilize cells.
3. Wash the cells two times in 1× Perm/Wash™ solution, pellet cells, and remove the supernatant.
4. Resuspend the cells in Staining Buffer with fluorochrome-conjugated specific antibodies. Incubate on ice for 30 min in the dark.

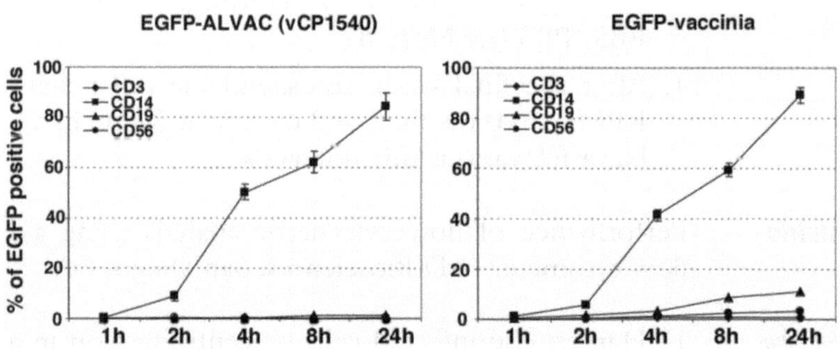

Fig. 1. Preferential infection of peripheral monocytes in PBMCs with ALVAC or vaccinia virus. Human PBMCs were exposed with a multiplicity of infection (*MOI*) of 10 of EGFP-expressing ALVAC recombinant virus vCP1540 or EGFP-Vaccinia virus for 1 h at 37 °C in a 5% CO_2 incubator. After washing three times with ice-chilled PBS plus 2% FBS, the cells were incubated for various time intervals. The infected cells were stained with fluorochrome-conjugated antibodies of $CD3^{APC}$/$CD4^{PerCP}$/$CD8^{PE}$, $CD14^{APC}$/$CD19^{PE}$, or $CD56^{PE}$, and then subjected to flow cytometric analysis. In (**A**) are kinetic data of ALVAC or vaccinia virus infection in PBMCs subsets from a healthy blood donor; a representative of six experiments is shown. *Y*-axis shows the percentage of EGFP-positive cells out of the total T cells (CD3), monocytes (CD14), B cells (CD19), or NK cells (CD56) as indicated. Infections are performed in triplicate and *bars* show standard error on the mean. In (**B**) one representative experiment shows the infection of T cells, B cells, monocytes, and NK cells with vCP1540 or EGFP-Vaccinia virus at 4 h postinfection. *Numbers* in each plot indicates the percentage of EGFP-positive cells within the indicated PBMC subsets. In (**C**) activation of $CD3^{+}$ T cells, by plate bound anti-human CD3/CD28/CD49d mAbs, induced minor changes in permissiveness to infection by ALVAC but a modest enhancement of infection of T cells with vaccinia virus. Representative data from one of three individuals are shown. Values in *upper left* quadrants indicate % EGFP-expressing cells of CD69-negative (unactivated) T cells and values in *upper right* quadrants indicate % EGFP-expressing cells of CD69-positive (activated) fraction.

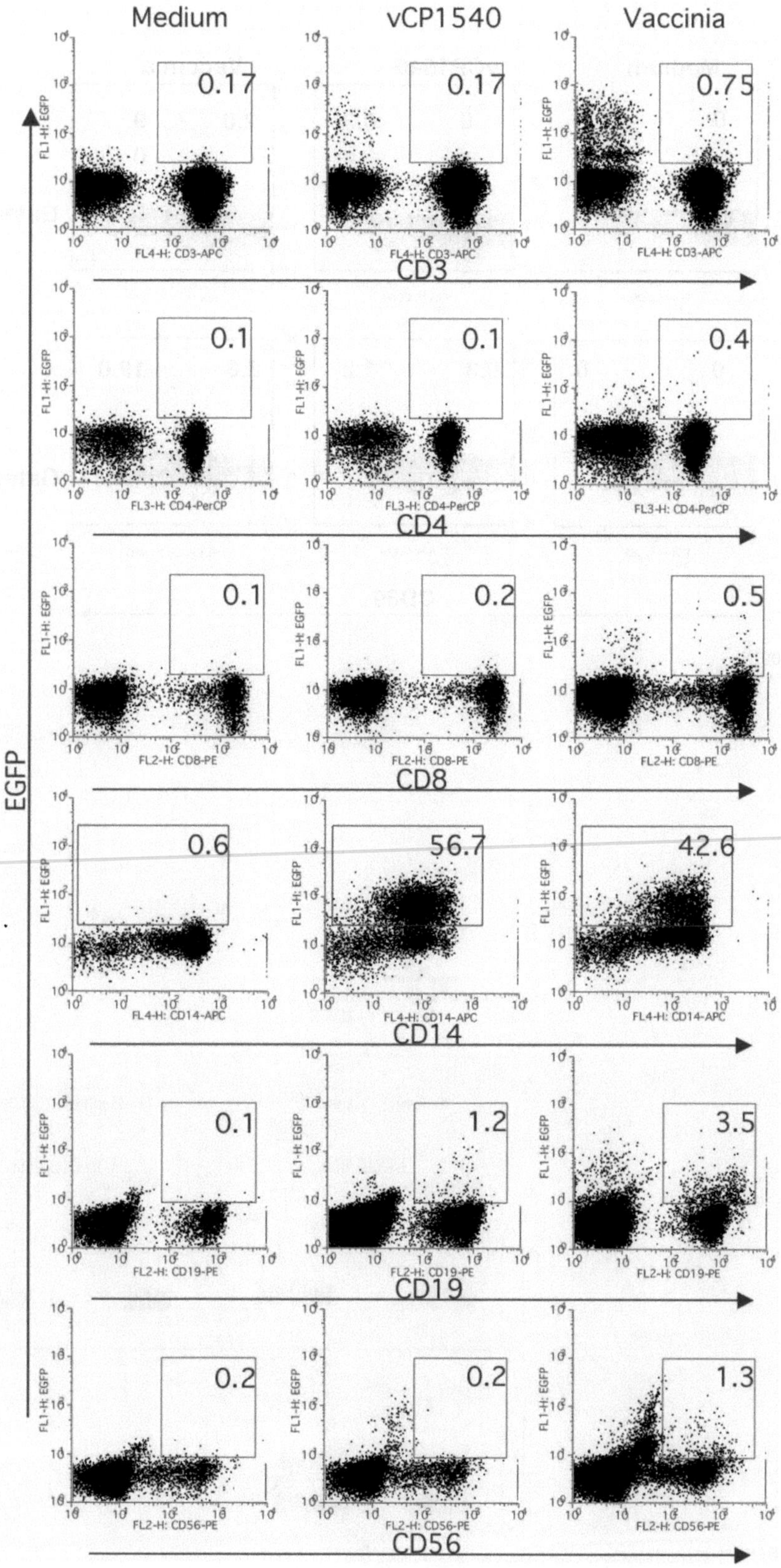

Fig. 1. (continued)

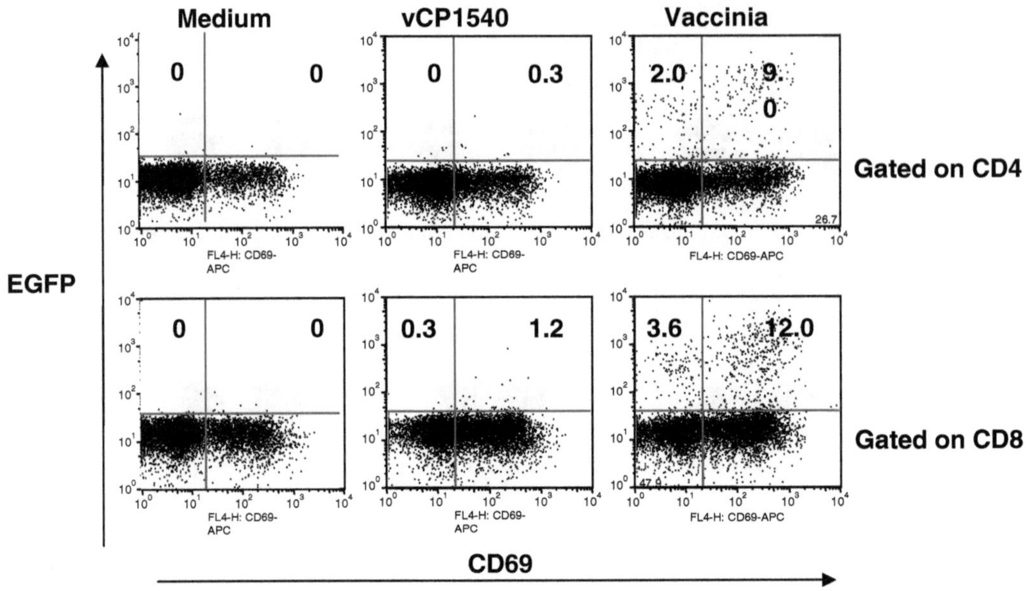

Fig. 1. (continued)

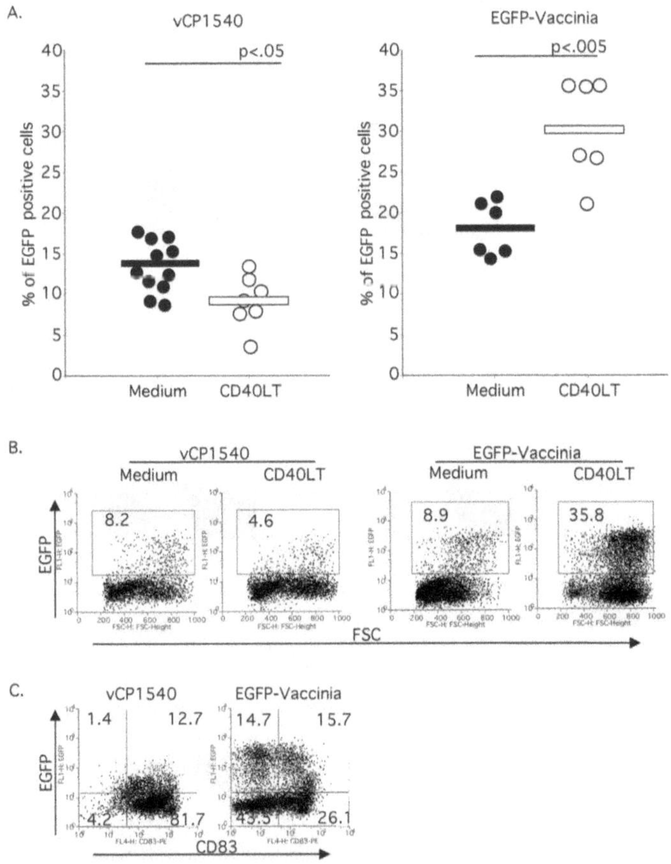

Fig. 2. ALVAC and vaccinia viruses demonstrate different infection patterns to CD40L-matured human MDDCs. For the infections, immature MDDCs and CD40LT-matured

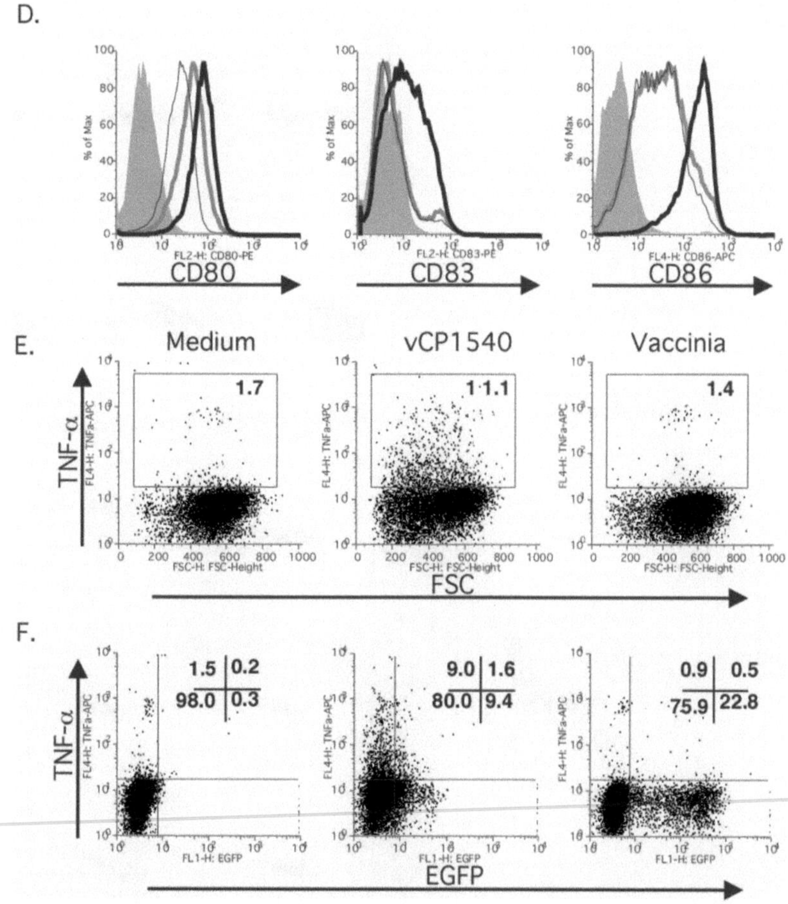

(continued) MDDCs were exposed to medium alone (no virus condition), vCP1540, or EGFP-Vaccinia virus at a multiplicity of infection (MOI) of 10 for 1 h at 37 °C. Free virus particles were removed by three washes with ice-chilled 2% FBS/PBS. Cells were cultured for 24 h and cell infection was determined by EGFP expression using flow cytometric analysis. Part (**A**) demonstrates the infection patterns of immature (medium treated) and CD40L-matured MDDCs to vCP1540 and EGFP-Vaccinia viruses. Part (**B**) is one representative experiment from at least six experiments, which shows the different infection patterns in immature MDDC (*left* plot) and mature MDDC (*right* plot). The number within each plot shows the percentage of the EGFP-positive MDDCs out of the total MDDCs. In (**C**) is a representative experiment showing expression of CD83 and infectivity of immature MDDC infected with vCP1540 (*left* plot) or EGFP-Vaccinia (*right* plot). In (**D**) surface staining for analyzing regulation of DC maturation markers CD80, CD83, and CD86 in response to ALVAC or vaccinia virus infection. Isotype control, *solid histogram*; medium control, *thin line*; ALVAC, *thick black line*; vaccinia virus, *thick gray line*. Shown is one representative experiment of six from six different donors. In (**E**) intracellular staining for TNF-α production after 24 h infection of immature MDDC with ALVAC or EGFP-Vaccinia virus, or medium control, and in (**F**), showing distribution of TNF-α in EGFP-positive and EGFP-negative MDDCs in the same experiments. Shown is one representative experiment of six from six different donors.

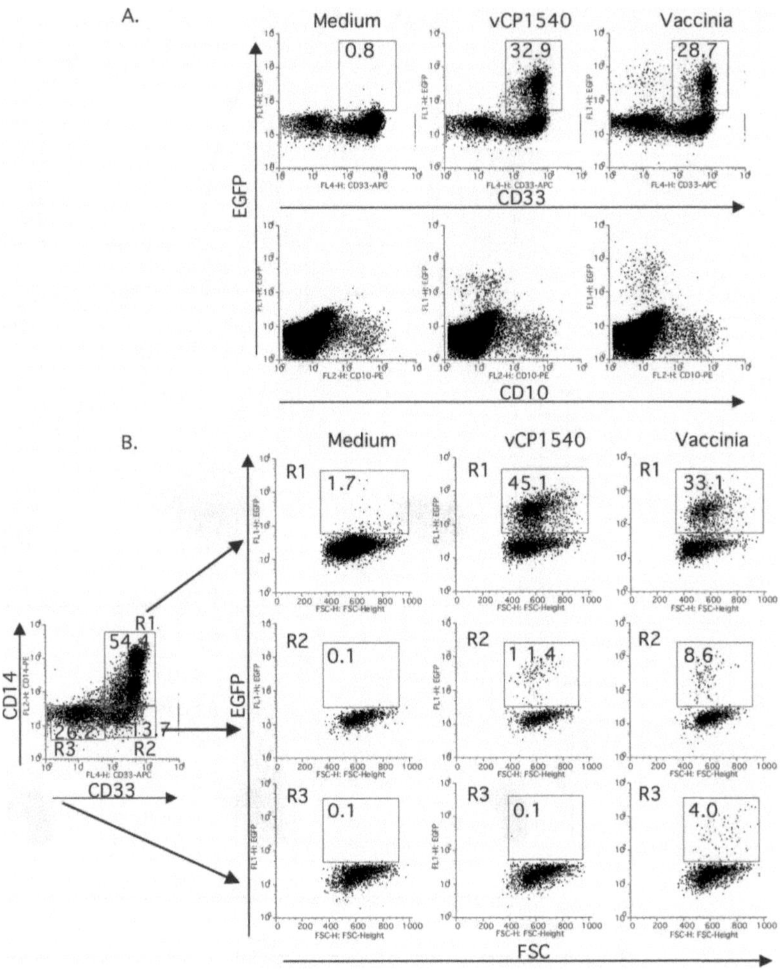

Fig. 3. ALVAC or vaccinia virus infects myeloid lineage cells but not lymphoid lineage cells in human bone marrow cells. Freshly isolated human bone marrow cells were infected with vCP1540 or EGFP-Vaccinia virus at a multiplicity of infection (*MOI*) of 10 for 8 h at 37 °C. The susceptibility and degree of myeloid lineage cells and lymphoid lineage cells to ALVAC or vaccinia virus infection were monitored using flow cytometric analysis with fluorochrome-conjugated mAbs specific for myeloid cell-specific transmembrane glycoprotein CD33^{APC} *(30, 31)* and lymphoid cell-specific surface antigen CD10^{PE} *(32)* (**A**). Infected bone marrow cells were also stained with CD33^{APC}/CD14^{PE} in order to study the susceptibility of myeloid progenitor cells to ALVAC or vaccinia virus infection (**B**). The *number* within each plot shows the percentage of the EGFP-positive monocytes in the cells expressing specific cell markers. One representative experiment shows the infection patterns of bone marrow cells from the three healthy bone marrow donors examined.

5. Wash two times with ice-cold PBS + 2% FBS at 4 °C to remove unbound antibody.
6. Gently resuspend the pellet and add cold 2% paraformaldehyde in 1× PBS (200 μL per sample) to fix the stained cells.
7. Store in the dark at 4 °C.
8. Acquire at least 50,000 events per sample using BD FACSCalibur flow cytometer.
9. Collect data for analysis using FlowJo software (**Fig. 2e, f**).

3.3.3. Fluorescence-Activated Cell Sorting

1. Perform fluorescence-activated cell sorting using a FACStar Plus (Becton-Dickinson) equipped with 5 W argon and 30 mW helium neon lasers, or a comparable instrument.
2. Stain PBMCs (exposed or unexposed to parental ALVAC or EGFP-Vaccinia virus) at 37 °C for 1 h with a combination of mAbs of $CD3^{APC}/CD19^{PE}$ or $CD14^{APC}/CD56^{PE}$.
3. Use a FACStar Plus to sort CD3, CD14, CD19, and CD56 fractions for DNA extraction.

3.4. DNA Extraction and PCR

1. Wash cell fractions sorted by CD3, CD14, CD19, and CD56 markers from PBMCs exposed or unexposed to parental ALVAC or EGFP-Vaccinia virus twice with ice-chilled PBS (*see* **Note 10**).
2. Resuspend cells in 200 μL of PBS for DNA extraction using QIAamp DNA Blood Mini Kit according to the manufacturer's guidelines.
3. Quantify the purified DNA using a spectrophotometer.
4. Perform PCR or dot blot hybridization (*see* **Note 11**).
5. A pair of oligonucleotides CNPV136F and CNPV136R, or VACV F4L.F389-409 and VACV F4L.R774-754 is used as PCR primers to detect ALVAC and vaccinia virus sequences, respectively, from DNA extracted from $CD3^+$ T cells, $CD14^+$ monocytes, $CD19^+$ B cells, and $CD56^+$ NK cells sorted from ALVAC- or vaccinia virus-exposed PBMCs.
6. Perform amplification of the human housekeeping gene glyceraldehyde-3-phosphate dehydrogenase (GAPDH) as a PCR control (*see* **Note 12**).
7. For all PCR, 10 ng of DNA was subjected to PCR with PLATINUM Taq according to the manufacturer's instruction. The PCR mixture contains 200 nM sense and antisense primers, nucleotide mix, plus 2 U of PLATINUM Taq polymerase. Start by denaturing at 95 °C for 2 min, and then perform 30 cycles of 94 °C for 30 s, 52 °C for 30 s, and 72 °C for 1 min.
8. Identify PCR products using ethidium-stained 1.5% agarose gel electrophoresis. The expected lengths of PCR amplicons were 525, 386, and 509 bp for ALVAC, vaccinia virus, and human GAPDH, respectively (**Fig. 22.4a, b**, upper panel).
9. Identities of PCR products were confirmed by DNA sequencing.

3.5. Dot Blot Hybridization

3.5.1. Preparation of DNA Samples

1. Wash the dot blot apparatus thoroughly with 1% SDS, and then with sterile water prior to use.
2. Cut a piece of nylon membrane and filter paper to cover the wells of the HYBRI-SLOT filtration manifold.

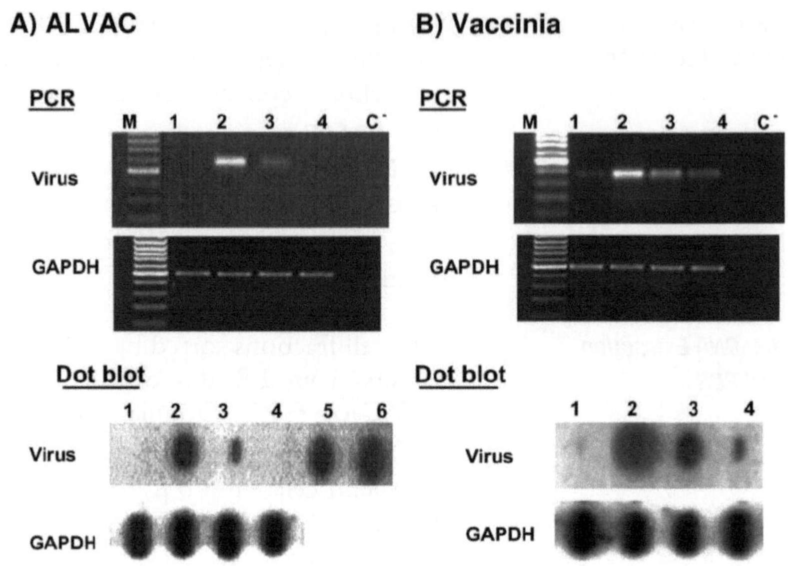

Fig. 4. Preferential infection of ALVAC or vaccinia virus to monocytes correlates with the level of virus binding. vCP1540 or EGFP-Vaccinia virus was absorbed to PBMCs at a multiplicity of infection (*MOI*) of 10 for 1 h at 37 °C in an environment of 5% CO_2. After removing unbound virus particles by washing three times with ice-chilled 2% FBS/PBS, virus-exposed PBMCs were stained with fluorochrome-conjugated antibodies of CD3, CD14, CD19, or CD56, and then subjected to fluorescence-activated cell sorting. Sorted $CD3^+$ T cells, $CD14^+$ monocytes, $CD19^+$ B cells, and $CD56^+$ NK cells were subjected to DNA isolation for PCR and dot blot hybridization. **a** Ethidium bromide-stained PCR-amplified ALVAC or vaccinia virus sequences from DNA samples prepared from sorted PBMC subsets. *M*: GeneRuler 100 bp DNA ladder plus; *1*, *2*, *3*, *4*: sorted $CD3^+$ T cells, $CD14^+$ monocytes, $CD19^+$ B lymphocytes, and $CD56^+$ NK cells, respectively; *C*−: PCR negative control. The specificity of PCR for both ALVAC and vaccinia viruses was confirmed by DNA sequences of the PCR products. Shown is one representative experiment of three from three different donors. **b** Dot blot analysis of ALVAC or vaccinia virus DNA in sorted PBMC subsets. *1*, *2*, *3*, *4*: sorted $CD3^+$ T cells, $CD14^+$ monocytes, $CD19^+$ B lymphocytes, and $CD56^+$ NK cells, respectively; *5*, *6*: 10 ng of DNA of purified PCR products or plasmid DNA of cloned PCR products generated from purified ALVAC virion served as positive controls. Oligonucleotide GAPDH.For was labeled with DIG-ddUTP and used as a probe for detecting human housekeeping gene GAPDH as the control of the DNA quality and quantity. Shown is one representative experiment of three from three different donors.

3. Soak both the nylon membrane and the filter paper in 2× SSC for 5 min prior to use.
4. Place the filter paper and membrane into the HYBRI-SLOT filtration manifold apparatus with the nylon membrane on top. Clamp the entire apparatus together.
5. Dilute 100 ng of each DNA sample prepared from sorted PBMCs subsets, which were exposed or unexposed to ALVAC or vaccinia virus, in 50 μL H_2O.
6. Add an equal volume of freshly made 0.7N NaOH to each diluted DNA sample.

7. Incubate the DNA solution on ice for 30 min.
8. Add 100 μL of 2 M ammonium acetate to the DNA solution and incubate on ice for 5 min.
9. Apply the DNA solutions to the dot blot apparatus according to the manufacture's instructions. Allow the DNA to incubate with the membrane at room temperature for 10 min.
10. Apply a vacuum to the dot blot apparatus to draw the DNA solution through the membrane.
11. After the DNA solution has been drawn through, remove the membrane from the apparatus.

3.5.2. Cross-Linking of DNA Sample to Membrane

1. Place the membrane with the bound DNA facing the lamps on a UV transilluminator (*see* **Note 13**).
2. Immobilize the DNA onto the membrane by irradiating (254 nm wavelength) for 5 min.
3. Rinse the membrane in 2× SSC and air-dry before use in the hybridization step (*see* **Note 14**).

3.5.3. Hybridization

1. Place the blotted membrane in a hybridization tube containing 6 mL prehybridization solution (the PerfectHyb™ Plus hybridization buffer without the labeled probe) per tube. Prehybridize at 60 °C for 10 min.
2. Boil the DIG-labeled probe for 10 min to denature and break secondary structures, and then chill on ice.
3. Replace the prehybridization solution with 6 mL of fresh hybridization solution containing 20 ng/mL of labeled probe (*see* **Note 15**). Hybridize overnight at 52 °C (*see* **Note 16**).
4. At the end of the hybridization, remove the membrane from the hybridization tube and save the hybridization solution at −20 °C for next use (*see* **Note 17**).
5. Wash the membrane twice with Low Stringency Wash Buffer (2× SSC, 0.1% SDS) plus 0.3% Tween-20 at room temperature, 5 min per wash (*see* **Note 18**). These washes remove unbound probe, which otherwise could lead to high background.
6. Wash the membrane twice with Ultra-High Stringency Wash Buffer (0.1× SSC, 0.1% SDS) plus 0.3% Tween-20 at room temperature, 15 min per wash.

3.5.4. Detection of Dig-Labeled Nucleotides

Perform all the incubations at room temperature with agitation.

1. After hybridization and posthybridization washes, equilibrate the membrane in Low Stringency Wash Buffer (2× SSC, 0.1% SDS) plus 0.3% Tween-20 for 1 min.

2. Block the membrane by gently agitating it in Low Stringency Wash Buffer containing 5% dry milk for 60 min.
3. Dilute the Anti-Dig-alkaline phosphatase 1:10,000 (75 mU/mL) in Low Stringency Wash Buffer containing 5% dry milk and incubate for 30 min.
4. Remove the antibody solution. Gently wash the membrane twice, 15 min per wash with Low Stringency Wash Buffer + 0.5% Tween-20.
5. Transfer the membrane to Ultra-High Stringency Wash Buffer 0.5% Tween-20 for 2 min.
6. Prewarm alkaline phosphatase chemiluminescent reagent, CSPD Ready-to-Use, to room temperature.
7. Transfer the blotted membrane carefully from Ultra-High Stringency Wash Buffer 0.5% Tween-20 to a plastic tray.
8. In the darkroom, incubate the blotted membrane with 5 mL CSPD Ready-to-Use for 1 min.
9. Drag blotted membrane along the edge of the tray to remove excess liquid, and place blotted membrane onto a plastic report cover.
10. Blot around edges with a tissue and then "close" the plastic cover. Rub gently to remove bubbles.
11. Expose to Kodak X-ray film in a cassette at room temperature. Exposures vary from blot to blot (1 min–2 h) (**Fig. 4a, b**, lower panel).

3.6. Statistical Analysis

Data were compared using the Wilcoxon signed rank test for paired samples.

Notes

1. Upon arrival, do not freeze these cells, but instead store them at 4 °C. These cells are best if used within 24–48 h, and the viability declines on a daily basis. The recommended cell concentration upon seeding is 4×10^6 per 25-cm^2 to achieve 80% confluency overnight.
2. The lyophilized cytokines are stable only for a few weeks at room temperature. Upon arrival, store at –20 °C. Reconstituted recombinant human GM-CSF and recombinant human IL-4 should be stored in working aliquots at –20 °C. Cytokines should be added to culture medium prior to use.

3. Blood and BM samples were obtained according to guidelines established by Institutional Review Boards for Human Research at University of Toronto (Toronto, Canada).
4. For activation of T cells, anti-human CD3 mAb OKT3 was immobilized in 96-well plates at a concentration of 50 ng/mL, and anti-human CD28 mAb and anti-human CD49d were immobilized at a concentration of 400 ng/mL unless otherwise noted.
5. Monocytes have a limited life span. Primary human monocytes cultured in the absence of appropriate exogenous stimuli undergo apoptosis that is enhanced by serum removal. The onset of apoptosis can be prevented by handling cells at low temperature (4 °C) or adding activating factors such as LPS and CD40 ligand. The whole procedure of monocyte isolation was carried out at 4 °C.
6. Monocytes express abundance of Fc receptors on cell surface, which can nonspecifically bind to antibody Fc. To block the nonspecific binding, Fc receptor blocking reagents should be used prior to addition of specific antibodies.
7. After differentiation for 7 days, cells with typical DC morphology will develop and can be observed under light microscopy: greatly increase in size, develop irregular shape, and typical dendritic cytoplasmatic extensions. The average yield of DCs in this condition is 41.2 ± 7.6% ($n = 9$) out of total monocytes.
8. Unbound virus can be detected from the first and second times of wash solutions (2% FBS/PBS) by PCR, but not from third time of washing solution. Therefore, three washes are required to completely remove free virus particles.
9. To accumulate cytokines in stimulated cells, addition of GolgiStop Protein Transport Inhibitor (containing monensin) at 4 μL of GolgiStop every 6 mL of cell culture is required. Keep GolgiStop in cell cultures for at least 6 h, but not longer than 12 h.
10. Equal numbers of sorted cells (2×10^6 cells), that is, CD3, CD14, CD19, and CD56 cells, are subjected to DNA isolation for PCR or dot blot hybridization.
11. DNA extracted from sorted CD3, CD14, CD19, or CD56 cell fractions is eluted in a final volume of 50 μL of TE elution buffer.
12. The human GAPDH gene consists of nine exons and eight introns with eukaryotic signals necessary for the transcription and translation of GAPDH mRNA. The oligonucleotides for GAPDH.For and GAPDH.Rev were derived from GAPDH

exons 7 and 8, respectively, which flank 193 bp length of intron 7 *(26)*.

13. Unless specified otherwise, membranes were dried prior to cross-linking. The side of the membrane with the bound DNA faced the lamps.
14. An alternative method for binding DNA to nylon membrane: the DNA dotted membrane was washed, air-dried, and baked for 2 h at 80 °C, and then used for hybridization.
15. Sufficient volumes of hybridization solution must be used to ensure complete coverage of the membrane.
16. Hybridization temperature may vary considerably from probe to probe, according to the GC content of the probes and homology to the target. To determine the optimal hybridization temperature (T_{opt}), first calculate the melting temperature (T_m) of the probe–target hybridization, and then set to a value that is 20–25 °C below the calculated T_{opt}.
17. The hybridization solution can be reused for up to three times.
18. Temperature for washing can be increased up to the temperature of hybridization if nonspecific binding of the probe to target nucleic acids is observed.

Acknowledgments

This work was carried out with funding from CFAR at University of Washington (AI 27757), R21 from NIAID/NIH (AI073250), Research Facilities Improvement Program Grant Number C06 RR015481-01 from NIH, Hawaii Community Foundation, OHTN, CIHR, Amfar, and Sanofi-UTRP program. Text and figures were modified from *Vaccine* (2006) **24**, 6376–6391 *(33)* with permission from the Elsevier Corporate.

References

1. Moss, B. (1992) Vaccinia and other poxvirus expression vectors. *Curr Opin Biotechnol* 3, 518–22.
2. Moss, B., Fuerst, T.R., Flexner, C., Hugin, A. (1988) Roles of vaccinia virus in the development of new vaccines. *Vaccine* 6, 161–3.
3. Moss, B., Flexner, C. (1987) Vaccinia virus expression vectors. *Annu Rev Immunol* 5, 305–24.
4. Bostock, C.J. (1990) Viruses as vectors. *Vet Microbiol* 23, 55–71.
5. Fries, L.F., Tartaglia, J., Taylor, J., Kauffman, E.K., Meignier, B., Paoletti, E., Plotkin, S. (1996) Human safety and immunogenicity of a canarypox-rabies glycoprotein recombinant vaccine: an alternative poxvirus vector system. *Vaccine* 14, 428–34.
6. McFadden, G. (2005) Poxvirus tropism. *Nat Rev Microbiol* 3, 201–13.
7. Krugman, S., Katz, S.L. (1969) Smallpox vaccination. *N Engl J Med* 281, 1241–2.
8. Welter, J., Taylor, J., Tartaglia, J., Paoletti, E., Stephensen, C.B. (2000) Vaccination against

canine distemper virus infection in infant ferrets with and without maternal antibody protection, using recombinant attenuated poxvirus vaccines. *J Virol* 74, 6358–67.

9. Tartaglia, J., Jarrett, O., Neil, J.C., Desmettre, P., Paoletti, E. (1993) Protection of cats against feline leukemia virus by vaccination with a canarypox virus recombinant, ALVAC-FL. *J Virol* 67, 2370–5.
10. Salmon-Ceron, D., Excler, J.L., Finkielsztejn, L., Autran, B., Gluckman, J.C., Sicard, D., Matthews, T.J., Meignier, B., Valentin, C., El Habib, R., Blondeau, C., Raux, M., Moog, C., Tartaglia, J., Chong, P., Klein, M., Milcamps, B., Heshmati, F., Plotkin, S. (1999) Safety and immunogenicity of a live recombinant canarypox virus expressing HIV type 1 gp120 MN MN tm/gag/protease LAI (ALVAC-HIV, vCP205) followed by a p24E-V3 MN synthetic peptide (CLTB-36) administered in healthy volunteers at low risk for HIV infection. AGIS Group and L'Agence Nationale de Recherches sur Le Sida. *AIDS Res Hum Retroviruses* 15, 633–45.
11. Karnasuta, C., Paris, R.M., Cox, J.H., Nitayaphan, S., Pitisuttithum, P., Thongcharoen, P., Brown, A.E., Gurunathan, S., Tartaglia, J., Heyward, W.L., McNeil, J.G., Birx, D.L., de Souza, M.S. (2005) Antibody-dependent cell-mediated cytotoxic responses in participants enrolled in a phase I/II ALVAC-HIV/AIDSVAX B/E prime-boost HIV-1 vaccine trial in Thailand. *Vaccine* 23, 2522–9.
12. Tartaglia, J., Perkus, M.E., Taylor, J., Norton, E.K., Audonnet, J.C., Cox, W.I., Davis, S.W., van der Hoeven, J., Meignier, B., Riviere, M., et al. (1992) NYVAC: a highly attenuated strain of vaccinia virus. *Virology* 188, 217–32.
13. Marovich, M.A., Mascola, J.R., Eller, M.A., Louder, M.K., Caudrelier, P.A., El-Habib, R., Ratto-Kim, S., Cox, J.H., Currier, J.R., Levine, B.L., June, C.H., Bernstein, W.B., Robb, M.L., Schuler-Thurner, B., Steinman, R.M., Birx, D.L., Schlesinger-Frankel, S. (2002) Preparation of clinical-grade recombinant canarypox-human immunodeficiency virus vaccine-loaded human dendritic cells. *J Infect Dis* 186, 1242–52.
14. Marovich, M.A. (2004) ALVAC-HIV vaccines: clinical trial experience focusing on progress in vaccine development. *Expert Rev Vaccines* 3, S99–S104.
15. Tartaglia, J., Cox, W.I., Taylor, J., Perkus, M., Riviere, M., Meignier, B., Paoletti, E. (1992) Highly attenuated poxvirus vectors. *AIDS Res Hum Retroviruses* 8, 1445–7.
16. Cox, W.I., Tartaglia, J., Paoletti, E. (1993) Induction of cytotoxic T lymphocytes by recombinant canarypox (ALVAC) and attenuated vaccinia (NYVAC) viruses expressing the HIV-1 envelope glycoprotein. *Virology* 195, 845–50.
17. Santra, S., Schmitz, J.E., Kuroda, M.J., Lifton, M.A., Nickerson, C.E., Lord, C.I., Pal, R., Franchini, G., Letvin, N.L. (2002) Recombinant canarypox vaccine-elicited CTL specific for dominant and subdominant simian immunodeficiency virus epitopes in rhesus monkeys. *J Immunol* 168, 1847–53.
18. Pal, R., Venzon, D., Letvin, N.L., Santra, S., Montefiori, D.C., Miller, N.R., Tryniszewska, E., Lewis, M.G., VanCott, T.C., Hirsch, V., Woodward, R., Gibson, A., Grace, M., Dobratz, E., Markham, P.D., Hel, Z., Nacsa, J., Klein, M., Tartaglia, J., Franchini, G. (2002) ALVAC-SIV-gag-pol-env-based vaccination and macaque major histocompatibility complex class I (A*01) delay simian immunodeficiency virus SIVmac-induced immunodeficiency. *J Virol* 76, 292–302.
19. Adler, S.P., Plotkin, S.A., Gonczol, E., Cadoz, M., Meric, C., Wang, J.B., Dellamonica, P., Best, A.M., Zahradnik, J., Pincus, S., Berencsi, K., Cox, W.I., Gyulai, Z. (1999) A canarypox vector expressing cytomegalovirus (CMV) glycoprotein B primes for antibody responses to a live attenuated CMV vaccine (Towne). *J Infect Dis* 180, 843–6.
20. Ghose, A., Iakhnina, E., Spaner, D., Tartaglia, J., Berinstein, N.L. (2000) Immunogenicity of whole-cell tumor preparations infected with the ALVAC viral vector. *Hum Gene Ther* 11, 1289–301.
21. Konishi, E., Kurane, I., Mason, P.W., Shope, R.E., Kanesa-Thasan, N., Smucny, J.J., Hoke, C.H., Jr., Ennis, F.A. (1998) Induction of Japanese encephalitis virus-specific cytotoxic T lymphocytes in humans by poxvirus-based JE vaccine candidates. *Vaccine* 16, 842–9.
22. Ignatius, R., Marovich, M., Mehlhop, E., Villamide, L., Mahnke, K., Cox, W.I., Isdell, F., Frankel, S.S., Mascola, J.R., Steinman, R.M., Pope, M. (2000) Canarypox virus-induced maturation of dendritic cells is mediated by apoptotic cell death and tumor necrosis factor alpha secretion. *J Virol* 74, 11329–38.
23. Chahroudi, A., Chavan, R., Kozyr, N., Waller, E.K., Silvestri, G., Feinberg, M.B. (2005) Vaccinia virus tropism for primary hematolymphoid cells is determined by restricted expression of a unique virus receptor. *J Virol* 79, 10397–407.
24. Sanchez-Puig, J.M., Sanchez, L., Roy, G., Blasco, R. (2004) Susceptibility of different leukocyte cell types to Vaccinia virus infection. *Virol J* 1, 10.

25. Norbury, C.C., Malide, D., Gibbs, J.S., Bennink, J.R., Yewdell, J.W. (2002) Visualizing priming of virus-specific CD8+ T cells by infected dendritic cells in vivo. *Nat Immunol* 3, 265–71.
26. Ercolani, L., Florence, B., Denaro, M., Alexander, M. (1988) Isolation and complete sequence of a functional human glyceraldehyde-3-phosphate dehydrogenase gene. *J Biol Chem* 263, 15335–41.
27. Yu, Q., Kovacs, C., Yue, F.Y., Ostrowski, M.A. (2004) The role of the p38 mitogen-activated protein kinase, extracellular signal-regulated kinase, and phosphoinositide-3-OH kinase signal transduction pathways in CD40 ligand-induced dendritic cell activation and expansion of virus-specific CD8+ T cell memory responses. *J Immunol* 172, 6047–56.
28. Yu, Q., Gu, J.X., Kovacs, C., Freedman, J., Thomas, E.K., Ostrowski, M.A. (2003) Cooperation of TNF family members CD40 ligand, receptor activator of NF-kappa B ligand, and TNF-alpha in the activation of dendritic cells and the expansion of viral specific CD8+ T cell memory responses in HIV-1-infected and HIV-1-uninfected individuals. *J Immunol* 170, 1797–805.
29. Ostrowski, M.A., Justement, S.J., Ehler, L., Mizell, S.B., Lui, S., Mican, J., Walker, B.D., Thomas, E.K., Seder, R., Fauci, A.S. (2000) The role of CD4 + T cell help and CD40 ligand in the in vitro expansion of HIV-1-specific memory cytotoxic CD8+ T cell responses. *J Immunol* 165, 6133–41.
30. Brinkman-Van der Linden, E.C., Angata, T., Reynolds, S.A., Powell, L.D., Hedrick, S.M., Varki, A. (2003) CD33/Siglec-3 binding specificity, expression pattern, and consequences of gene deletion in mice. *Mol Cell Biol* 23, 4199–206.
31. Freeman, S.D., Kelm, S., Barber, E.K., Crocker, P.R. (1995) Characterization of CD33 as a new member of the sialoadhesin family of cellular interaction molecules. *Blood* 85, 2005–12.
32. Imamura, R., Miyamoto, T., Yoshimoto, G., Kamezaki, K., Ishikawa, F., Henzan, H., Kato, K., Takase, K., Numata, A., Nagafuji, K., Okamura, T., Sata, M., Harada, M., Inaba, S. (2005) Mobilization of human lymphoid progenitors after treatment with granulocyte colony-stimulating factor. *J Immunol* 175, 2647–54.
33. Yu, Q., Jones, B., Hu, N., Chang, H., Ahmad, S., Liu, J., Parrington, M., Yewdell, J., Ostrowski, M.A. (2006) Comparative analysis of tropism between Canarypox (ALVAC) and Vacciniaviruses reveals a more restricted and preferential tropism of ALVAC for human cells of the monocytic lineage. *Vaccine* 24, 6376–91.

Chapter 23

Reverse Genetics of Nipah Virus to Probe Viral Pathogenicity

Misako Yoneda, Kentaro Fujita, Hiroki Sato, and Chieko Kai

Summary

Enhanced green fluorescent protein (EGFP) is a useful marker protein which enables the tracing of virus infection. Recombinant viruses expressing EGFP are useful for the investigation of the underlying mechanism of viral infection in vitro and in vivo. Using EGFP-expressing recombinant Nipah virus (NiV) and canine distemper virus (CDV), we tested the susceptibility of a variety of cells to infection. Receptor usage in CDV infection was also investigated.

Key words: Nipah virus, Canine distemper virus, Enhanced green fluorescent protein (EGFP), Viral susceptibility.

1. Introduction

Nipah virus (NiV) was first identified in Malaysia in 1998 during an outbreak of infection in pigs and pig farmers. The NiV infection resulted in more than 40% fatality in humans in Malaysia and Singapore *(1)*. Encephalitis was the most prominent clinical symptoms observed in humans. Respiratory disorders were the signs most frequently observed in pigs. The fruit bat (suborder Megachiroptera) was found to be the natural host and the NiV proved to have a broad host range *(2)*.

NiV has been assigned to genus *Henipavirus*, family *Paramyxoviridae (3)*. Genetically, the genus *Henipavirus* is most closely related to the genus *Morbillivirus*, family *Paramyxoviridae.*

Barry W. Hicks (ed.), *Methods in Molecular Biology, Viral Applications of Green Fluorescent Protein, vol. 515*

DOI: 10.1007/978-1-59745-559-6_23

Canine distemper virus (CDV), which belongs to the genus *Morbillivirus*, had initially been thought to affect animals in the family *Canidae*. However, increasing evidence suggests that seals (*Phoca sibirica*), lions (*Panthera leo*), tigers (*Panthera tigris*), jaguars (*Panthera onca*), and leopards (*Panthera pardus*) can also be infected *(4–6)*. Given the broad range of host animals and cross-species infections, we examined the susceptibility of various cell types and molecules that are involved in NiV and CDV infection. The enhanced green fluorescent protein (EGFP)-expressing recombinant viruses were useful for this investigation, since viral infection can be monitored with ease. In this chapter, we describe the construction of recombinant NiV and CDV expressing EGFP (rNiV-EGFP, rCDV-EGFP), analysis of their infection using various cells, and infection inhibition assays for rCDV-EGFP, using a monoclonal antibody against one of the receptors for CDV, SLAM *(7, 8)*.

2. Materials

2.1. Construction of Plasmids and Recovery of Viruses

1. Plasmid-encoding NiV full genome (pNiV) and supporting plasmids encoding the N, P, and L proteins of NiV (pKS-N, pKS-P and pGEM-L), which were constructed in-house *(7)*.
2. Plasmid encoding the EGFP gene (pEGFP-N1) (Clontech, CA, USA).
3. LA-Taq DNA polymerase (Takara, Japan).
4. pGEM-T Easy subcloning vector (Promega, WI, USA).
5. Standard reagents and apparatus for PCR.
6. Primers for amplifying the EGFP gene.

   EGFP-F; 5'-GAGCTCATGGTGAGCAAGGGCGAGGA-3'

   (*Sac*I site underlined) and EGFP-R; 5'-GGCCGGCC**TATAT**C-TACTTGTACAGCTCGTCCATGCC-3' (*Fse*I site underlined, five additional nucleotides for rule of six in bold).
7. Primers for amplifying the intergenic region between the N and P junction. NP-F; 5'-GGCCGGCCTCCAATAT-TCTA-3' (*Fse*I site underlined) and NP-R; 5'-GAGCTC-CATTGGATGAATTGTTATTA-3' (*Sac*I site underlined).
8. STBL2 competent cells for cloning and plasmid DNA.
9. Plasmid DNA Isolation Midi Kit (Qiagen, CA, USA).
10. Target cell lines: CV-1 (monkey kidney), Vero (monkey kidney), HeLa (human uterus), 293 (human kidney), P815 (mouse mast cells), CHO (hamster ovary), L2 (rat lung), MDCD MDCK (canine kidney), CRFK (feline kidney),

RK-13 (rabbit kidney), BHK-21 (hamster kidney), MDBK (bovine kidney), CPK (porcine kidney), and B95a (monkey B cells), 3132 (canine lymphoma), and NIH-3T3 (mouse embryonic fibroblast). Organisms from which these cells were derived can be seen in **Table 1**.

11. Chicken embryonic fibroblast cells.
12. Dulbecco's modified Eagle's medium (DMEM) (GIBCO BRL, Invitrogen, CA, USA) supplemented with 5 or 10% foetal calf serum (FCS) (Sigma, MO, USA).
13. RPMI 1640 medium (Sigma, MO, USA) supplemented with 1.5, 5, or 10% FCS.
14. Opti-MEM media (Invitrogen, CA, USA).
15. The MVAGKT7 recombinant vaccinia virus strain that expresses the bacteriophage T7 RNA polymerase gene. Expand this virus in chicken embryonic fibroblast cells and use the culture supernatant as virus source.
16. Fugene 6 transfection reagent (Roche, Switzerland).
17. Normal 6-well plates and collagen I-coated 6-well plates (Iwaki, Japan).

Table 1
Comparison of rCDV-EGFP infectivity among various cell lines of different species

Cell line	Infectivity (%)
HeLa (human)	38.0
293 (human)	40.0
Vero (African green monkey)	8.4
MDCK (dog)	20.3
CRFK (cat)	50.5
RK-13 (rabbit)	22.6
BHK-21 (baby hamster)	2.0
MDBK (cattle)	21.4
CPK (pig)	15.3
NIH-3T3 (mouse)	17.5
B95a (marmoset)	92.6
3132 (dog)	32.2

Cells were infected with rCDV-EGFP at an MOI of 2 $TCID_{50}$/cell (determined in B95a cells) and analysed by flow cytometry at 40 hpi

18. Monoclonal antibody against human SLAM (clone IPO-3) (Kamiya Industry, Biomedical, WA, USA).
19. Fluorescence microscopy.

3. Methods

3.1. Plasmid Construction and Virus Rescue for rNiV-EGFP

1. To construct the full-length genome plasmid to express the EGFP, the EGFP gene can be amplified from pEGFP-N1 by LA-Taq with EGFP-F and EGFP-R primers. The intergenic region between the N and P junction of NiV can be amplified from the original full-length genome plasmid by LA-Taq with NP-F and NP-R primers. The PCR products are then cloned into a pGEM-T Easy vector. The intergenic region and EGFP gene can be connected using the *Sac*I recognition site. Finally, the plasmid, which was inserted into the NP intergenic region and EGFP gene, are digested with *Fse*I and ligated into the full-length plasmid (pNiV-EGFP), then check the sequences using the BigDye Terminator Cycle Sequencing Kit (*see* **Note 1**).
2. For transfection, the pNiV-EGFP and supporting plasmids for expression of N, P, and L proteins were transformed into STBL2 competent cells (*see* **Note 2**) and purified using the QIAGEN Plasmid Midi Kit.
3. Seed CV-1 cells into 6-well plates (2×10^5 cells/well) 1 day before infection and transfection. Infect the cells for 1 h before transfection with replication-deficient MVAGKT7 vaccinia virus from medium diluted 200–400 times in DMEM supplemented with 2% FCS (*see* **Note 3, 4**).
4. Add 28.4 μL of Fugene 6 to 300 μL of serum-free DMEM and incubate for 5 min. During that time, mix plasmids (7 μg of pNiV and the plasmids encoding transacting proteins, 1.25 μg pKS-N, 0.8 μg pKS-P, and 0.4 μg pGEM-L) with 100 μL of serum-free DMEM.
5. Carefully pipette the plasmid mixtures into the diluted Fugene 6.
6. Remove the MVAGKT7-containing medium from the cells and replace with 2 mL of DMEM containing 2% FCS. Add the mixture of plasmids and Fugene 6. The cells were cultured at 37 °C.
7. Three days later, add 1.5×10^5 CV-1 cells/well (*see* **Note 5**).
8. Observe the cells by fluorescence microscopy until the fluorescence of EGFP can be detected in the cells.
9. Collect the supernatants for virus solution when EGFP expression and cytopathic effect (CPE) appear (*see* **Note 6**).

3.2. Plasmid Construction and Virus Rescue for rCDV-EGFP

1. Perform a PCR reaction using LA Taq DNA polymerase, pEGFP-N1, and the following primers: 5′-TAAGGCCGGCCAAACTCATTATAAAAAACTTAGGGCTCAGGTAGTCCAACAATGGTGAGCAAGGGCGAGGA-3', 5′-TCGAGGCCGGCCTTACTTGTACAGCTCGTCCA-3' (*Fse*I sites underlined).
2. Gel-purify the PCR product, digest it with *Fse*I, and ligate it to the dephosphorylated *Fse*I-digested pCDV vector, which is constructed in-house *(8)* (*see* **Note 1**).
3. Transform ligated DNA into STBL2 competent cells (*see* **Note 2**).
4. Purify the plasmid using the QIAGEN Plasmid Midi Kit.
5. Check the sequences using the BigDye Terminator Cycle Sequencing Kit or comparable method.
6. Seed 293 cells in collagen-coated 6-well plates (2×10^5 cells/well) in 2 mL DMEM/10% FCS 2 days before transfection (*see* **Note 7**).
7. On the day of transfection, remove the DMEM media and replace with 200–300 μL Opti-MEM, containing an appropriate amount of MVAGKT7 supernatant (usually make a 1:100 to 1:300 dilution; *see* **Notes 3, 4**).
8. Incubate for 30–60 min with gentle rocking every 10–15 min.
9. Meanwhile, prepare the transfection mixture as follows: mix 20 μL Fugene 6 with 175 μL Opti-MEM, and allow to stand for 5 min at room temperature. Then mix 1 μg of full-genomic plasmid, 1 μg of pKSN1, 1 μg of pKSP, and 0.1 μg of pGEML. Combine the Fugene 6/Opti-MEM mixture and the DNA mixture and leave for 15 min at room temperature.
10. Remove the MVAGKT7 inoculum, and add 2 mL of DMEM/5% FCS. Add MVAGKT7 DNA/Fugene 6 mixture in drop-wise manner, and swirl the plate gently.
11. Incubate at the cells at 37°C for 3 days.
12. Remove the media and add 2×10^6 B95a cells in 2 mL of RPMI/1.5% FCS.
13. Incubate until extensive syncytia are observed (*see* **Note 8**).
14. Collect the culture supernatant containing the virions. Store at −80 °C.

3.3. Infection of Various Cells with rNiV-EGFP

1. Count and seed adherent cells into 6-well plates. Nonadherent cells are transferred into 15 mL plastic tubes and gently pelleted (at ~1,000 rpm).

2. Remove media 1 h after preparation of the cells and add an appropriate quantity of rNiV-EGFP (0.01 pfu/cell is typical). Incubate for 1 h with gentle mixing every 20 min.
3. Observe the cells using a fluorescence microscope 48 h after virus infection.
4. Examples of infections are shown in **Fig. 1**.

3.4. Infection of Various Cells with rCDV-EGFP

1. Detach cells from dishes using PBS/0.02% EDTA.
2. Transfer 1×10^5 cells to sterile 1.5 mL Eppendorf tubes and spin cells down at 3,000 rpm for 5 min.
3. After washing twice with the PBS/0.02% EDTA solution, resuspend the cells in 200 μL of virus solution containing 2×10^5 $TCID_{50}$ (50% tissue culture infective dose) rCDV-EGFP.
4. Transfer the cells to 24-well plates and incubate at 37 °C for 40 h.
5. Harvest the cells in PBS/0.02% EDTA.
6. Spin down and fix the cells in PBS containing 4% paraformaldehyde for 10 min.
7. Wash the cells twice in PBS and suspend them in 500 μL 3% FCS/0.01% NaN_3 in PBS after the final wash.
8. Analyse the cells by flow cytometric analysis using FACS Calibur, in accordance with the manufacturer's instructions.
9. Calculate the positive percentage using CellQuest software, in accordance with the manufacturer's instructions.
10. A summary of results is presented in **Table 1**.

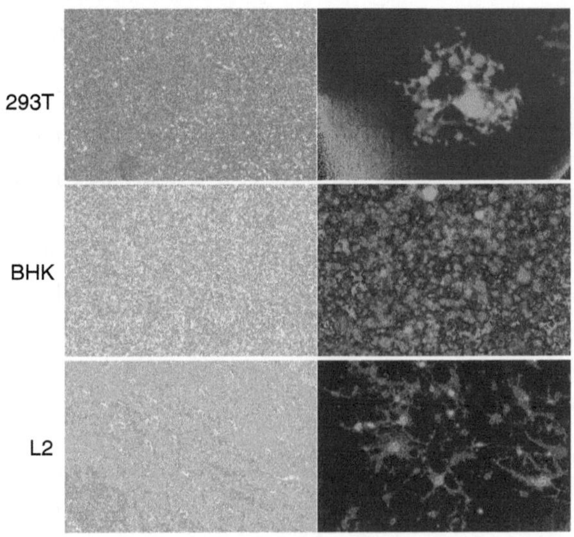

Fig. 1. Cells were infected with rNiV-EGFP at a MOI of 0.01 pfu/cell and observed 48 h later, either by light microscopy or by fluorescence. rNiV-GFP produced large multi-nucleate giant cells in 293T, but the CPE was reduced in BHK and L2 cells. The expression of the EGFP, observed by fluorescence, appeared simultaneously and paralleled the CPE. Using the above CPE and fluorescence criteria, no infection was observed in 208f, CHO, or P815 cells (data not shown)

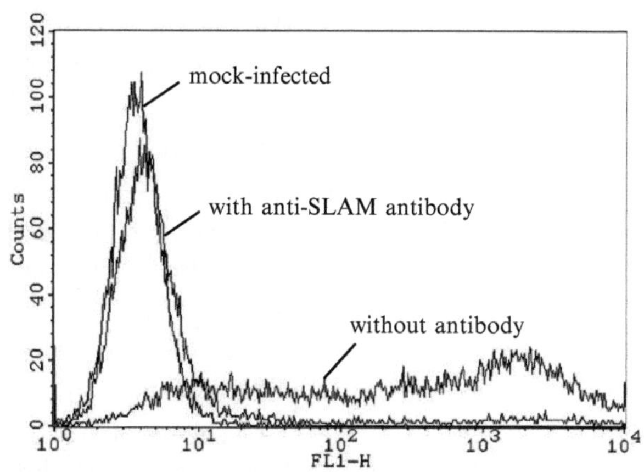

Fig. 2. B95a cells were infected with rCDV-EGFP in the absence or presence of a monoclonal antibody against SLAM (a receptor for CDV).

3.5. Inhibition of Infection by a Monoclonal Antibody Against SLAM

1. Seed B95a cells (1×10^5 cells) in 24-well tissue plates and keep for 1–2 h, until the cells attach to the base.
2. Remove media and add 500 μL of RPMI/1.5% FCS containing 5 μg/mL of anti-SLAM monoclonal antibody. Incubate at 37 °C for 1 h.
3. Remove media and add virus solution containing 2×10^5 $TCID_{50}$ of rCDV-EGFP and 5 μg/mL of anti-SLAM monoclonal antibody. Incubate at 37 °C for 1 h.
4. Remove inoculum and wash cells twice in RPMI/1.5% FCS. Incubate for 40 h.
5. Conduct flow cytometric analysis as described above.
6. An example of the results obtained is shown in **Fig. 2**.

Notes

1. The entire cDNA of NiV or CDV is flanked by T7 promoter and HDV ribozyme sequence followed by T7 terminator. This plasmid produces full-length positive-sense RNA of viral genome by T7 RNA polymerase in MVAGKT7-infected cells. Unique restriction enzyme sites were introduced between each structural gene to facilitate manipulation of the genome. An *Fse*I site was introduced downstream of the N gene.
2. STBL2 competent cells are suitable for the host of full-genomic cDNA in our hands. We noticed that some plasmids containing a morbillivirus sequence produced deletion or insertion in other host cells, such as DH5α.

3. We use the T7 promoter to rescue this virus, which can act in cytoplasm, because the transfected full genome plasmid should be transcribed in cytoplasm.
4. The optimal dilution should be determined practically.
5. After the transfection of plasmids, many of the CV-1 cells often show severe cell fusion and death. Therefore, it is better to add fresh CV-1 cells to keep the cells in a good condition.
6. Although it takes ~7–8 days normally after transfection to appear GFP-expressing cells, it would vary with each recombinant virus.
7. 293 cells tend to detach easily from the base of the dishes. Collagen-coated dishes prevent such undesired cell loss to some degree.
8. Usually, syncytia are observed at 2–5 days after the addition of B95a cells. However, it sometimes takes 10–15 days to observe syncytia. The time for observation of extensive syncytia (~30–70% of the cells are present as syncytia consisting of 10–50 nuclei per gigantic cell) varies dependent on the viruses, cell condition, or else. Recovery of the recombinant CDV expressing EGFP took 2–3 weeks to observe the syncytia. The reason of the delayed time is not known, probably depends on virus type, inserted genes, or compatibility of virus with cells. When we recover recombinant CDVs expressing firefly luciferase, we usually observe syncytia at 2–3 days after co-cultivation. Therefore, even if no detectable syncytium is observed after 4–5 days of incubation, it is worth splitting the cells every 4–5 days until 2 weeks after the addition of B95a cells.

Acknowledgments

The authors gratefully acknowledge the assistance of P4 laboratory staffs for the treatment of live NiV in the INSERM Jean Mérieux P4 laboratory. This study was supported by grants-in-aid from the Ministry of Education, Science, Culture and Sports of Japan.

References

1. Chua, K. B., Goh, K. J., Wong, K. T., Kamarulzaman, A., Tan, P. S., Ksiazek, T. G., Zaki, S. R., Paul, G., Lam, S. K., and Tan, C. T. (1999) Fatal encephalitis due to Nipah virus among pig-farmers in Malaysia. *Lancet* 354, 1257–1259.
2. Hooper, P. S., Zaki, P. D., and Middleton, D. (2001) Comparative pathology of the diseases caused by Hendra and Nipah viruses. *Microbes Infect.* 3, 315–322.
3. Büchen-Osmond, C. (2004) A summary of taxonomic changes recently approved by ICTV. *Arch. Virol.* 147, 1655–1656.
4. Appel, M. J., Yates, R. A., Foley, G. L., Bernstein, J. J., Santinelli, S., Spelman, L. H., Miller, L. D.,

Arp, L. H., Anderson, M., Barr, M., Pearce-Kelling, S., and Summers, B. (1994) Canine distemper epizootic in lions, tigers, and leopards in North America. *J. Vet. Diagn. Invest.* 6, 277–288.

5. Grachev, M. A., Kumarev, V. P., Mamaev, L. V., Zorin, V. L., Baranova, L. V., Denikina, N. N., Belikov, S. I., Petrov, E. A., Kolesnik, V. S., Kolesnik, R. S., Dorofeev, V. M., Beim, A. M., Kudelin, V. N., Nagieva, F. G., and Sidorv, V. N. (1989) Distemper virus in Baikal seals. *Nature* 338, 209–210.
6. Roelke-Parker, M. E., Munson, L., Packer, C., Kock, R., Cleaveland, S., Carpenter, M., O'Brien, S. J., Pospischil, A., Hofmann-Lehmann, R., and Lutz, H. (1996) A canine distemper virus epidemic in Serengeti lions (*Panthera leo*). *Nature* 379, 441–445.
7. Yoneda, M., Guillaume, V., Ikeda, F., Sakuma, Y., Sato, H., Wild, T. F., and Kai, C. (2006) Establishment of Nipah virus rescue system. *Proc. Natl. Acad. Sci. USA* 103, 16508–16513.
8. Fujita, K., Miura, R., Yoneda, M., Shimizu, F., Sato, H., Muto, Y., Endo, Y., Tsukiyama-Kohara, K., and Kai, C. (2006) Host range and receptor utilization of canine distemper virus analyzed by recombinant viruses; Involvement of heparin-like molecule in CDV infection. *Virology* 359, 324–335.

Chapter 24

Baculovirus-Mediated Gene Transfer into Mesenchymal Stem Cells

Chien-Tai Tsai, Ching-Kuang Chuang, and Yu-Chen Hu

Summary

Mesenchymal stem cells (MSCs) have drawn considerable attention as vehicles for cell- and gene-based therapies. Additionally, baculovirus has emerged as a novel gene therapy vector because of its large cloning capacity for insertion of multiple genes, its minimal cytotoxic effects, and its inability to replicate in mammalian cells. These features have prompted efforts to employ baculovirus vectors carrying mammalian expression cassettes for gene delivery into MSCs. This chapter demonstrates the use of GFP expression to monitor baculovirus-mediated gene transfer into MSCs.

Key words: Baculovirus, Transduction, Mesenchymal stem cell, Gene therapy, GFP.

1. Introduction

In recent years, mesenchymal stem cells (MSCs) have become a promising source for cell therapy, tissue engineering, and regenerative medicine. This is because of their extensive capability for self-renewal, and because they can differentiate into a variety of cell types including adipocytes, chondrocytes, osteoblasts, tenocytes, and even neural cells under appropriate induction conditions *(1)*. Furthermore, MSCs can be genetically modified using various viral and nonviral vectors to serve as a platform for cell-based gene therapy. After genetic modification, the expressed therapeutic gene products can promote or modulate cellular differentiation and accelerate tissue regeneration.

Barry W. Hicks (ed.), *Methods in Molecular Biology, Viral Applications of Green Fluorescent Protein, vol. 515*

DOI: 10.1007/978-1-59745-559-6_24

In contrast to common viral vectors (e.g., retrovirus, lentivirus, adenovirus, or adeno-associated virus), baculovirus (*Autographa californica* multiple nucleopolyhedrovirus) is an insect virus widely employed for recombinant protein production in insect cells *(2)*. Since the finding that it is capable of transducing liver cells in 1995 *(3)*, baculovirus has emerged as a novel gene delivery vector thanks to its high efficiency for gene delivery for many cell types, its large cloning capacity, its inability to replicate, and its low cytotoxicity in mammalian cells (for reviews, *see 2, 4, 5*). Furthermore, baculovirus is easy to construct and produce in high titers. These attributes have triggered growing interests in employing baculovirus vectors harboring mammalian expression cassette(s) for in vitro and in vivo gene delivery, development of cell-based assays, surface display of eukaryotic proteins, study of gene functions, and even delivery of vaccine immunogens *(6)*.

Given the potentials of MSCs for cell therapy and of baculovirus for gene therapy, we have demonstrated that baculovirus can efficiently transduce human MSCs derived from bone marrow and cord blood. Baculovirus transduction does not obstruct MSCs proliferation *(7)* nor subsequent adipogenic, chondrogenic, and osteogenic differentiation *(8)*. To facilitate the monitoring of transduction, we constructed a recombinant baculovirus (Bac-CE) harboring the gene for enhanced green fluorescent protein (EGFP) under the control of cytomegalovirus immediate-early (CMV) promoter *(9)*. This chapter discusses preparation and titration of the Bac-CE virus, and its use in transduction of MSCs. The transduction, along with subsequent EGFP expression in mammalian cells, allows one to monitor the transduction efficiency and transgene expression by flow cytometry. An additional benefit of EGFP is that it enables simple and rapid quantification of the baculovirus titer.

2. Materials

2.1. Culture of Mesenchymal Stem Cells

1. Human bone marrow MSCs (bMSCs): The bMSCs were derived and characterized as described previously *(7)* and stored in the liquid nitrogen (1×10^6 cells/vial).
2. bMSC culture medium: Alpha-modified minimal essential medium (α-MEM, Hyclone, Ogden, UT) supplemented with 20% fetal bovine serum (FBS, Hyclone), 4 ng/mL basic fibroblast growth factor (b-FGF, RD Systems, Minneapolis, MN), 100 U/mL penicillin (Sigma, St. Louis, MO), and 100 μg/mL streptomycin (Sigma).
3. Trypsin–EDTA (0.05% (w/w), Sigma).

4. Tissue culture flasks: 75 cm^2 (T-75) flasks (Corning Inc., Corning, NY).
5. Humidified incubator controlled at 37 °C and 5% CO_2.
6. Water bath controlled at 37 °C.

2.2. Preparation of Recombinant Baculovirus Bac-CE

1. Tissue culture dishes and plates: 35-mm dishes, 6- and 96-well plates (Corning).
2. Tissue culture flasks: T-25 (Corning) or spinner flasks (Bellco Biotechnology, Vineland, NJ).
3. pEGFP-C1 plasmid (Clontech, Mountain View, CA).
4. pFastBac™ DUAL (Invitrogen, Carlsbad, CA).
5. Bac-to-Bac®; System (Invitrogen).
6. DH10Bac *E. coli* (Invitrogen).
7. Thermocycler (Thermo Scientific, Waltham, MA).
8. Insect cells: Sf-9 (ATCC, CRL 1711).
9. Insect cell medium: TNM-FH medium (Sigma) supplemented with 10% FBS or serum-free medium (e.g., Sf-900 II, Invitrogen).
10. Sucrose cushion solution: 25% (w/w) sucrose containing 5 mM NaCl and 10 mM EDTA.
11. Ultracentrifuge (e.g., L8-70M, Beckman, Fullerton, CA) and rotor (e.g., Beckman SW28).
12. Ultracentrifuge tubes (Beckman SW28 tubes or equivalent).
13. Phosphate-buffered saline (PBS): 1 mM $Na_2HPO_4•7H_2O$, 10.5 mM KH_2PO_4, 140 mM NaCl, 40 mM KCl, pH 6.2.

2.3. Baculovirus Transduction

1. Bac-CE: Either the unconcentrated virus supernatant directly harvested from the infected cell culture or the virus that is concentrated by ultracentrifugation (*see* **Subheading 3.2.2**).
2. Human bMSCs.
3. bMSC culture medium.
4. Six-well plates.
5. Dulbecco's phosphate-buffered saline (D-PBS, Gibco): 10 mM Na_2HPO_4, 1.8 mM KH_2PO_4, 137 mM NaCl, and 2.7 mM KCl, pH 7.4.
6. Trypsin–EDTA (0.05%, Sigma).
7. Rocking plate (Midwest Scientific, St. Louis, MO).

2.4. Measurement of Transduction Efficiency and Protein Expression Levels

1. Fluorescence microscope (Nikon, Kanagawa, Japan) equipped with a digital camera (CoolSNAP, Media Cybernetics).
2. D-PBS: The D-PBS used in flow cytometry needs to be filtered through 0.22 μm.

3. Flow cytometer (FACSCalibur, BD Biosciences, La Jolla, CA).
4. CellQuest software (BD Biosciences).

3. Methods

3.1. Culture of Mesenchymal Stem Cells

MSCs can be derived from a variety of sources (e.g., bone marrow or cord blood). The derivation and characterization methods are beyond the scope of this chapter and can be found in previous protocols *(10)*. This chapter only describes the culture and expansion of bMSCs from frozen stocks.

1. Rapidly thaw the frozen bMSCs (1.0×10^6 cells in each cryovial) at a 37 °C in a water bath for 1–2 min. Transfer the cells to a 15-mL centrifuge tube and add 5 mL bMSC medium and gently centrifuge at 250 × g for 5 min. Discard the medium and resuspend the cells in 10 mL bMSC medium.
2. Seed the cells to a T-75 flask and incubate in a 37 °C humidified CO_2 incubator.
3. When the cells reach ~80–90% confluency (about 3 days later), detach the bMSCs with trypsin–EDTA. Replenish the flasks with 30 mL bMSC medium. Split the cells to three T-75 flasks and culture at 37 °C.
4. Expand the bMSCs by repeating step 3 until sufficient cells are available for transduction experiments (*see* **Note 1**).

3.2. Preparation of Recombinant Baculovirus Bac-CE

Recombinant baculovirus can be generated using different commercially available systems (for review, *see ref. (11)*). One popular system for constructing the EGFP-expressing virus is Bac-to-Bac® System, which relies on site-specific transposition of the expression cassette cloned in the donor plasmid (e.g., pFastBac™ DUAL) to a baculovirus shuttle vector (bacmid) propagated within the *E. coli* DH10Bac™. The transposition results in the recombinant bacmid consisting of baculoviral genome and the expression cassette. The recombinant bacmid is selected, purified, and subsequently transfected into insect cells in which the baculoviral genes are expressed and initiate ensuing virus replication (**Fig. 1**). Note, however, that the baculovirus promoter (e.g., polyhedrin or p10) provided in the donor plasmid is not active in mammalian cells, thus the EGFP gene must be placed under the control of a mammalian promoter such as CMV *(9)*, RSV (Rous sarcoma virus) *(12)*, CAG *(13)*, or another mammalian promoter *(14)*.

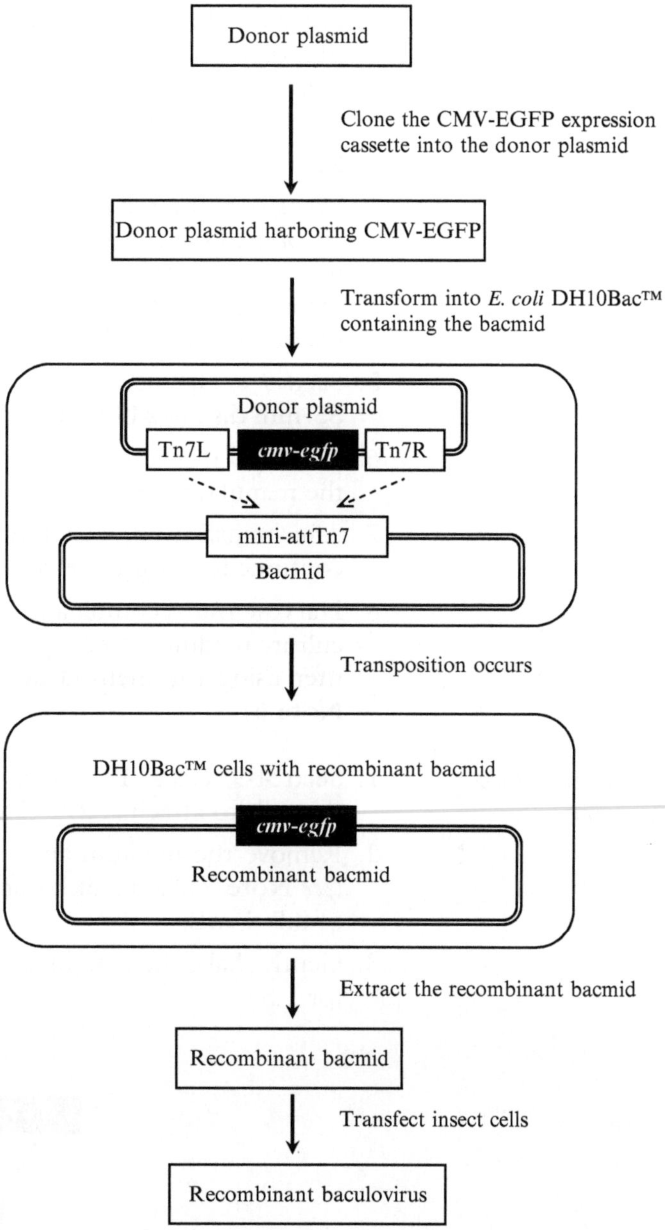

Fig. 1. Flow chart of recombinant baculovirus construction using Bac-to-Bac® System. The DH10Bac™ cells contain the bacmid and a helper plasmid (not shown) that expresses the transposase. The transposase mediates the transposition of the expression cassette flanked by Tn7R and Tn7L elements on the donor plasmid to the attachment site on the bacmid.

3.2.1. Generation of Bac-CE that Harbors CMV-EGFP Expression Cassette (9)

1. Amplify the CMV-EGFP expression cassette from pEGFP-C1 plasmid (*see* **Note 2**) by 30 cycles of polymerase chain reaction (PCR) using the following conditions: 45 s at 95 °C, 30 s at 55 °C, and then 180 s at 72 °C. The primers are designed as follows: 5′-CGCG <u>AGATCT</u> TAG TTA TTA ATA GTA ATC AAT TA-3′ and 5′-<u>AAGCTT</u>TTA CTT GTA CAG CTC GTC CAT GCC G-3′ (the underlined nucleotides encode the enzyme sites *BglII* and *HindIII*, respectively).

2. Subclone the PCR product into the BamHI/HindIII sites of pFastBac™ DUAL (*see* **Note 3**) so that the expression cassette is in the multiple cloning site (MCS I) downstream of the polyhedrin promoter as shown in **Fig. 2** (*see* **Note 4**).
3. Select and isolate the recombinant donor plasmid *(15)*.
4. Transform the recombinant donor plasmid into DH10Bac™ *E. coli* and select the recombinant bacmid following the manufacturer's instructions. Confirm the insertion of CMV-EGFP cassette by PCR and DNA sequencing according to the manufacturer's protocols.
5. Seed 9 × 10^5 Sf-9 cells in 2 mL insect cell medium to each 35-mm dish (*see* **Note 5**).
6. Transfect the recombinant bacmid into Sf-9 cells following the manufacturer's instructions.
7. After transfection, wash the Sf-9 cells with 2 mL medium and continue to culture the cells at 27 °C.
8. Harvest the recombinant virus (designated Bac-CE) in the culture medium at 72 h posttransfection. Determine the virus titer using the method described in **Subheading 3.2.3** (*see* **Note 6**).

3.2.2. Amplification of Baculovirus

1. Seed Sf-9 cells to T-25 flasks at ~50–70% confluency and allow the cells to attach at 27 °C for at least 1 h.
2. Remove the medium and inoculate the virus at a low MOI (*see* **Note 7**). Replenish complete medium to a final volume of 0.5–1 mL.
3. Gently shake the flask on a rocking plate at room temperature for 1 h.

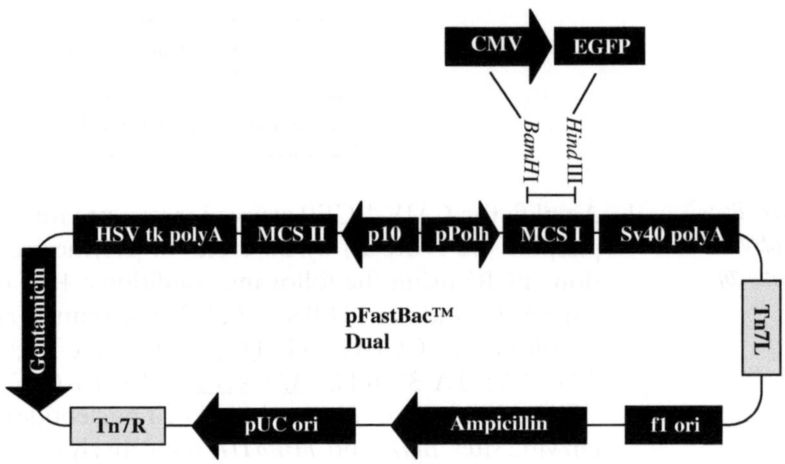

Fig. 2. Schematic illustration of the recombinant donor plasmid harboring CMV-EGFP which is subcloned into MCS I downstream of the polyhedrin promoter.

4. Add complete medium to a final volume of 5–7 mL. Incubate in a 27 °C humidified incubator.
5. Collect the medium at 72–96 h postinfection (*see* **Note 8**) and centrifuge at 1,000 × g for 5 min.
6. Collect and store the virus supernatant at 4 °C, protected from light. Store at –80 °C for long-term storage (*see* **Note 9**).
7. To amplify the virus stock to a larger scale, the process can be repeated by infecting the cells cultured in larger spinner flasks at a low MOI.
8. If concentrated baculovirus is required, load 33 mL virus supernatant in each of six 38 mL ultracentrifuge tubes.
9. Underlay with 3 mL sucrose cushion solution to each tube.
10. Centrifuge at 80,000 × *g* (24,000 rpm in an SW28 rotor) for 75–90 min at 4 °C.
11. Decant the supernatant and carefully remove traces of sucrose. The viral pellet appears translucent white.
12. Resuspend the pellets in 1–2 mL PBS (pH 6.2).
13. Filter the concentrated virus through 0.22 μm filters and determine the virus titer (*see* **Subheading 3.2.3**). Store at 4 °C.

3.2.3. Determination of Baculovirus Titers by End-Point Dilution Method (16)

1. Serially dilute the virus by tenfold (10^{-1} to 10^{-10}) in the microfuge tubes with complete insect cell medium.
2. Dilute Sf-9 cells that are in log phase with complete insect cell medium to 1×10^5 cells/mL.
3. For each dilution, mix 150 μL virus with 1.5 mL cell suspension.
4. For each dilution, inoculate 110 μL virus/cell mixture to each of the 12 wells (i.e., 10 μL virus/well) on the same row of the 96-well plate.
5. Seed uninfected cells to four wells as the negative control. Seed the cells infected by virus of 10^{-1} dilution to another four wells as the positive control.
6. Incubate the plates at 27 °C with wet tissue paper in a sealed bag to avoid dehydration.
7. Observe the cells under the fluorescence microscope 6–7 days postinfection (*see* **Note 10**). Score the wells that contain green fluorescent cells as positive.
8. For the following example (*see* **Table 1**), the numbers of infected wells and uninfected wells for each dilution are recorded (**Columns 2 and 3**), which can be used to calculate the cumulative infected and uninfected wells (**Columns**

Table 1
Example of infectious titer result

Dilution	Infected wells	Uninfected wells	Cumulative infected wells	Cumulative uninfected wells	% infected
10^{-6}	12	0	20	0	100.0 (20/20)
10^{-7}	7	5	8	5	61.5 (8/13)
10^{-8}	1	11	1	16	5.9 (1/17)
10^{-9}	0	12	0	28	0.0 (0/28)

4 and 5). Then calculate the percentage of infected wells (% infected).

In this case, the dilution that would yield 50% of infection (% infected) lies between 10^{-7} and 10^{-8}. The proportionate distance (PD) of 50% response from the response above 50% is:

$$PD = \frac{(\text{The \% responseabove 50\%}) - 50}{(\text{The \% responseabove 50\%}) - (\text{The \% responsebelow 50\%})}$$

Because the % response above 50% is 61.5 (at dilution 10^{-7}) and the % response below 50% is 5.9 (at dilution 10^{-8}), PD is calculated as follows:

$$PD = (61.5 - 50)/(61.5 - 5.9) = 0.206$$

$TCID_{50}$, the Tissue Culture Infectious $Dose_{50}$, is calculated using the following formula:

Log $TCID_{50}$ = (log of the dilution giving a response greater than 50%) – PD = – 7 – 0.206 = – 7.206

Therefore, $TCID_{50} = 10^{-7.206}$.

The virus titer is the reciprocal of $TCID_{50}$ ($1/10^{-7.206}$) and is converted to $TCID_{50}$/mL, thus the titer is $(1/10^{-7.206}) \times (1{,}000/10) = 1.6 \times 10^{9}$ ($TCID_{50}$/mL), where 10 denotes the 10 μL virus used for infection in each well, and 1,000/10 converts the unit from per 10 μL to per mL.

Usually, the baculovirus titer is expressed as plaque forming units (pfu)/mL (1 pfu = 0.69 × $TCID_{50}$), thus the titer is $1.6 \times 10^{9} \times 0.69 = 1.1 \times 10^{9}$ pfu/mL.

3.3. Baculovirus Transduction

For gene delivery into mammalian cells, baculovirus is typically concentrated and resuspended in PBS as described in **Subheading 3.2.2**. Virus transduction is then performed by incubating the cells with the concentrated virus for 1 h at 37 °C, using the culture medium (e.g., α-MEM) as the surrounding solution to adjust the final volume. Alternatively, baculovirus transduction can be performed by incubating the cells with unconcentrated virus for 4 h at 25 °C, using D-PBS as the surrounding solution

(7, 8, 17–18). The second method obviates the need for ultracentrifugation which is time consuming and impairs virus stability, thus represents a simpler and efficient approach.

3.3.1. Protocol A: Transduction Using Concentrated Bac-CE

1. Seed bMSCs onto 6-well plates at 5 × 10^5 cells/well and incubate in a 37 °C humidified CO_2 incubator overnight (*see* **Note 11**).
2. Prior to transduction, remove the medium and wash the cells with D-PBS.
3. Prepare the transduction solution by mixing the concentrated virus with bMSC medium. The volume of concentrated virus depends on the virus titer and the required MOI.
4. Add 0.5–1 mL transduction solution to triplicated wells. In parallel, add 0.5–1 mL bMSC medium to triplicated wells as the mock-transduction control.
5. Incubate the plates for 1 h at 37 °C.
6. Aspirate the transduction solution, and wash the cells with D-PBS.
7. Replenish 2 mL fresh bMSC medium to each well and incubate the plates at 37 °C for 24 h or a longer period of time.

3.3.2. Protocol B: Transduction Using Unconcentrated Bac-CE

1. Repeat steps 1 and 2 in Protocol A.
2. Prepare the transduction solution by mixing the unconcentrated virus with D-PBS (*see* **Note 12**). Depending on the MOI required, the volumetric ratio of D-PBS to virus can be varied (*see* **Note 13**).
3. Add 0.5–1 mL transduction solution to triplicated wells. In parallel, add 0.5–1 mL D-PBS to triplicated wells as the mock-transduction control (*see* **Note 14**).
4. Shake the plates on a rocking plate at 25–27 °C for 4 h (*see* **Note 15**).
5. Repeat steps 6 and 7 in Protocol A.

3.4. Assessment of Transduction Efficiency and Protein Expression Level

1. Observe the green fluorescence emitted by the transduced bMSCs under the fluorescence microscope (**Fig. 3A**).
2. At the selected time points, detach the transduced (and mock-transduced) cells by trypsin–EDTA. Wash and resuspend the cells in 1 mL D-PBS.
3. Analyze the mock-transduced cells using the flow cytometer. Gate the range of mock-transduced cells in the FL1 detector as the background by CellQuest software.
4. Analyze the transduced cells using the flow cytometer. Count 10,000 cells in each measurement and analyze the data by CellQuest software (*see* **Fig. 3B**) to obtain the percentage of cells emitting fluorescence (% GFP^+ cells) and mean fluorescence

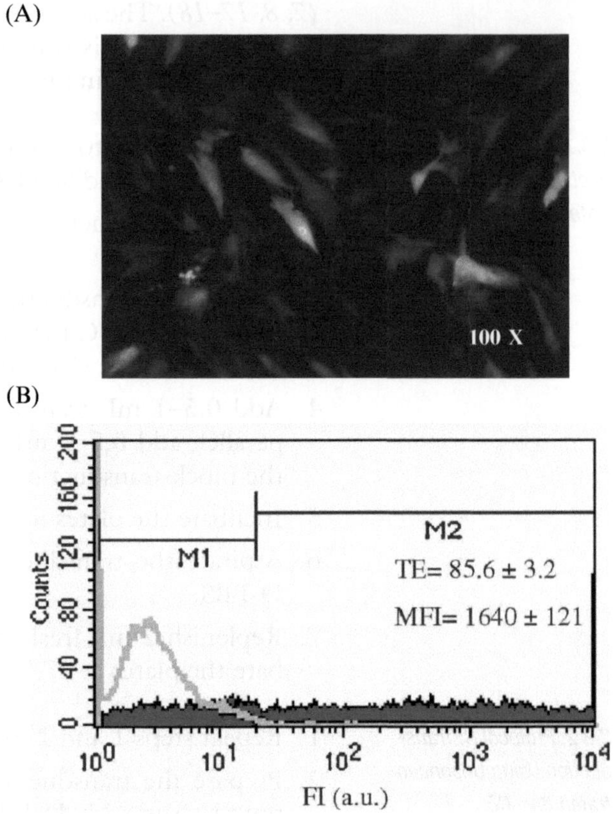

Fig. 3. Baculovirus-mediated EGFP expression in the bMSCs. a Fluorescence micrograph; b the flow cytometry data. The cells were seeded onto 6-well plates at 5.0×10^5 cells/well and were transduced with 100 μL of unconcentrated virus (MOI 60) with 400 μL PBS as the surrounding solution for 4 h at 25 °C. The fluorescence micrographs and the flow cytometry data were obtained at 24 h posttransduction. The transduction efficiency (TE) and mean fluorescence intensity (MFI) are shown as means plusmn; SD (standard deviation) of three independent experiments. The fluorescence intensity (FI) is expressed in arbitrary units (a.u.). Magnification, 100×.

intensity (MFI). The % GFP^+ cells represents the transduction efficiency (TE) (*see* **Note 16**).

5. To evaluate the transgene expression level, multiply the MFI by TE and 10,000 to yield the total fluorescence intensity (TFI) (*see* **Note 17**).

$$TFI = MFI \times TE \times 10{,}000$$

Notes

1. The bMSCs can maintain its ability to differentiate after serial passage, but it is recommended that the cells are not passed more than 20 times. We usually used to pass cells 10 times (P10) for experiments.
2. The CMV-EGFP expression cassette is available in a variety of commercial vectors. The readers can find any vector of choice for cloning.
3. The restriction enzymes sites for *BglII* and *BamHI* share the same sticky end sequence.

4. This arrangement places both polyhedrin and CMV promoter in tandem and upstream of EGFP gene. The resultant baculovirus not only expresses strong EGFP in mammalian cells, but expresses weak EGFP in insect cells.
5. Sf-9 cell is anchorage independent and can be routinely maintained in spinner flasks or T-flasks at 27 °C without CO_2. It is recommended to use the cells from a 3- to 4-day-old suspension culture in mid-log phase with a viability of >97%.
6. Baculovirus infectious titer can be determined based on the ability of baculovirus to infect insect cells by plaque assay or end-point dilution method. The corresponding virus dosage for infection is expressed as multiplicity of infection (MOI).
7. Deletion of foreign gene and emergence of defective virus particle could occur during the serial passage; it is thus critical to amplify the virus by infecting cells at a low MOI (0.01–0.1) and to avoid excessive passage.
8. The virus infection eventually kills the insect cells and results in cell lysis in 3–5 days, and the virus titer generally reaches the maximum at day 3 or 4.
9. Do not store the virus at –20 °C as the virus titer decays faster at this temperature.
10. The CMV promoter is weakly active in insect cells, thus the infected cells can emit weak EGFP that is observable under the fluorescence microscope and could distinguish them from uninfected cells.
11. Do not seed the cells at a density exceeding 90–100% confluency. Transduction of over-confluent cells would result in sharply decreased transduction efficiency.
12. Compared with **Protocol A,** this protocol can achieve comparable or superior transduction efficiencies when using a lower MOI. One key to the high efficiency is that D-PBS does not contain $NaHCO_3$, a common buffering agent supplemented to α-MEM or other mammalian cell culture media. $NaHCO_3$ considerably hinders baculovirus-mediated gene expression in mammalian cells (19). Alternatively, one can use $NaHCO_3$-deficient α-MEM as the surrounding solution which results in comparably high transduction efficiency (unpublished data).
13. The volumetric ratio of PBS to virus supernatant should be at least 2–4 because excess spent insect cell medium in the transduction solution may hinder baculovirus transduction.
14. For 6-well plates, we usually add 0.5 mL of solution to each well. The volume can be increased to up to 1 mL if a higher MOI is required. For transduction occurring in larger vessels, the liquid volume is adjusted proportionally.

15. Compared with **Protocol A,** this protocol increases the incubation time to 4 h because a longer incubation time enhances the transduction efficiency and transgene expression. Incubation of bMSCs with transduction solution for a prolonged period of time (e.g., 8 h), however, results in the detachment of cells. This protocol also decreases the temperature to 25–27 °C because incubation of baculovirus at 37 °C results in the rapid decay of virus titer.
16. The transduction efficiency depends on the MOI and the source of MSCs. The transduction efficiency can be up to 80–90% for bMSCs and exceed 95% for MSCs derived from cord blood (unpublished data).
17. Baculovirus mediates transient expression, thus the protein expression decays with time. However, in some cases, the % GFP+ cells may remain high while the mean FI decreases rapidly. Therefore, the measurement of total fluorescence intensity is more indicative of the protein expression than the % GFP+ cells. It should also be noted that EGFP is stable within the cells, thus caution should be used in data interpretation.

Acknowledgments

The authors gratefully acknowledge the financial support from the National Health Research Institutes (NHRI-EX96-9412EI) and National Science Council (NSC 95-2221-E-007-215), Taiwan.

References

1. Caplan A.I., and Bruder S. (2001) Mesenchymal stem cells: building blocks for molecular medicine in the 21st century. *Trends Mol Med* 7, 259–264.
2. Kost T.A., Condreay J.P., and Jarvis D.L. (2005) Baculovirus as versatile vectors for protein expression in insect and mammalian cells. *Nat Biotechnol* 23, 567–575.
3. Hofmann C., Sandig V., Jennings G., Rudolph M., Schlag P., and Strauss M. (1995) Efficient gene-transfer into human hepatocytes by baculovirus vectors. *Proc Natl Acad Sci USA* 92, 10099–10103.
4. Hu Y.-C. (2005) Baculovirus as a highly efficient expression vector in insect and mammalian cells. *Acta Pharmacol Sin* 26, 405–416.
5. Hu Y.-C. (2006) Insect Viruses: Biotechnological Applications (Bonning, B. C., Ed.), pp. 287–320. Elsevier, New York.
6. Yang D.-G., Chung Y.-C., Lai Y.-K., Lai C.-W., Liu H.-J., and Hu Y.-C. (2007) Avian influenza virus hemagglutinin display on baculovirus envelope: cytoplasmic domain affects virus properties and vaccine potential. *Mol Ther* 15, 989–996.
7. Ho Y.-C., Chung Y.-C., Hwang S.-M., Wang K.-C., and Hu Y.-C. (2005) Transgene expression and differentiation of baculovirus-transduced human mesenchymal stem cells. *J Gene Med* 7, 860–868.
8. Ho Y.-C., Lee H.-P., Hwang S.-M., Lo W.-H., Chen H.-C., Chung C.-K., et al. (2006) Baculovirus transduction of human mesenchymal stem cell-derived progenitor cells: variation of transgene expression with cellular differentiation states. *Gene Ther* 13, 1471–1479.
9. Hu Y.-C., Tsai C.-T., Chang Y.-J., and Huang J.-H. (2003) Enhancement and prolongation

of baculovirus-mediated expression in mammalian cells: focuses on strategic infection and feeding. *Biotechnol Prog* 19, 373–379.

10. Lennon D.P., and Caplan A.I. (2006) Culture of Cells for Tissue Engineering (Vunjak-Novakovic, G., and Freshney, R. I., Eds.). Wiley-Liss, Hoboken.
11. Chung Y.-C., and Hu Y.-C. (2007) Virus Expression Vectors (Hefferon, K. L., Ed.). Transworld Research Network, Kerala, India.
12. Boyce F.M., and Bucher N.L.R. (1996) Baculovirus-mediated gene transfer into mammalian cells. *Proc Natl Acad Sci USA* 93, 2348–2352.
13. Shoji I., Aizaki H., Tani H., Ishii K., Chiba T., Saito I.et al., (1997) Efficient gene transfer into various mammalian cells, including non-hepatic cells, by baculovirus vectors. *J Gen Virol* 78, 2657–2664.
14. Ong S.T., Li F., Du J., Tan Y.W., and Wang S. (2005) Hybrid cytomegalovirus enhancer-H1 promoter-based plasmid and baculovirus vectors mediate effective RNA interference. *Hum Gene Ther* 16, 1404–1412.
15. Ausubel F.M., Brent R., Kingston R.E., Moore D.D., Seidman J.G., Smith J.A., et al. (2002) Short Protocols in Molecular Biology. Wiley, New York.
16. O'Reilly D., Miller L., and Luckow V. (1992) Baculovirus Expression Vectors: A Laboratory Manual. WH Freeman and Co., New York.
17. Ho Y.-C., Chen H.-C., Wang K.-C., and Hu Y.-C. (2004) Highly efficient baculovirus-mediated gene transfer into rat chondrocytes. *Biotechnol Bioeng* 88, 643–651.
18. Hsu C.-S., Ho Y.-C., Wang K.-C., and Hu Y.-C. (2004) Investigation of optimal transduction conditions for baculovirus-mediated gene delivery into mammalian cells. *Biotechnol Bioeng* 88, 42–51.
19. Shen H.-C., Lee H.-P., Lo W.-H., Yang D.-G., and Hu Y.-C. (2007) Baculovirus-mediated gene transfer is attenuated by sodium bicarbonate. *J Gene Med* 9, 470–478.

INDEX

A

AAV, *See* adeno-associated virus
actin........ 2, 147
polymerization........ 2
promoter........ 47, 52, 78, 94
tail........ 2
AccuPrime DNA polymerase........ 219
AdEasy........ 109, 133
adeno-associated virus (AAV)........ 64, 340
Alliance HIV-1 p24 ELISA Kit........ 173
allophycocyanin........ 161, 171
alphaherpesvirus........ 227–228
anesthetic........ 51, 75–76, 222, 304
antiangiogenesis drugs........ 45
antiretroviral therapy........ 166, 177
Antisedan anti-anesthetic........ 217
avertin anesthetic........ 193
avoiding viral exposure........ 125
axon........ 63–65, 92
guidance........ 69

B

BAC clone containing human promoters and genes........ 219
Baculovirus........ 339–350
barley stripe mosaic virus........ 287–288
BBB, *See* blood-brain barrier
bevacizumab........ 46
bicystronic vector........ 64, 90, 139
biolistic bombardment........ 287–288, 296
bioluminescence........ 13–30
biopsy........ 56
biosafety level two........ 100, 134, 139
biotin–streptavidin........ 79–80
bFGF........ 21
bleeding........ 58, 87, 192
blood-brain barrier (BBB)........ 177–178
transmigration by virally infected cells........ 192
blue LED flashlight........ 51
bovine herpesvirus........ 249
brain........ 49, 63, 166, 199, 215
adult........ 68
healthy........ 98
macrophages........ 177, 193
sections........ 193, 199, 222
slices........ 208
buprenorphine analgesia........ 88

C

C57BL/6 mice........ 52–56, 91
canarypox........ 309–311
canine distemper virus........ 310, 329–330
carboxy terminus........ 19, 275
cardiac myocytes........ 107–113, 117
CCD camera........ 16, 39, 52, 130, 182, 242
cell detachment........ 35, 42, 122, 166, 196, 350
cell lysis........ 42, 349
cell monolayer........ 4, 24, 173, 195, 234
cell sorting........ 14, 29, 171, 321–322
cell specific promoters........ 199–200
cell transplantation........ 13, 14
Cellquest Pro software........ 131, 144, 146, 344, 347
cellufine sulfate........ 77
central nervous system (CNS)........ 63, 97, 177, 215
CNS gene therapy........ 178
selective expression within CNS........ 215
chemiluminescent detection........ 148, 324
chemotherapy........ 17, 46, 315
chimera........ 4, 191
Chinese Hamster Ovary cells........ 35, 113, 121, 330, 334
Chloroplasts........ 287, 291–292, 296
cholecystokinin........ 199, 206
chrome-alum coated slides........ 224
clinical isolates........ 33–43
clinical trials........ 17, 46, 64, 310
CNS, *See* central nervous system
co-transfection........ 67, 154, 187, 227–233
conditioned medium........ 29, 54, 187
corneal protection........ 71
coverslips........ 9, 79, 91, 241–245
CMV, *See* cytomegalovirus
Cre, *See* cre-recombinase
Cre-recombinase (Cre)........ 68, 69, 76, 99, 211
cryoprotection
CryoStor........ 174
OCT compound........ 72
-30% sucrose........ 101
cryostat........ 51, 78–79, 183, 193
CsCl gradient for adenoviral isolation........ 102
cytomegalovirus (CMV)
promoter........ 20–24, 50, 68, 78, 94, 199, 204–206, 247, 340, 349
human........ 33–43

cytopathic.. 42, 166, 232,
cytotoxicity 20, 98, 340

D

DAPI 43, 194
dendritic spine 70, 99
dendritic cells 57, 165, 299–300, 309–311
dentate gyrus 69, 209–210
diafiltration 90
diaminobenzidine 57, 192, 218
dissociated neuron cultures 91
DNA
  chromosomal 20
  maxiprep 202, 207, 212
  midiprep 5, 293
  miniprep 201, 205, 256, 265
doxorubicin 49
DPX mounting media 80, 224
drug susceptibility 33, 41
dual color imaging 45–58
dura mater 103, 223

E

EGFP, Enhanced GFP, *See* green fluorescent protein
electrophysiology 199–200
electroporation 19, 63–67, 75–94, 114–116, 121, 201, 267, 294
electroporator 67, 75, 121, 201, 263, 291
encephalomyocarditis virus 137–139
endostatin 46
endothelial cells 46–49, 85, 126
enhanced expression from weak promoters 224
ensheathing glia 63–66, 74, 94
envelope 269, 296
  protein. 29, 140, 152, 159, 190, 300
  vector 90, 180, 207, 216
  viral 1, 53, 250
essential replication protein ICP8 239–245
ethics 52, 109
euthanasia 72, 74
explants 72, 82–83
eye vitreous 78, 92

F

FACS, *See* fluorescent activated cell sorting
fast green dye 75, 89
feeder plate 27
FlowJo software 29, 316
fluorescent activated cell sorting (FACS) 20, 27, 85, 141–146, 161, 303, 315, 321, 334
fibroblast 21, 35, 82, 85, 93, 311
fibronectin 36, 40
fibrosarcoma cells
  RFP-expressing 48
  human 49, 76
firefly luciferase 13–18, 336
foot-and-mouth disease virus cleavage factor 137–139
frozen sections 29, 30, 57, 79
furin 249–253
fusion construct 17–19, 220

G

G418 36, 53, 54, 58, 241
gB, *See* glycoprotein B
Gag, pol, tat and rev proteins of HIV 152–153
gancyclovir 17, 35
granulocyte-macrophage colony stimulating factor 301, 311
green fluorescent protein (GFP/EGFP)
  antibody 30, 76, 101, 147, 218, 223
  expressing blood vessels 57
  expression for evaluation of viral production 160
  fusion protein with TNF-α receptor
  kinetics of expression156, 258
  RT-PCR 118, 190, 292
  silencing in myocytes by shRNA 118
  tagged protein 68, 69, 288
  viral supernatant assay 151
gelatin-coated plates 26
GFP, *See* green fluorescent protein
gene silencing 20, 107–122
glial cells 63, 75, 77
glutamatergic neurons 98
glycoprotein 120 (gp120) 261–262, 265–273
glycoprotein B (gB) 249–259
gp120, *See* glycoprotein 120
Golgi network 2, 249–250, 296
Golgi staining 92, 97, 104
GolgiStop 312, 325

H

HEK 293T cells 70, 77, 101, 116, 135, 206–207
HeLa cells 5, 140, 145, 331
hemocytometer 25, 75, 84, 94, 142, 184
hepatoma cell line Hep3B 126–132
Herpesviridae 227, 249
herpes simplex virus 239,
hippocampus 63, 90, 208, 210
HIV-1 packaging vector 21, 24
Hoechst nuclear dye 5, 72, 87, 96, 210
horseradish peroxidase 57, 182
HT1080 cells 90
human host-cell factor 36
HuH7 cells 154, 157–159

I

ImageJ software 39, 81
Immunopurification of Schwann Cells 82–83, 93
immunostain 14, 33, 195
infrared light 19
infection risk of laboratory personnel 126
internal ribosome entry site (IRES) 64, 92, 137–140, 145–149, 180, 211
intradermal injection 54–55
intracranial injection 215, 222
in utero 63, 66–67, 75, 88
inverted confocal laser-scanning microscopy 281
iodixanol gradient for viral isolation 72, 77
IRES, *See* internal ribosome entry site
isofluorane 82

J

judging by colony PCR 268

K

K562 cells 140

L

laminectomy 87
Langendorff perfusion 117
laser ablation 168, 172–173
latent infection of alphaherpersvirus 228
length of transgene expression 149, 199
lentiviral packaging proteins 29
lentiviral primers 221
leucocytes 167
line of Bregma 104, 207
lipofectamine 7, 53, 116, 206
live cell imaging 69, 77, 91, 241
loxP 68
luciferin 16, 28
Lucifer yellow dye 65, 73, 81
lymphomas 47

M

macrophage 57, 165–174, 177–180, 184, 194
magnetic beads for monocyte isolation 168, 314
Madin-Darby bovine kidney (MDBK) cells 253, 256–258, 331
matrigel 29
microcircuitry 97
Microcon concentrators 266, 272
microdissection 165
microelectrode 73
microinfusion pump 203, 210, 212
micromanipulators 81
microtome 91
  for frozen sections 30, 104, 193, 223
mitogen
  for Schwann cells 73–74, 83–84
  for T cells 166
mitomycin C 29
MOI, *See* multiplicity of infection
Moloney murine leukemia virus 50
Moloney murine sarcoma virus 50
moncyte-derived cells 165, 186–187, 309–314
morphogenesis 1–9, 36
mowiol mounting media 7, 76
multigene 17–19, 137–138
mutagenesis
  insertional 20
  site-directed 17, 277–280, 287, 290
  truncation 7
multicistronic lentiviral vectors 137
multiplicity of infection (MOI)
  baculovirus 349
  bovine herpes virus 257
  determination by titration 86
  increased MOI 178
  lentiviral transduction 27
  poxvirus 315–322
muscarinic receptor 110, 119
myeloid cell 309, 320

N

Nanofil syringe 203, 212
neocortex 63, 98–99
nestin 48–52
neuroAIDS 177
neurobasal medium 77, 91
Neurolucida software 82
Nipah virus 329
nitrocellulose 182, 284
NP-40 detergent 35
Nuclear
  cryptic localization signals 159
  domain 33, 240
  export 216
  expression 69, 282
  geneome 275–277
  localization 17, 22, 68, 239
nucleofection 20
nude mouse 45–50

O

olfactory bulb 84
optic nerves 78–79
origin-binding protein 240

P

P4 microPET scanner 28
packaging cells 50, 53

paraformaldehyde 7, 30, 43, 72, 91, 104, 217, 223, 312
percoll gradient 169, 291–292, 314
peptide foldability 262
perineurium 82, 93
phycoerythrin 185
picospritzer 88–89
PK15 cells 229, 232–235
plant bioreactors 275, 277
plant transformation 275, 277
plaque 4, 233, 257
  green fluorescing 234, 250, 257
  plaque forming units 6, 53, 127, 346
  purification 227, 237, 250, 253, 257
peroxisome 17, 22
polybrene 50, 54, 161, 188, 195
polyethylene glycol for viral storage 153, 155
portal vein 54
positron emission tomography 13–30
posttranscriptional gene silencing 107
prereplicative sites 239–245
primary antibody 79, 148, 192
primary cells 68, 114, 122, 126
progenitor cells 14, 48, 300–307, 32
promoter
  CamKII 94
  CCK 206
  CMV 20–24, 50, 68, 78, 94, 187, 199, 213, 247, 340, 349
    excision 204
  GFAP 215
  H5 3
  nestin 52
  NSE 215
  PGK 139
  Thy1 68–70
  T7 .... 9, 264, 335
  U6 .... 111, 117
  ubiquitin 20
  viral 6, 23, 166
promoter length 199
promyelocytic leukemia body 33
propidium iodide 130, 172
protamine sulfate 144–145
protease 9, 249, 262
proviral polyprotein cleavage 138
pseudo-transfection 162
pseudorabies virus 227–228
PT67 cells 53–54

Q

qualitative examination of viral expression 27, 47, 113

R

radiation source 22
ras oncogene 46
RCG, *See* retinal ganglion cells
recombinant adenoviruses 100, 108, 113–116, 126, 135
real-time monitoring 1, 15, 33, 50, 69, 220
receptor 46, 119, 171, 261, 325, 330
red fluorescent protein (RFP) 13–29, 45–58, 238, 288, 296
retinal ganglion cells (RGC) 64, 65
retrograde transport 97–98
RFP, *See* red fluorescent protein
RNA interference 97, 107–108, 120

S

SARS, *See* severe acute respiratory syndrome coronavirus
S phase-dependent sites 240, 243
Seakam Agarose 253
sendai virus 299–300
Schwann cells 65–66, 73–74, 84
SE/15 cells 33–43
semaphorin 68, 69
severe acute respiratory syndrome (SARS) 261–262, 265–273
shRNA 107–122
  oligonucleotide design 110
    software 110
siRNA 107–112, 120
skull 103, 193, 207, 222
  fetal 88–89
Slide-A-Lyzer dialysis 102–103
small animal imaging system 51
smallpox 1, 310
solid tumor 45, 307
sonicator 5, 230
spinal cord 49, 63, 82, 87–88
stem cells
  CNS 48
  human embryonic 13–30
  hematopoietic 157, 178
  mesenchymal 339–350
stereotaxic frame 75, 88, 103, 222
striatum 67, 89, 218
subcellular 2, 52
  localization 8, 152, 275, 279, 287
subcutaneous injection 18, 48, 54, 304, 306
Sudan Black 193
SUMOylation 34
SuperScript II RT-PCR kit 182
surface glyoproteins 261
synaptic structure 67, 98, 100, 228

T

targeted gene delivery........ 199
terrific broth........ 202
thymidine kinase........ 13–23
Thy 1.1 antibody........ 74, 93
Time-lapse microscopy........ 71, 91, 245–246
tissue trituration........ 84–85
titration assay of viral titer........ 156–157
TOPO TA Cloning........ 112
transduced cells........ 20, 29
transfected cells........ 6, 15, 159, 247, 259
transcription factors........ 23, 139
transcription start sequence........ 203
transfection
  calcium phosphate........ 24, 77, 90–91, 142–143, 187, 206, 256
  efficiency........ 19, 152, 159–160, 195, 236
  polyethylenimine........ 153–154
transgenic mice........ 45–47, 76, 211, 219
transmembrane........ 3, 320
transplantation
  CNS........ 63–66, 73, 75, 87
  orthoptic........ 45, 49
  therapy........ 13–17
transport........ 68, 276
  inhibitor........ 312
  intracellular........ 8, 250
  vesicles........ 69–71
  retrograde........ 97–98
trans-synaptic infection........ 98, 100
tricistronic lentiviral vectors........ 137–140
trimethoprim sulfamethoxazole antibiotic........ 196
TRIzol........ 128, 190
tumor suppressor gene........ 46
tumor volume formula........ 306
tweezertrode........ 75, 89

U

UCSC Genome Bioinformatics Genome Browser........ 203
ultracentrifugation for viral isolation........ 90, 143, 188, 196, 220, 225, 347
Use of Laboratory Animals........ 57
uterine wall........ 67, 98

V

vaccinia........ 1, 309
vaccine........ 1, 249, 275–284
  development........ 261–273
  EGFP........ 311, 316
  virus infection........ 310
  recombinant virus........ 4, 311, 331
Vero cells........ 241, 245, 279
vesicular stomatitis virus........ 21, 24, 152, 180, 216
vesicle........ 71, 91, 289, 291
Virakit........ 100
viral decontamination........ 125–126
viral genes6,........ 137, 342
viral load........ 162, 166
viral peptide fragments........ 261–262
viral replication compartments........ 239–243, 287
viral resivoirs........ 165
viral susceptibility........ 329
viral titer
  determination........ 103, 146, 158–159, 346
  for cell specific promoters........ 220–221
  overestimation........ 105
virion........ 2, , 125, 300
  extracellular........ 2, 69, 250
  infectious........ 2, 129
  quantification........ 167
  maturation........ 2, 240
  recovered........ 125
  transport........ 4
  wrapped........ 2
virotherapy........ 299–301
virus
  trypsinized

W

western blot........ 146, 189
woodchuck hepatitis posttranscriptional regulatory element........ 94, 139, 216

X

xenograft........ 47
xylazine........ 54, 75

Y

yellow autofluorescence from dead cells........ 134
yellow fluorescent protein (YFP)........ 8, 15
YFP, *See* yellow fluorescent protein

Z

zeocin........ 34, 36, 39

Springer

MIX
Papier aus verantwortungsvollen Quellen
Paper from responsible sources
FSC® C105338

If you have any concerns about our products,
you can contact us on
ProductSafety@springernature.com

In case Publisher is established outside the EU,
the EU authorized representative is:
Springer Nature Customer Service Center GmbH
Europaplatz 3, 69115 Heidelberg, Germany

Printed by Libri Plureos GmbH
in Hamburg, Germany